HANDBOOK OF
RADIO AND
WIRELESS
TECHNOLOGY

Handbook of Radio and Wireless Technology

Stan Gibilisco

McGraw-Hill

New York · San Francisco · Washington, D.C. · Auckland · Bogotá
Caracas · Lisbon · London · Madrid · Mexico City · Milan
Montreal · New Delhi · San Juan · Singapore
Sydney · Tokyo · Toronto

Library of Congress Cataloging-in-Publication Data

Gibilisco, Stan.
    Handbook of radio and wireless technology / Stan Gibilisco.
        p.   cm.
    Includes index.
    ISBN 0-07-023024-2
    1. Radio.   2. Wireless communication systems.   I. Title.
TK6550.G515   1998
621.382—dc21   98-8471
                                                                CIP

McGraw-Hill

A Division of The McGraw-Hill Companies

1 2 3 4 5 6 7 8 9 0   DOC/DOC   9 0 3 2 1 0 9 8

ISBN 0-07-023024-2

The sponsoring editor for this book was Scott Grillo, the editing supervisor was Stephen M. Smith, and the production supervisor was Pamela A. Pelton. It was set in Vendome ICG by Joanne Morbit and Michele Zito of McGraw-Hill's Hightstown, N.J., Professional Book Group composition unit.

Printed and bound by R. R. Donnelley & Sons Company.

McGraw-Hill books are available at special quantity discounts to use as premiums and sales promotions, or for use in corporate training programs. For more information, please write to the Director of Special Sales, McGraw-Hill, 11 West 19th Street, New York, NY 10011. Or contact your local bookstore.

 This book is printed on recycled, acid-free paper containing a minimum of 50% recycled de-inked fiber.

To Tony, Tim, and Samuel
from Uncle Stan

CONTENTS

Contents

Contents

Contents

PREFACE

When scientists and inventors first toyed with the idea of conveying messages via electromagnetic (EM) fields, they called the new mode "wireless telegraph." Video and data EM signaling became common after World War II, and the modes were given names such as *radiotelegraphy, radiotelephony, radar, telemetry,* and *television.*

In the late 1980s, the term "wireless" was resurrected. Today it refers to any and all communications, networking, control devices, and security systems in which signals travel without direct electrical connections.

This book begins with a discussion of fundamental concepts, devices, and systems: EM fields and waves, electronic components, antennas, power supplies, receivers, and transmitters. From there, the topics become more specific, including satellite communications, radionavigation, radiolocation, the Internet, measurement and control systems, privacy, and security. The book concludes with a discussion of personal and hobby communications systems.

Suggestions for future editions are welcome. I can be contacted on the Internet through my Web site at

http://members.aol.com/stangib

Acknowledgments

Some of the drawings in this book were done using *CorelDRAW* graphics software. Some clip art in illustrations is courtesy of Corel Corporation, 1600 Carling Avenue, Ottawa, Ontario, Canada K1Z 8R7.

— STAN GIBILISCO

Electromagnetic Waves

Most wireless devices work by transmitting and receiving *electromagnetic (EM) waves*. These waves can range in frequency from a fraction of 1 cycle per second, or *hertz*, to many trillions of hertz. All EM waves can travel through a complete vacuum, and sometimes they exert effects over vast distances. These properties make EM waves useful and, to some scientists, fascinating.

Electric Fields

Any electrically charged object is surrounded by an *electric field*. This field has effects on other charged objects nearby. The electric field around a charged object can be depicted by *lines of flux*.

The intensity of an electric field is measured in *volts per meter*. Two opposite charges with a potential difference of 1 volt, separated by 1 meter (m), produce a field of 1 volt per meter. The greater the electric charge on an object, the greater the intensity of the electric field surrounding the object, and the greater the force exerted on other charged objects in the vicinity. Lines of electric flux are not actual, visible lines. They are theoretical, each "line" representing a certain amount of electric charge. Electric flux density is measured in coulombs per square meter.

A single charged object produces lines of flux that emanate directly outward from, or inward toward, the charge center (Fig. 1-1A). The direction of the field is considered to be outward from the positive pole, or inward toward the negative pole. When two charged objects are brought near each other, their electric fields interact. If the charges are both positive or both negative, repulsion occurs. If the charges are opposite, attraction takes place, and the flux lines appear as shown in Fig. 1-1B.

Figure 1-1
At A, flux lines around a single charged object. At B, flux lines in the vicinity of oppositely charged objects that are near each other.

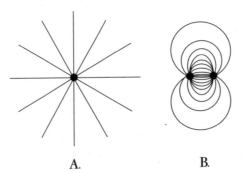

A.   B.

Electric *flux density* diminishes with increasing distance from a charged object, according to the *inverse-square law.* If the distance from a charge center is doubled, the electric flux density is reduced to one-quarter its previous value. If the distance increases by a factor of 10, the electric flux density becomes 1/100 as great. This can be envisioned by imagining spheres centered on the charged particle shown at A in the illustration. No matter what the radius of the sphere, all of the electric flux around the charged particle passes through the surface of the sphere. As the sphere radius is increased, the surface area of the sphere grows according to the square of the radius. A region of the sphere with a fixed surface area, such as 1 square meter, decreases in relative size according to the square of the radius of the sphere; therefore, proportionately fewer electric lines of flux pass through the region.

Magnetic Fields

Whenever an electric current flows (that is, when charge carriers move), a *magnetic field* accompanies the current. A magnetic field can also occur because of atomic alignment in magnetic substances.

In the case of a straight, current-carrying conductor such as a length of wire, magnetic lines of flux surround the wire in circles, with the wire at the center. This is shown in Fig. 1-2A. The current is represented by the arrows. When a current-carrying wire is coiled up, two well-defined magnetic poles develop, as shown in Fig. 1-2B. You might sometimes hear of a certain number of flux lines per unit cross-sectional area, such as 100 lines per square centimeter. This is a relative way of talking about the intensity of the magnetic field. By convention, the lines of flux are said to emerge from the *magnetic north pole* and to run inward toward the *magnetic south pole.*

Figure 1-2
At A, magnetic flux lines surrounding a straight, current-carrying wire. At B, magnetic flux lines near a current-carrying coil.

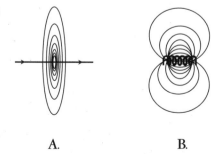

A.   B.

Magnetic fields are produced when the atoms of certain materials align themselves. Iron is the most common metal with this property. In the core of the earth, atoms of iron are aligned to some extent, making the planet a gigantic magnet. This is why a magnetic compass can be used for navigation, always pointing toward the geomagnetic poles. The geomagnetic field is in part responsible for the concentration of charged particles that creates *aurora* (northern lights and southern lights) following solar eruptions.

A piece of magnetic material such as iron, nickel, or steel, whose atoms are aligned, is always surrounded by a more or less stable magnetic field. Such an object is called a *permanent magnet*. Gradually, the atoms in a permanent magnet become less well aligned, and the strength of the magnet weakens. High temperatures and/or physical shock will speed up the process of deterioration. The ability of some metals to maintain their alignment of atoms is the basis for the operation of *magnetic media* such as recording tapes and computer diskettes.

The intensity of a magnetic field can be greatly increased by placing a special core inside of a coil. The core must consist of iron or some other material that can be readily magnetized. Such substances are called *ferromagnetic materials*. A core of this kind cannot actually increase the total quantity of magnetism in and around a coil, but it will cause the lines of flux to be much closer together inside the material. This is the principle by which an *electromagnet* works. It also makes possible the operation of electrical transformers for utility current. *Powdered iron* and *ferrite* concentrate magnetic flux at high alternating-current frequencies, and allow for compact inductors and transformers in radio communications equipment.

Electromagnetic Fields

EM fields were discovered in the nineteenth century by physicists who noticed that alternating-current (AC) energy fields can exert effects at a distance. The earliest practical use of EM fields was in wireless communication and broadcasting, which later became known as *radio* and *television*. Today, human-made EM fields permeate the environment, conveying information via many types of systems, everywhere accompanied by a steady, 60-hertz (Hz) EM field from utility lines and house wiring.

An EM field is produced when charged objects, such as electrons in a wire, accelerate and decelerate. All EM fields display properties of *wavelength* and *frequency*. The higher the frequency, the shorter the wavelength. Radio waves in the standard amplitude-modulation (AM) broadcast band are several hundred meters long. In the frequency-modulation (FM) radio broadcast band, EM waves are about 3 m long. At one extreme, some radio waves measure kilometers in length; at the other extreme, they can be as short as a few millimeters.

All EM fields contain electric lines of flux and magnetic lines of flux. The electric lines of flux are perpendicular to the magnetic lines of flux at every point in space. The direction in which the EM field propagates (travels) is perpendicular to both sets of flux lines.

The controlled acceleration and deceleration of electrons is responsible for *radio waves* having frequencies ranging from about 3000 Hz (3×10^3 Hz) to 3000 gigahertz (GHz; 3×10^{12} Hz). The 60-Hz AC utility produces an EM field whose frequency is lower (and whose wavelength is much longer) than that of radio waves. *Infrared, visible light, ultraviolet, X rays,* and *gamma rays* are forms of EM energy with frequencies higher (and wavelengths shorter) than those of radio waves.

The range of EM frequencies or wavelengths is known as the *EM spectrum*. Theoretically, there is no limit to how low or high the frequency can be, nor, correspondingly, to how long or short the wavelength can be. The most common EM wavelengths range from about 1,000,000 m (10^6 m) to around a billionth of a millimeter (10^{-9} mm).

Scientists use a logarithmic scale to depict the EM spectrum. A simplified rendition is shown in Fig. 1-3A, labeled for wavelength in meters. To find the frequencies in megahertz (MHz), divide 300 by the wavelength shown. For frequencies in hertz, use 300,000,000 (3.00×10^8) instead of 300. For kilohertz (kHz), use 300,000 (3.00×10^5); for gigahertz, use 0.300. Radio waves fall into a subset of this spectrum, at frequencies between approximately 9 kHz and 1000 GHz. This corresponds to wavelengths of 33 kilometers (km) and 0.3 mm. The *radio spectrum,* which includes television and microwaves, is "blown up" in Fig. 1-3B, and is labeled for frequency. To find wavelengths in meters, divide 300 by the frequency in megahertz. For frequencies in hertz, use 3.00×10^8; for kilohertz, use 3.00×10^5; for gigahertz, use 0.300.

Frequency

Frequency is the rate at which the cycle of a periodic disturbance repeats. The fundamental unit of frequency is the *hertz,* equivalent to 1 cycle per second. Frequencies are often specified in larger units: kilohertz or

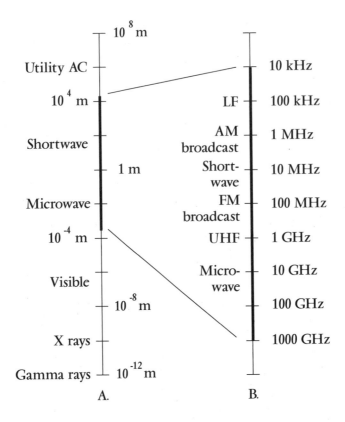

thousands of hertz, megahertz or millions of hertz, gigahertz or billions of hertz, and terahertz (THz) or trillions of hertz. A frequency of 1 kHz=1000 Hz; 1 MHz=10^6 Hz; 1 GHz=10^9 Hz; 1 THz=10^{12} Hz. If the *period,* or the time required for one complete cycle, is equal to T seconds, then the frequency f, in hertz, is equal to $1/T$.

Wave disturbances in the EM spectrum can have frequencies ranging from less than 1 Hz to trillions of terahertz. The *audio-frequency (AF)* range, at which the human ear can detect acoustic energy, ranges from about 20 Hz to 20 kHz. The radio spectrum, or *radio-frequency (RF)* range, extends from a few kilohertz up to many gigahertz.

Frequency can be measured in various ways. The most direct method is to have a digital device count the cycles within a given interval of time; this is done with a *frequency counter.* Frequency counters provide extremely accurate indications of the frequency of a periodic wave. In some instruments the resolution is 10 or even 12 significant digits.

An *absorption wavemeter* uses a tuned circuit in conjunction with an indicating meter to give an approximate measurement of the frequency of a radio signal. A capacitor or inductor in a tuned circuit is adjustable, and has a calibrated scale showing the resonant frequency over the tuning range. When the device is placed near the signal source, the meter shows the energy transferred to the tuned circuit. The coupling is maximum, resulting in a peak reading of the meter, when the tuned circuit is set to the same frequency as the signal source.

A *gate-dip meter* operates on a principle similar to that of the absorption wavemeter, except the gate-dip meter contains a variable-frequency oscillator. An indicating meter shows when the oscillator is set to the same frequency as a tuned circuit under test. Gate-dip meters are commonly used to determine unknown resonant frequencies in tuned amplifiers and antenna systems. The name of this device comes from the fact that it uses the gate circuit of a field-effect transistor to obtain readings.

A *heterodyne frequency meter* contains a variable-frequency oscillator, a mixer, and an indicator such as an analog meter. The oscillator frequency is adjusted until zero beat is reached with the signal source. This condition is shown by a dip in the meter indication. An audio amplifier might be coupled to the output of the device instead of a meter; in this case the heterodyne appears as an audible tone whose frequency drops to zero when the oscillator frequency is equal to the signal frequency being measured.

A *cavity frequency meter* is often used to determine resonant frequencies in the very-high-frequency (VHF), ultra-high-frequency (UHF), and microwave parts of the radio spectrum. The principle of operation is similar to that of the absorption wavemeter, except that an adjustable resonant cavity is used rather than a tuned circuit.

The *Lecher wire* method of frequency measurement employs a tunable section of RF transmission line to determine the wavelength of an EM signal in the VHF, UHF, and microwave regions. The system consists of two parallel wires with a movable shorting bar. The shorting bar is set until the system reaches resonance, as shown by an indicating meter. A correction is provided to take the *velocity factor* of the transmission line into account. The velocity factor, usually denoted v, is the fraction (or percentage) of the speed of light at which an EM disturbance travels through a particular medium. This factor is equal to 1 (or 100 percent) in a perfect vacuum, and is close to this value in air. In some media, such as polyethylene, v is much less than 1. It is never greater than 1.

A *spectrum analyzer* can be used for the purpose of approximate frequency measurement. This instrument can display the frequencies of several signals at the same time. General-coverage radio receivers can be used for determining the frequencies of signals. An *oscilloscope* can also be used for approximate frequency measurements.

Free-Space Wavelength

All wave disturbances have a certain physical length, regardless of the medium in which they travel. This length, called the *wavelength*, depends on two factors: the frequency of the disturbance, and the speed with which it travels.

Wavelength is inversely proportional to frequency. That is, as the frequency increases, the wavelength decreases. Doubling the frequency will cut the wavelength in half, all other things being equal. If the frequency is reduced to one-tenth of its former value, the wavelength will become 10 times as great, if nothing else changes.

Wavelength is directly proportional to the speed with which a disturbance propagates. Thus, doubling the speed will double the wavelength, and reducing the propagation speed by a factor of 10 will reduce the wavelength by a factor of 10, assuming nothing else changes.

In a periodic disturbance, the wavelength is defined as the distance between identical points in two adjacent waves. These points can be wave *crests* (maxima), *troughs* (minima), or any other specifically defined points. For an EM field, the wavelength can range from less than the diameter of an atom to more than the diameter of a galaxy. In radio communications, common EM wavelengths are measured in meters, centimeters, or millimeters. A radio station broadcasting at a frequency of 1 MHz creates an EM wave 300 m long in free space; a station at 100 MHz creates a free-space wave measuring 3 m; a communications link at 100 GHz has a signal whose free-space wavelength is only 3 mm.

The general formula for the relationship between wavelength s, frequency f, and speed c for a wave disturbance is

$$s = c/f$$

For an EM wave in free space, c is about 300,000,000 (3.00×10^8) meters per second, the same as the speed of light. Thus,

$$s = 3.00 \times 10^8 / f$$

Wavelength is an important consideration in the design of antenna systems at radio frequencies. Wavelength is also of interest to the designers of optical devices such as fiberoptic cables.

Electrical Wavelength

In any medium carrying RF signals, the *electrical wavelength* is the distance between one point in the cycle and the next identical point. This is usually expressed as the separation between points where the instantaneous amplitude of the electric field is zero and increasing positively (Fig. 1-4). The electrical wavelength depends on the velocity factor of the medium through, or along, which the field propagates. The electrical wavelength also depends on the frequency of the energy.

Along a single wire conductor, the electrical wavelength at a given frequency is somewhat shorter than the wavelength in free space. The velocity factor in a single wire is approximately 0.95, or 95 percent. Therefore, the above formula for free-space wavelength must be modified to approximately

$$s = 3.00 \times 10^8 \times 0.95/f = 2.85 \times 10^8/f$$

In general, in a transmission line with a velocity factor v (given as a fraction rather than as a percentage):

$$s = 3.00 \times 10^8 v/f$$

Often, this formula is modified for values of f expressed in megahertz rather than in hertz. The equation in this case becomes

Figure 1-4
Wavelength is the distance between identical points on two adjacent waves.

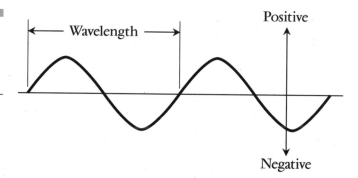

$$s = 300\,v/f$$

The electrical wavelength of a signal in an RF transmission line is always less than the wavelength in free space. When designing *impedance-matching networks* or *resonant circuits* using lengths of transmission line, it is important that the line be cut to the proper length in terms of the electrical wavelength, taking the velocity factor into account.

Polarization

The term *polarization* refers to the orientation of the flux lines in an electric, EM, or magnetic field. Polarization is of special importance in radio and television communications and broadcasting. It is also significant in some optical communications systems.

EM Wave Polarization

The polarization of an EM field is defined as the orientation of the electric flux lines. The polarization is generally parallel with the active element of a radio transmitting or receiving antenna. Thus, a vertical antenna radiates and receives fields with *vertical polarization,* and a horizontal antenna radiates and receives fields having *horizontal polarization.* These are special forms of *linear polarization,* in which the electric flux lines maintain a constant orientation.

Some antennas radiate and receive an EM field whose polarization continually and rapidly rotates. This is called *elliptical polarization.* If the rate of rotation is constant, it is known as *circular polarization.*

Polarization effects occur at all wavelengths, from the very low frequencies to the gamma-ray spectrum. These phenomena are noticeable in the visible-light range. Light having horizontal polarization, for example, will reflect well from a horizontal surface (such as a pool of water), while vertically polarized light reflects poorly off the same surface. Simple experiments, conducted with polarized sunglasses, illustrate the effects of visible-light polarization.

When radio waves are propagated via the ionosphere, the polarization is almost always affected. For example, if a vertical antenna is used at a shortwave transmitting station and this signal is received at a distant place after one or more "hops" via the ionosphere, the polarization will likely fluctuate because of the nonuniform nature of the ionized layers. However, in the immediate vicinity of the transmitting station (a radius of a few miles), the polarization will be constant and vertical.

Horizontal Polarization

In radio communications, horizontal polarization has certain advantages and disadvantages at various wavelengths.

At the low and very low frequencies (below about 300 kHz), horizontal polarization is not often used. This is because the *surface wave*, an important factor in propagation at these frequencies, is more effectively transferred when the electric field is oriented vertically. Most standard AM broadcast stations, operating between 535 and 1605 kHz, also employ vertical rather than horizontal polarization.

In the high-frequency part of the electromagnetic spectrum (3 to 30 MHz), horizontal polarization becomes practical. The polarization is always parallel to the orientation of the radiating antenna element. Horizontal wire antennas are simple to install for use at frequencies above 3 MHz. The surface wave is of lesser importance at high frequencies than at low and very low frequencies; the *sky wave* is the primary mode of propagation above 3 MHz. This becomes increasingly true as the wavelength gets shorter. Horizontal polarization is just as effective as vertical polarization in the sky-wave mode.

In the VHF and UHF range (above 30 MHz), either vertical or horizontal polarization can be used. Horizontal polarization generally provides better noise immunity and less fading than vertical polarization in this part of the spectrum.

Vertical Polarization

Vertical polarization is a condition in which the electric lines of flux of an EM wave are vertical, or perpendicular, to the surface of the earth. In radio and television communications and broadcasting, vertical polarization has certain assets and limitations, depending on the application and the wavelength.

At low and very low frequencies, vertical polarization is ideal, because surface-wave propagation, the major mode of propagation at these wavelengths, requires a vertically polarized field. Surface-wave propagation is effective in the standard AM broadcast band as well; most AM broadcast antennas are vertical. Vertical polarization is also common at high frequencies, especially among electronics hobbyists. This is mainly because self-supporting vertical antennas are practical at these wavelengths, and such structures require a minimal amount of real estate. At very high and ultra-high frequencies, vertical polarization is used mainly for mobile communications, and also in repeater installations.

The main disadvantage of vertical polarization is that most human-made noise tends to be vertically polarized. Thus, a vertical antenna picks up more of this interference than would a horizontal antenna in most cases. At very high and ultra-high frequencies, vertical polarization results in more "flutter" in mobile communications, compared with horizontal polarization.

Elliptical and Circular Polarization

The polarization of an EM wave can be deliberately made to change as the wave propagates through space. If the orientation of the electric lines of flux rotates as the signal is propagated from the transmitting antenna, the signal is said to have elliptical polarization. An elliptically polarized EM field can rotate either clockwise or counterclockwise as it moves through space. The intensity of the signal might remain constant as the wave rotates, although this need not necessarily be the case.

Elliptical polarization is useful because it allows the reception of signals having variable polarization with a minimum of fading and loss. Ideally, the transmitting and receiving antennas should both have elliptical polarization, although signals with linear polarization (usually either vertical or horizontal) can be received with an elliptically polarized antenna. If the transmitted signal has elliptical polarization in a sense contrary to that of a receiving antenna, there is substantial loss. For example, if a receiving antenna is designed for signals that rotate counterclockwise as they approach, a signal that rotates counterclockwise as it approaches will be received well, but a signal that rotates clockwise will be attenuated greatly.

Uniformly rotating elliptical polarization is called circular polarization. The orientation of the electric-field lines of flux completes one rotation for every cycle of the wave, with constant angular speed corresponding to the frequency of the signal. The flux rotation is accomplished by electrical means. For example, two *Yagi antennas*, oriented at right angles to each other, can be fed 90 degrees out of phase. The signals from the antennas add vectorially to create a rotating field. The direction, or *sense*, of the rotation can be reversed by changing the phase at one of the antennas by 180 degrees (1/2 cycle). If the antennas receive equal power, and if they have the same forward gain, then the resulting field will have circular polarization. If the *effective radiated power* (ERP) from one antenna is greater than that from the other, the resulting field will have noncircular elliptical polarization.

Circular polarization is compatible with linear polarization, but there is a 3-decibel (dB) power loss. When one station is communicating with another station also using circular polarization, the senses must be in agreement (clockwise with clockwise, or counterclockwise with counter-clockwise). If a circularly polarized signal arrives with opposite sense from that of the receiving antenna, the loss is approximately 30 dB compared with matched rotational sense.

Circular polarization is generally used at ground stations for satellite communication. For signal reception, the use of a circularly polarized antenna reduces the fading caused by changing satellite orientation. For signal transmission, the use of circular polarization ensures that the satellite will always receive a good signal for retransmission.

Propagation Modes

Radio waves behave in diverse ways as they travel over the earth, through the atmosphere, and through outer space. At very long wavelengths, signals can travel for many miles in direct contact with the earth's surface. The lowest part of the atmosphere causes radio waves to be bent, reflected, and/or scattered at short wavelengths. The upper atmosphere returns radio waves to the earth at some frequencies. In outer space, radio waves generally propagate in straight lines regardless of the wavelength.

The Ground Wave

In radio communication, the *ground wave* consists of three distinct components: the *direct wave* (also called the *line-of-sight wave*), the *reflected wave*, and the *surface wave*.

The direct wave travels in a straight line. It plays a significant role only when the transmitting and receiving antennas are connected by a straight geometric line entirely above the earth's surface. At most radio frequencies, EM fields pass through objects such as trees and frame houses with little attenuation. Concrete-and-steel structures cause some loss in the direct wave at higher frequencies. Earth barriers such as hills and mountains block the direct wave.

A radio signal can be reflected from the earth or from certain structures such as concrete-and-steel buildings. The reflected wave combines with the direct wave (if any) at the receiving antenna. Sometimes the two are exactly out of phase, in which case the received signal is extremely weak. This effect occurs primarily at frequencies above 30 MHz (wavelengths less than 10 m).

The surface wave travels in contact with the earth, and the earth forms part of the circuit. This happens only with vertically polarized fields at frequencies below about 15 MHz. Above 15 MHz, there is essentially no surface wave. Below about 300 kHz, the surface wave propagates for hundreds or even thousands of miles. Sometimes the surface wave is called the "ground wave," although technically this is a misnomer.

The Ionosphere

The atmosphere of our planet becomes less dense with increasing altitude. Because of this, the energy received from the sun is much greater at high altitudes than it is at the surface. High-speed subatomic particles, UV rays, and X rays cause ionization of the rarefied gases in the upper atmosphere. Ionized regions occur at specific altitudes and comprise the so-called *ionosphere*.

The ionosphere causes absorption and refraction of radio waves. This makes long-distance communication or reception possible on some frequencies. The highest level of the ionosphere, the *F layer*, regularly returns radio waves to the earth at frequencies up to approximately 10 MHz. This layer is responsible for worldwide "shortwave" signal propagation. (Ionospheric layers are discussed in more detail below.) F-layer conditions vary between about 10 and 30 MHz, depending on the time of day, the time of year, and the level of solar activity. Some F-layer effects are observed above 30 MHz, but their extent depends greatly on the recurring sunspot cycle. When sunspot levels are at their peak, a situation that takes place every 11 years, the F layer sometimes returns radio waves at frequencies up to about 70 MHz. At frequencies higher than that, the F layer almost never returns radio signals to the earth.

Ionization in the upper atmosphere actually occurs at four different layers. The lowest region is called the *D layer*. It exists at an altitude of about 30 miles (mi; 50 km), and is ordinarily present only on the daylight side of the planet. This layer does not contribute to long-distance communication, and in fact sometimes impedes it. The *E layer,* about 50 mi (80 km) above the surface, also exists mainly during the day, although nighttime ionization is sometimes observed. The E layer can facilitate medium-range radio communication at certain frequencies. The uppermost layers are called the *F1 layer* and the *F2 layer.* The F1 layer, normally present only on the daylight side of the earth, forms at about 125 mi (200 km) altitude; the F2 layer exists at about 180 mi (300 km) more or less around the clock. Sometimes the distinction between the F1 and F2 layers is ignored, and they are spoken of together as the F layer.

Under normal conditions, communication by means of F-layer propagation can usually be accomplished between any two points on the earth, at some frequency or frequencies between 5 and 30 MHz. Moderate to high transmitter power, and large antennas, improve reliability of communications. But elaborate installations are not always necessary. Amateur radio operators have made two-way contacts with small antennas and transmitter outputs of less than 1 watt under ideal conditions at 21, 24, and 28 MHz.

The four ionospheric layers and their relative distances above the surface are illustrated in Fig. 1-5. These altitudes can vary slightly; also, the layers are not sharply defined, but are "fuzzy." Sometimes ionization

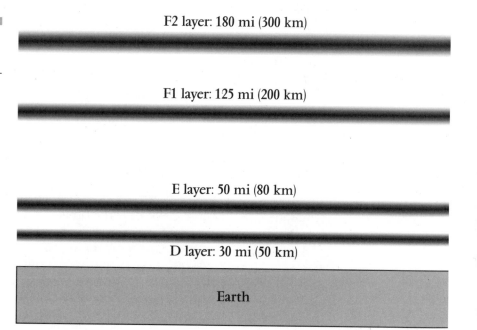

Figure 1-5
Layers of the ionosphere.

F2 layer: 180 mi (300 km)

F1 layer: 125 mi (200 km)

E layer: 50 mi (80 km)

D layer: 30 mi (50 km)

Earth

occurs nonuniformly in patches or "clouds." This phenomenon is especially common in the E layer.

Sporadic-E

At certain radio frequencies, the ionospheric E layer occasionally returns signals to the earth. This kind of propagation tends to be intermittent, and conditions can change rapidly. For this reason, it is known as *sporadic-E propagation*. It is most likely to occur at frequencies between approximately 20 and 150 MHz. Occasionally it is observed at frequencies as high as 200 MHz. The propagation range is on the order of several hundred miles, but occasionally communication is observed over distances of 1000 to 1500 mi.

The standard FM broadcast band is sometimes affected by sporadic-E propagation. The same is true of the lowest television (TV) broadcast channels, especially channels 2 and 3. Sporadic-E propagation often occurs on the amateur bands at 21 through 148 MHz. Sporadic-E propagation is sometimes mistaken for effects that take place in the lower atmosphere, independently of the ionosphere.

Solar Activity

The number of sunspots is not constant, but changes from year to year. The variation is periodic and dramatic. This fluctuation of sunspot numbers is called the *sunspot cycle*. It has a period of approximately 11 years. The rise in the number of sunspots is generally more rapid than the decline, and the maximum and minimum sunspot counts vary from cycle to cycle.

The sunspot cycle affects propagation conditions at frequencies up to about 70 MHz for F-layer propagation and 150 to 200 MHz for sporadic-E propagation. When there are not many sunspots, the *maximum usable frequency (MUF)* is comparatively low, because the ionization of the upper atmosphere is not dense. At or near the time of a sunspot peak, the MUF is higher because the upper atmosphere is more ionized.

A *solar flare* is a violent storm on the surface of the sun. Solar flares can be seen with astronomical telescopes equipped with projecting devices to protect the eyes of observers. A solar flare appears as a bright spot on the solar disk, thousands of miles across and thousands of miles high. Solar flares cause an increase in the level of radio noise that comes from the sun, and cause the sun to emit an increased number of high-speed subatomic particles. These particles travel through space and arrive at the earth a few hours after the first appearance of the flare. Because the particles are electrically charged, they are accelerated by the earth's magnetic field. Sometimes a *geomagnetic storm* results. Then we see the "northern lights" or "southern lights" (*aurora borealis* or *aurora australis*, often called simply the *aurora*) at high latitudes during the night, and experience a sudden deterioration of ionospheric radio-propagation conditions. At some frequencies, communications can be cut off within seconds. Even wire communications circuits are sometimes affected.

Solar flares can occur at any time, but they seem to take place most often near the peak of the 11-year sunspot cycle. Scientists do not know exactly what causes solar flares, but the events seem to be correlated with the relative number of sunspots.

Aurora and Meteor Trails

In the presence of unusual solar activity, the aurora often return EM energy to earth. This is called *auroral propagation*.

The aurora occur in the ionosphere, at altitudes of about 40 to 250 mi above the surface of the earth. Theoretically, auroral propagation is possible, when the aurora are active, between any two points on the earth's surface from which the same part of the aurora lie on a line of sight. Auroral propagation seldom occurs when one end of the circuit is at a latitude less than 35 degrees north or south of the equator.

Auroral propagation is characterized by rapid and deep fading. This almost always renders voice signals unreadable. Radiotelegraphy is the most effective mode for communication via auroral propagation, but the carrier is often spread out ("smeared") over several hundred hertz as a result of phase modulation induced by auroral motion. Auroral propagation can take place at frequencies well above 30 MHz, and is often accompanied by deterioration in ionospheric propagation via the E and F layers.

When a meteor from space enters the upper part of the atmosphere, an ionized trail is produced because of the heat of friction. Such an ionized region reflects EM energy at certain wavelengths. This phenomenon, known as *meteor-scatter propagation*, can result in over-the-horizon radio communication or reception.

A meteor produces a trail that persists for a few tenths of a second up to several seconds, depending on the size of the meteor, its speed, and the angle at which it enters the atmosphere. This amount of time is not sufficient for the transmission of very much information, but during a meteor shower, ionization can be almost continuous. Meteor-scatter propagation has been observed at frequencies considerably above 30 MHz.

Meteor-scatter propagation is mainly of interest to experimenters and radio amateurs. This mode is sometimes used in amateur *packet radio* because message elements (packets) are of short duration. Meteor-scatter propagation occurs over distances ranging from just beyond the horizon up to about 1500 mi, depending on the altitude of the ionized trail, and on the relative positions of the trail, the transmitting station, and the receiving station.

Propagation Forecasting

Ionospheric conditions vary considerably from day to day, month to month, and year to year at frequencies below 70 MHz. To some extent, these changes are predictable.

The sun is constantly monitored during the daylight hours for fluctuations in the intensity of radiation at various EM frequencies, especially 2800 MHz. The *2800-MHz solar flux,* as it is called, provides a good indicator of propagation conditions for the ensuing several hours. A sudden increase in the solar flux means that a solar flare is occurring. Within a short time, propagation conditions usually become disturbed following a flare. Flares sometimes wipe out F-layer communications completely, but the E layer might become sufficiently ionized to allow propagation over moderate distances. At higher latitudes, auroral propagation can usually be observed after a solar flare; signals are reflected whether it is nighttime or daytime.

Regular variations in propagation conditions take place. At frequencies below about 10 MHz, ionospheric propagation is generally better at night than during the day. Above about 10 MHz, this situation is reversed. The winter months are generally better for propagation conditions below 10 MHz, while the summer months bring better conditions above 10 MHz.

The 11-year sunspot cycle produces regular changes in the MUF. The MUF is highest at the time of the sunspot maximum. Occasionally the MUF can rise to about 70 MHz. When the sunspot level is at or near a minimum, the MUF can fall below 10 MHz.

Propagation-forecast bulletins are regularly transmitted by the National Bureau of Standards time-and-frequency stations, WWV and WWVH. Long-term propagation forecasts are also issued each month in amateur-radio magazines and other publications for the electronics hobbyist and professional.

Tropospheric Bending

The lowest 10 or 12 mi of the earth's atmosphere comprise the *troposphere*. This region has an effect on radio-wave propagation at certain frequencies. At wavelengths shorter than about 15 m (frequencies above 20 MHz), refraction and reflection can take place within and between air masses of different density. The air also produces some scattering of EM energy at wavelengths shorter than about 3 m (frequencies above 100 MHz). All of these effects are generally known as *tropospheric propagation,* which can result in communication over distances of hundreds of miles.

A common type of tropospheric propagation takes place when radio waves are refracted in the lower atmosphere. This happens to some extent all the time, but is most dramatic near weather fronts, where warm, relatively light air lies above cool, denser air. The cooler air has a higher *index of refraction* than the warm air, causing EM fields to be bent downward at a considerable distance from the transmitter. This is known as *tropospheric bending.* It is often responsible for anomalies in reception of FM and TV broadcast signals. On TV, a station hundreds of miles away might suddenly appear on a channel that is normally vacant. Unfamiliar stations might be received on a high-fidelity (hi-fi) FM radio tuner. Sometimes two or more distant stations come in on a single channel, interfering with each other. In the case of TV, the effects are often mistaken for *electromagnetic interference (EMI)*.

Duct Effect

The *duct effect* is a form of tropospheric propagation that takes place at approximately the same frequencies as bending. Also called *ducting,* this form of propagation is most common very close to the surface, sometimes at altitudes of less than 1000 feet.

A *duct* forms when a layer of cool air becomes sandwiched between two layers of warmer air. This is common along and near weather fronts in temperate latitudes. It also takes place frequently above water surfaces during the daylight hours, and over land surfaces at night. Cool air, being denser than warmer air at the same humidity level, exhibits a higher index of refraction for radio waves of certain frequencies. *Total internal reflection* takes place inside the region of cooler air, in much the same way as light waves are trapped inside an optical fiber.

For the duct effect to provide communications, both the transmitting and receiving antennas must be located within the same duct, and this duct must be unbroken and unobstructed between the two locations. A duct might measure only a few feet from top to bottom, but nevertheless cover many thousands of square miles parallel to the surface. Ducting often allows over-the-horizon communication of exceptional quality, over distances of hundreds of miles, at VHF and UHF.

Troposcatter

At frequencies above about 100 MHz, the atmosphere has a scattering effect on radio waves. The scattering allows over-the-horizon communication at VHF, UHF, and microwave frequencies. This mode of propagation is called *tropospheric scatter* or *troposcatter*. Dust and clouds in the air increase the scattering effect, but some troposcatter occurs regardless of the weather. Troposcatter takes place mostly at low altitudes where the air is the most dense. Some effects occur at altitudes up to about 10 mi.

Troposcatter propagation can provide reliable communication over distances of several hundred miles when the appropriate equipment is used. Communication via troposcatter requires the use of high-gain antennas. The physical dimensions of high-gain directive arrays are reasonable at UHF and microwave frequencies. The transmitting and receiving antennas are aimed at the same parcel of air, which is ideally located midway between the two stations. The maximum obtainable range depends not only on the gain of the antennas used for transmitting and receiving, but on their height above the ground: the higher the antennas, the greater the range. The terrain also affects the range; flat terrain is best, while mountains impede troposcatter propagation. High-power transmitters and sensitive receivers are also advantageous.

Tropospheric scatter, while occurring for different reasons than bending or ducting, is often observed along with other modes of tropospheric propagation. While communicating via troposcatter, a sudden improvement in conditions can take place because of bending and/or ducting, if a weather front happens to pass both the transmitting and receiving locations at about the same time.

Figure 1-6 shows typical tropospheric scatter and bending effects. The transmitting station is at the lower left. There is a temperature inversion in this example; it exaggerates the bending effect. If the boundary

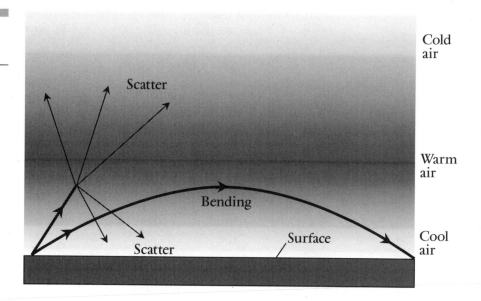

Figure 1-6
Tropospheric scatter
and bending.

between the cool air near the surface and the warm air above is well-defined enough, reflection will occur in addition to the bending. If the inversion covers a large geographic area, signals can bounce repeatedly between the inversion boundary and the surface, providing exceptional long-range communication, especially if the surface is salt water. The Gulf of Mexico, for example, is known among radio amateurs for "openings" of this kind.

Propagation at Various Frequencies

The following several sections discuss the nature of wave propagation in the earth's atmosphere throughout the EM spectrum, from the lowest frequencies (longest wavelengths) to the highest frequencies (shortest wavelengths).

Very Low Frequencies (below 30 kHz)

At *very low frequencies (VLF)*, propagation takes place mainly via a wave-guide-like effect between the earth and the ionosphere. Surface-wave propagation also occurs for considerable distances. With high-power transmitters and large antennas, communications can be realized over distances of several thousand miles. The earth-ionosphere waveguide has a low-end cutoff frequency of approximately 9 kHz. For this reason, signals much below 9 kHz suffer severe attenuation and do not propagate well.

Antennas for VLF transmitting must be vertically polarized. Otherwise, surface-wave propagation will not take place, because the earth tends to short-circuit horizontally polarized electromagnetic fields.

Propagation at VLF is remarkably stable; there is very little fading. Solar flares occasionally disrupt communication in this frequency range by making the ionosphere absorptive, and by raising the cutoff frequency of the earth-ionosphere waveguide.

Because the surface-wave mode is very efficient at frequencies below 30 kHz, it is possible that the VLF band might be used for over-the-horizon communications on planets that lack ionized layers concentrated enough to return signals at higher frequencies. The moon and Mars are examples of such planets.

Low Frequencies (30 to 300 kHz)

At *low frequencies (LF)*, propagation takes place in the surface-wave mode, and also as a result of ionospheric effects.

Toward the low end of the LF band, wave propagation is similar to that in the VLF range. As the frequency increases, surface-wave propagation becomes less efficient. A surface-wave range of more than 3000 mi is common at 30 kHz, but it is unusual for the range to be greater than a few hundred miles at 300 kHz.

Ionospheric propagation at LF usually occurs via the E layer. This increases the useful range during the nighttime hours, especially toward the upper end of the band. Intercontinental communication is possible with high-power transmitters.

Solar flares disrupt conditions at LF. Following a flare, the D layer becomes highly absorptive, preventing ionospheric propagation.

Medium Frequencies (300 kHz to 3 MHz)

Propagation at *medium frequencies (MF)* occurs by means of the surface wave, and by E-layer and F-layer ionospheric modes.

Near the low end of the MF band, surface-wave communications can be had up to several hundred miles. As the frequency is raised, the surface-wave attenuation increases. At 3 MHz, the range of the surface wave is limited to about 150 or 200 mi.

Ionospheric propagation at MF is almost never observed during the daylight hours, because the D layer prevents EM waves from reaching the higher E and F layers. During the night, ionospheric propagation takes place mostly via the E layer in the lower portion of the band, and primarily via the F layer in the upper part of the band. The range increases as the frequency is raised. At 3 MHz, worldwide communication is sometimes possible over darkness paths.

The 11-year sunspot cycle and the season of the year affect propagation at MF. Propagation is usually better in the winter than in the summer. This is partly because darkness prevails over a greater proportion of the hemisphere in the winter than in the summer, and partly because there is less interference from *sferics* (thunderstorm-induced noise) in the winter than in the summer.

High Frequencies (3 to 30 MHz)

Propagation at *high frequencies (HF)* exhibits widely variable characteristics. Effects are much different in the lower-frequency part of this band than in the upper-frequency part. Consider the lower portion as the range 3 to 10 MHz, and the upper portion as the range 10 to 30 MHz.

Some surface-wave propagation occurs in the lower part of the HF band. At 3 MHz, the maximum range is 150 to 200 mi; at 15 MHz it decreases to roughly the radio-horizon distance, or about 15 mi. Above 15 MHz, surface-wave propagation is essentially nonexistent.

Ionospheric communication occurs mainly via the F layer. In the lower part of the band, there is very little daytime ionospheric propagation because of D-layer absorption; at night, worldwide communication can be had because the D layer disappears, allowing signals to reach the F layer. In the upper part of the band, signals penetrate the D layer, allowing worldwide communication during the day, but conditions often deteriorate at night because F-layer ionization is not dense enough to return the waves to the surface.

Communications in the lower part of the HF band are generally better during the winter months than during the summer months. In the upper part of the band, this situation is reversed. Some E-layer propagation is occasionally observed in the upper part of the HF band. This is usually of the sporadic type. It can occur even when the F-layer MUF is below the communication frequency. Solar flares cause dramatic changes in conditions in the HF band. Sometimes, ionospheric communications are almost totally wiped out within a matter of minutes by the effects of a solar flare.

At the extreme upper-frequency end of the HF band, near 30 MHz, there is often no ionospheric propagation. This is especially true when the sunspot cycle is at or near a minimum.

Very High Frequencies (30 to 300 MHz)

Propagation at *VHF* occurs most commonly along the line-of-sight path or via tropospheric modes. Ionospheric F-layer propagation is rarely observed, although it can occur at times of sunspot maxima at frequencies as high as about 70 MHz. Sporadic-E propagation is fairly common, and can occur up to 150 or 200 MHz.

Meteor-scatter and auroral propagation can sometimes be observed in the VHF band. The range for meteor scatter is typically a few hundred miles; auroral propagation can provide communications over distances of up to about 1500 mi. Slow-speed radiotelegraphy or data transmission must usually be employed for auroral communications, because the auroras cause phase-shifting effects that confuse receivers of high-speed data or voice signals.

Repeaters are used extensively at VHF to extend the range of mobile communications equipment. A repeater at a high-elevation site such as a hilltop can provide coverage over thousands of square miles. At VHF, active communications satellites are often used to provide worldwide coverage on a reliable basis. The satellites are actually orbiting wideband repeaters. The moon, a passive satellite, is sometimes used, mostly by amateur radio operators. This is known as *earth-moon-earth* mode or *moonbounce*.

Ultra-High Frequencies (300 MHz to 3 GHz)

Propagation at *UHF* occurs almost exclusively via the line-of-sight mode and via satellites and repeaters. "Shortwave"-type ionospheric propagation does not take place. Auroral, meteor-scatter, and tropospheric propagation are occasionally observed in the lower-frequency portion of this band. Ducting can result in propagation over distances of several hundred miles.

A major asset of the UHF band is its size: 2700 MHz of spectrum space. It has room for a large number of signals. Another advantage of this band is its relative immunity to the effects of most solar flares. Since UHF energy has nothing to do with the ionospheric D, E, and F layers, disruptions in the ionosphere have no effect. But an extremely intense solar flare can interfere indirectly with UHF circuits by disrupting the utility power grid.

Although the UHF band was once regarded as a practically infinite spectral resource, this situation is changing. Digital communications is being carried out among an ever-increasing number of people, at ever-faster speeds. The numbers of signals, and their bandwidths, are skyrocketing. As a result, competition for UHF spectrum space has become intense among various services.

Microwaves (above 3 GHz)

Microwaves travel essentially in straight lines through the atmosphere. Microwaves are rarely affected by such phenomena as temperature inversions or tropospheric scattering, and are not refracted or reflected by the ionosphere. The primary mode of propagation in the microwave range is line-of-sight. This facilitates repeater and satellite communications.

The microwave band is much more vast than the UHF band. The upper limit of the microwave range is the LF (long-wavelength) *infrared (IR)* band of the EM spectrum. But even at the highest microwave frequencies, the spectrum is teeming with signals in some large cities, and interference can occur.

At some microwave frequencies, atmospheric attenuation becomes a consideration. Rain, fog, and other weather effects can increase path loss as the wavelength becomes comparable with the diameter of water droplets and dust particles. Transmitter and receiver designs get critical as the frequency increases, especially above several hundred gigahertz. A multimegawatt transmitter can be built for use at VLF, LF, MF, or HF, but such a thing is unheard of in the microwave range. However, a device called the *maser*, which is in fact a microwave *laser*, can concentrate energy into narrow, coherent beams at microwave frequencies. This offers some possibilities for long-distance links, especially in outer space.

Beyond Microwaves

Electromagnetic waves whose frequencies are higher than those of radio microwaves always travel in straight lines through the atmosphere. The only factor that varies is the path loss.

Water in the air causes severe attenuation of IR between the wavelengths of approximately 4500 and 8000 nanometers (nm). (1 nm is equal to 0.000000001, or 10^{-9}, m.) Carbon dioxide gas interferes with the transmission of IR at wavelengths ranging from about 14,000 to 16,000 nm. Rain, snow, fog, and dust also interfere with the propagation of IR.

The visible light spectrum extends from about 750 nm (red light) to 390 nm (violet light). Visible light is transmitted fairly well through the atmosphere at all wavelengths. Scattering increases toward the short-wavelength end of the spectrum (blue and violet). For free-space laser communications, red is the preferred visible-light color. Helium-neon lasers produce red light, and are available at reasonable prices for experimenters. Rain, snow, fog, and dust interfere with the transmission of visible light through the air. A free-space laser communications link can be rendered useless in a storm.

Ultraviolet (UV) at the longer wavelengths can penetrate the air fairly well, although some scattering takes place. Ozone pollution increases the loss. At shorter wavelengths, attenuation increases. Rain, snow, fog, and dust interfere with UV propagation in the same way as they interfere with visible light.

X rays and gamma rays do not propagate well for long distances through the air. This is because the sheer mass of the air, over paths of any appreciable distance, is sufficient to block these types of radiation. In addition to this problem, effective transmitters are almost impossible to construct. The radiation at these wavelengths is dangerous because it ionizes living tissue. It is doubtful that X rays and gamma rays will ever be routinely employed for communications purposes, although they could conceivably be used in some short-range wireless control systems in outer space.

Electromagnetic Interference

Electromagnetic interference is a phenomenon in which electronic devices upset each other's operation. In recent years, this problem has been getting worse because consumer electronic devices are proliferating, and they have become more susceptible to EMI.

Computers, Hi-fi, and TV

Much EMI results from inferior equipment design. To some extent, faulty installation also contributes to the problem. A *personal computer*

(PC) produces wideband RF energy in the form of an EM field. Most of this energy comes from the cathode-ray-tube (CRT) monitor. The digital pulses in the main unit can also cause problems in some cases. The energy gets out of the computer via the interconnecting cables and power cords, because they act as miniature transmitting antennas unless *electromagnetic shielding* is employed. Figure 1-7A shows RF energy (zigzags) coming from computer interconnecting cables.

Computers, TV receivers, and hi-fi sound equipment can malfunction because of strong RF fields, such as those from a nearby radio or TV transmitter. This can, and often does, happen when the transmitter is working perfectly. In these cases, and also in cases involving cellular telephones, citizens band (CB) radios and amateur ("ham") radios, the transmitting equipment is usually not at fault. The problem is generally the result of improper or ineffective shielding in some part of the home-entertainment system or computer.

In a computer system, cables and cords can act as receiving antennas (Fig. 1-7B), thereby letting RF energy into the machine. In hi-fi sound equipment, RF can get in via the speaker wires, the power cord, the tuner antenna and feed line, and cables between an amplifier and externals such as a compact-disk (CD) player or tape deck (C). In TV receiving installations, RF can enter through the power cord, the cable system, a satellite antenna system, and cords between the TV set and peripheral equipment such as a videotape recorder (D).

In general, the more interconnecting cables there are in a home entertainment system, the greater the chance for EMI from an RF field of a given intensity and frequency. Also, as the interconnecting cables are made longer, the likelihood of EMI increases. It is always wise to use as few connecting cords as possible, and to keep them as short as possible. If there is excess cord or cable and you do not want to cut it shorter, coil it up and tape it in place. A good RF ground is important for hi-fi and TV systems; electrical grounding is imperative for all equipment. Proper grounding can help minimize the chance for EMI taking place.

In Amateur Radio

If you are an amateur radio operator or shortwave radio enthusiast, EMI will especially affect you. You will be more likely than the average person to experience EMI-related radio reception problems at some time or another. If you are a radio ham with a sophisticated station, you might

Figure 1-7
Electromagnetic inter-
ference: emitted by
a computer (A),
received by a com-
puter (B), received by
a hi-fi system (C), and
received by a home
TV system (D).

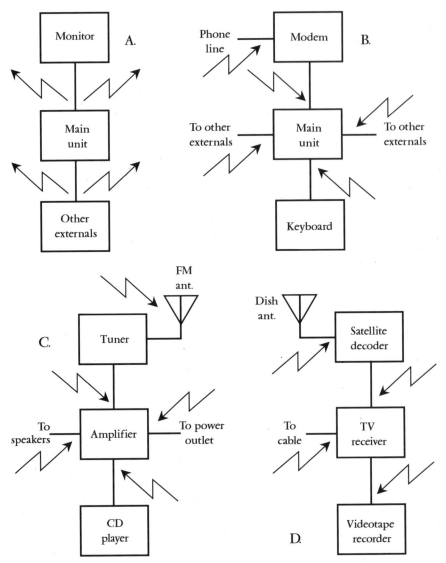

be blamed for interference to home entertainment equipment, whether
it is technically your fault or not.

Amateur-radio-induced EMI, also often called *radio-frequency interference (RFI)*, can sometimes be partly the fault of the transmitter and/or
radio operator. Radio amateurs should always use the minimum amount
of power necessary for the desired communication. Transmitter systems
should be properly aligned to ensure that they do not radiate excessive
harmonics, parasitic oscillations, or *spurious signals.* Antenna systems should

be installed in such a way that they radiate as little energy as possible into nearby homes and other buildings. These are things that radio amateurs can control, and they can sometimes reduce or eliminate RFI in mild cases.

Unfortunately, many RFI cases are caused by inadequate RF shielding of home entertainment equipment. Radio hams can still sometimes solve such a problem at their station by reducing power or by switching to another frequency band. Diplomacy, and a willingness to compromise, can go a long way toward securing the cooperation of neighbors. There's no sense making enemies because of faulty design or installation of electronic apparatus.

It is unwise to make any modifications to a neighbor's equipment yourself. The safest course of action is to contact the manufacturer of the equipment, or recommend that the neighbor contact a professional engineer.

EMI to Radio Receivers

Radio receivers, especially at low, medium, and high frequencies, often experience EMI from common appliances. Vacuum cleaners, hair driers, and TV sets are well-known culprits. The manufacturers of these appliances generally don't worry about whether their products will interfere with radio reception. Including filters to minimize such problems, such as capacitors and/or chokes in the AC line cord, would raise the cost of the appliances without providing a tangible return. Consumers would be more likely to complain about "that brick in the power cord" of a hair drier than appreciate the fact that it might allow them to receive shortwave radio signals more clearly.

Utility lines can radiate EM energy. This can cause EMI to sensitive equipment. Fortunately, these fields are rarely strong enough to interfere with consumer electronic systems directly, although they can cause trouble for shortwave radio listeners and amateur radio operators. This type of interference is caused by electric sparks. A malfunctioning transformer, a bad street light, or a salt-encrusted insulator can be responsible. Help can often be obtained by calling the utility company. Sometimes whole neighborhoods are affected. This can give additional clout to requests for help from utility companies. However, cooperation is not always immediately forthcoming. There are nightmare stories, told mostly by amateur-radio operators, about power-line EMI cases whose resolutions were immensely complicated.

Internal-combustion engines occasionally cause EMI to amateur radio equipment and shortwave receivers. This usually represents a less severe problem than power-line EMI because of its intermittent or infrequent nature. But in some locations, it can cause ongoing trouble. Utility-line filters won't get rid of noise from these sources; the only remedy in such cases is to alter the receiving antenna system and/or employ a *noise blanker*. A receiver with a well-designed noise blanker can reduce the effects of electrical and motor-generated EMI, especially when caused by devices that emit periodic EM "spikes," such as gasoline-powered engines. Directional antennas for receiving, such as small loops, can also work. Sometimes it is possible to find the source of the interference and modify or eliminate it.

In a Ham-Radio Shack

Personal computers have become increasingly common in households throughout the United States, and many radio hams now own them. It is possible for the circuitry of the computer to generate signals that interfere with ham radio or shortwave reception. Amateur transmissions can also sometimes upset the working of the computer.

Transient suppressors in the power lines to a computer and its peripherals are essential for any PC workstation. Bypassing the power lines to each individual piece of equipment, and/or installing RF chokes in series with all power-carrying wires, might help solve in-the-shack EMI problems. Keep the computer away from antenna tuners and feed lines. Be sure the station equipment is well grounded and that there are no ground loops. You might contact the manufacturer of the computer.

Sometimes, a radio transmitter's antenna feeders will radiate enough so that RF is present in the shack, and this can cause keyers and other peripheral equipment to malfunction. This can be an especially persistent and tricky problem. Good antenna balance, proper grounding and shielding, and sometimes a specific choice of feed-line length are necessary to minimize this. Reducing power can often prove a temporary solution until an electronic means is found to correct this type of EMI.

Other Examples

As the number and variety of consumer electronic devices expands, EMI problems are almost certain to become more frequent and troublesome. Here are some more examples of RFI that can originate in radio and wireless transmitting equipment.

Telephone interference can be especially bothersome for radio amateurs with high-power single-sideband stations. Filtering devices are available for telephone sets, but they do not always work. It is to the consumer's advantage to purchase a well-engineered telephone set. A bad phone line can sometimes cause an RFI problem. The telephone company is responsible for the upkeep of their lines, but not usually for telephone sets.

Unwanted mixing of radio signals can occur in the most unsuspected places. This is called *cross modulation*. It produces a kind of RFI that can be exceedingly difficult to track down and correct. Poor electrical connections in house wiring, plumbing, and even in exterior structures such as fences and rain gutters can generate harmonics and image signals affecting radio reception and producing "phantom" signals from a nearby transmitter.

Intermodulation is common in the downtown areas of large cities, where many radio transmitters are in operation simultaneously. It causes false signals in radio receivers, often sounding garbled or broken up. In the worst cases, FM receivers can be blocked by this interference.

Interference to television receivers is called *television interference (TVI)*. It, and hi-fi FM stereo interference, is sometimes caused by excessive harmonic emission from a ham transmitter. You should be certain that your transmitter's harmonic emissions are within legal limits, but this is not always sufficient to prevent TVI and FM stereo interference. Low-pass filters and antenna tuners can provide additional attenuation of harmonics. Linear amplifiers should not be overdriven or underloaded.

Radio transmitters sometimes malfunction in such a way that signals are transmitted on unintended (and often unknown) frequencies. If you suspect such a malfunction, write or call the manufacturer of the transmitting equipment. Technicians are usually cooperative concerning these types of problems.

Electromagnetic fields can upset the workings of certain medical electronic devices. If amateur radio operation causes such problems in any situation, the amateur must immediately cease operation, and not resume operation until a definite resolution is reached. For other classes of radio and wireless operation, the situation might be more complicated from a legal standpoint, but it is hard to imagine how any sort of radio transmission could take precedence over a human being's life.

Cures

There are methods of keeping RF from getting out of, or into, home entertainment equipment. A *choke* is a coil that blocks RF currents; a *bypass* is a capacitor that shorts RF currents to ground. They are installed in cords, cables, and/or cable connectors. You must be sure these devices do not interfere with the transmission of power, signals, or data through cables. For advice, consult the dealer or manufacturer of the equipment in question, or a competent engineer. If you try to install choke or bypass devices yourself, they might not work properly. You will also run the risk of voiding the warranty on the equipment.

Transient suppression, also called *surge suppression,* in the power cord is essential for reliable operation of a computer. A line filter, consisting of capacitors between each side of the power line and ground, can help prevent RF from getting into equipment via the utility lines. Some transient suppressors act as line filters; some do not.

Devices that react unfavorably to amateur radio transmissions are also frequently affected by commercial and government transmissions. As more and more transmitters are built, manufacturers will perhaps feel increasing pressure to provide EMI protection in electronic equipment. However, the problem of EMI will probably always exist to some extent, no matter how much the manufacturers might cooperate.

Passive Electronic Components

In radio and wireless systems, *passive components* are those that do not require an external source of power for their operation. The most common of these are *resistors, capacitors, inductors,* and most *diodes.*

Resistors

A *resistor* is a component deliberately placed in a circuit to reduce, control, or limit the flow of current. This property is called *resistance* and is symbolized by the letter *R.* The standard unit of resistance is the *ohm.*

Uses

Resistors play various roles in electrical and electronic equipment. Here are a few of the ways resistors are used.

Biasing In order to work efficiently, transistors or vacuum tubes require the correct *bias.* This means that the control electrode—the base, gate, or grid—must carry a certain voltage or current. Networks of resistors accomplish this. Different bias levels are needed for different types of circuits. A radio-transmitting amplifier will usually be biased differently than an oscillator or a low-level receiving amplifier.

Current Limiting Resistors interfere with the flow of electrons in a circuit. Sometimes this is essential to prevent damage to components. A good example is in a receiving amplifier. A resistor can keep the transistor from dissipating undue power as heat.

Power Dissipation A resistor can sometimes be used as a "dummy" component, so that a circuit "sees" the resistor as if it were something more complicated. In radio, for example, a resistor can be used to take the place of an antenna. A transmitter can then be tested in such a way that it does not interfere with communications on the airwaves.

Bleeding Off Charge In a high-voltage, direct-current (DC) power supply, capacitors are used to smooth out the fluctuations in the output. In high-voltage power supplies, these capacitors hold the full output voltage (750 volts, for example) even after the supply has been turned off, and even after it is unplugged from the wall outlet. Resistors, connected across the filter capacitors, drain the stored charge from the capacitors, so servicing the supply is not dangerous. (It is nevertheless a good idea to short out all filter capacitors, using a screwdriver with an insulated handle, before working on a high-voltage DC power supply.)

Impedance Matching   A more subtle use for resistors is in the coupling in a chain of amplifiers, or in the input and output circuits of amplifiers. In order to produce the greatest possible amplification, the impedances must agree between the output of a given amplifier and the input of the next. The same is true between a source of signal and the input of an amplifier. Also, this applies between an amplifier output and the load.

Attenuators   An *attenuator* is a network, usually constructed from resistors, designed to cause a specific reduction in the amplitude of a signal. Such circuits are useful for sensitivity measurements and other calibration purposes. In a well-designed attenuator, the amount of attenuation is constant over the entire range of frequencies in the system to be checked. Most attenuator circuits will function from the audio-frequency (AF) range up to several hundred megahertz. At higher frequencies, most resistors exhibit *inductive reactance* because the wavelength is so short that the leads are long electrically. Attenuators for ultra-high and microwave radio-frequency (RF) applications must be designed especially to suit the short wavelengths.

Types and Characteristics

Resistors can be manufactured in many ways. The best type of resistor for a given application depends on how much current it must carry, the frequency (if the current is alternating or contains a signal component), and the extent to which variations in resistance can be tolerated.

Carbon-Composition Resistor   One method of making a resistor is to mix up finely powdered carbon (a fair electrical conductor) with some nonconductive substance, press the resulting claylike stuff into a cylindrical shape, and insert wire leads in the ends. The resistance depends on the ratio of carbon to the nonconducting material, and on the physical distance between the wire leads. The nonconductive material is usually phenolic, similar to plastic. This results in a *carbon-composition* resistor. This kind of resistor is essentially nonreactive: it introduces almost pure resistance into a circuit, with minimal capacitance or inductance. Such resistors can be made to have almost any value, from very low to very high. This makes carbon-composition resistors useful in radio receivers and transmitters.

Wirewound Resistor   Another way to get resistance is to use a length of wire that is not a good conductor. The wire can be wound around a cylindrical form, like a coil. The resistance is determined by how well the wire metal conducts, by its diameter (gauge), and by its length. An asset of *wirewound resistors* is that they can be made to have values within a very close tolerance; that is, they are precision components. Another advantage is that wirewound resistors can be made to handle large amounts of power. A disadvantage of wirewound resistors, in some applications, is that they act like inductors. This makes them unsuitable for use in most RF circuits. Wirewound resistors typically have low to moderate values of resistance.

Film Resistor   A mixture of ceramic, carbon, and/or metal can be applied to a cylindrical form in a thin layer to obtain a desired value of resistance. This type of resistor is called a *carbon-film resistor* or *metal-film resistor*. The cylindrical form is made of an insulating substance, such as porcelain. The film can be deposited on this form by various methods, and the value tailored as desired. Metal-film units can be made to have nearly exact values. Film-type resistors usually have low to medium-high resistance. A major advantage of film-type resistors is that they, like carbon-composition units, do not have much inductance or capacitance. A disadvantage, in some applications, is that they cannot handle as much power as the more massive carbon-composition units, or as wirewound types.

IC Resistor   Increasingly, entire electronic circuits are being fabricated on semiconductor wafers known as *integrated circuits (ICs)*. Resistors can be etched onto the semiconductor chip that makes up an IC. The thickness, and the types and concentrations of impurities added, determine the resistance of the component. Integrated-circuit resistors can only handle a tiny amount of power because of their small size. But because IC circuits in general are designed to consume minimal power, this is not a problem. The small signals produced by ICs can be amplified, using circuits made from discrete components, if it is necessary to obtain higher signal power.

Rheostat   A variable resistor can be made from a wirewound element. This is called a *rheostat*. A rheostat can have either a rotary control or a sliding control. This depends on whether the wire is wound around a doughnut-shaped form (toroid) or a cylindrical form (solenoid). Rheostats always have inductance as well as resistance. They share the advantages and disadvantages of fixed wirewound resistors. Rheostats are not continuously adjustable because the movable contact slides along from turn to turn of the wire coil. The smallest possible increment is the resistance in one turn of the coil.

Potentiometer   A *potentiometer* is a form of variable resistor that uses a strip of resistive material rather than a coil of wire, across which a contact slides. Potentiometers are commonly used in electronic equipment for volume and tone adjustment, cathode-ray-tube controls, and other purposes. They are low-power devices.

Standard Values   In theory, a resistor can have any value, from the lowest possible (such as a shaft of solid silver) to the highest (open air). In practice, it is unusual to find resistors with values less than about 0.1 ohm, or more than about 100,000,000 ohms. Resistors are manufactured in multiples of 1.0, 1.2, 1.5, 1.8, 2.2, 2.7, 3.3, 3.9, 4.7, 5.6, 6.8, and 8.2. Units are commonly made with values derived from these values, multiplied by some power of 10. Thus you will see units of 47 ohms, 180 ohms, 6.8 kilohms (K), or 18 megohms (M), but not 380 ohms, 650 K, or 215 M. In addition to the above values, there are others that are used for resistors made with greater precision, or closer tolerance. These are power-of-10 multiples of 1.1, 1.3, 1.6, 2.0, 2.4, 3.0, 3.6, 4.3, 5.1, 6.2, 7.5, and 9.1.

Tolerance   The first set of numbers above represents standard resistance values available in tolerances of plus or minus 10 percent. The second set, along with the first set, of numbers represents standard resistance values available in tolerances of plus or minus 5 percent. Some resistors are available in tolerances tighter (less) than 5 percent. These precision units are employed in circuits where a little error can make a big difference. In most AF and RF oscillators and amplifiers, 10 or 5 percent tolerance is good enough.

Power Rating   All resistors are given a specification that determines how much power they can safely dissipate. Typical values are 1/4 watt (W) and 1/2 W. Units also exist with ratings of 1/8 W, 1 W, or 2 W. These dissipation ratings are for continuous duty. You can figure out how much current a given resistor can handle by using the formula for power (*P*) in terms of current (*I*) and resistance (*R*): $P = I^2 R$. Just work this formula backward, plugging in the power rating for *P* and the resistance of the unit for *R*, and solve for *I*. Or you can find the square root of *P/R*. Remember to always use amperes for current, ohms for resistance, and watts for power.

Temperature Compensation   All resistors change value somewhat when the temperature changes dramatically. Because resistors dissipate power, they can get hot as a result of the current they carry. There are various ways to approach problems of resistors changing value when they get hot. One method is to use specially manufactured resistors that do not appreciably change value when they heat up. Such units are called *temperature-compensated resistors*. Another approach is to use a power rating that is much higher than the actual dissipated power in the resistor. This will keep the resistor from getting very hot. Still another scheme is to use a series-parallel network of identical resistors to increase the power dissipation rating.

Color Coding

Many resistors have color bands that indicate their values and tolerances. You will usually see three, four, or five bands around carbon-composition resistors and film resistors. Other units are large enough so the values can be printed on them in numerals. Table 2-1 shows the numbers and multipliers of 10 that are assigned to the colors on resistors that have color bands. Figure 2-1 shows the arrangement of bands on a typical resistor.

Suppose you find a resistor whose first three bands are yellow, violet, and red, in that order. Then the resistance is 4700 ohms (4.7 K). Read as follows: yellow=4, violet=7, red=×100. As another example, suppose you reach into a bag and pull out a resistor with bands of blue, gray, orange. Referring to the table, determine: blue=6, gray=8, orange=×1000. Therefore, the value is 68,000 ohms (68 K).

The fourth band, if there is one, indicates tolerance. If it is silver, the resistor is rated at plus or minus 10 percent. If it is gold, the resistor is rated at plus or minus 5 percent. If there is no fourth band, the resistor is rated at plus or minus 20 percent.

The fifth band, if there is one, indicates the percentage that the value might change in 1000 hours of use. A brown band indicates a maximum change of 1 percent of the rated value. A red band indicates 0.1 percent; an orange band indicates 0.01 percent; a yellow band indicates 0.001 per-

TABLE 2-1

Resistor Color Code

Color of band	Numeral (bands 1 and 2)	Multiplier (band 3)
Black	0	1
Brown	1	10
Red	2	100
Orange	3	1 K
Yellow	4	10 K
Green	5	100 K
Blue	6	1 M
Violet	7	10 M
Gray	8	100 M
White	9	1000 M

See text for discussion of bands 4 and 5.

Figure 2-1
Placement of color-code bands on a typical resistor.

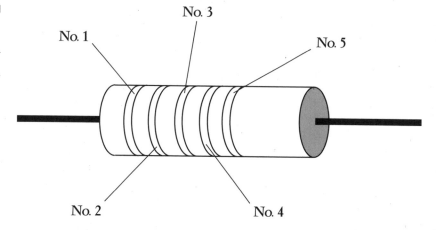

No. 3

No. 1

No. 5

No. 2

No. 4

cent. If there is no fifth band, it means that the resistor might deviate by more than 1 percent of the rated value after 1000 hours of use.

It is good engineering practice to test a resistor with an ohmmeter before installing it. If the unit happens to be labeled wrong, it is easy to identify and correct the error before placing the component in a circuit. If a circuit is assembled and it will not work because one of its resistors is mislabeled (this can and does happen), it is exceedingly difficult to find the problem, because it is not generally possible to accurately measure the value of a resistor when it is installed in a circuit.

Capacitors

Capacitance, symbolized by the letter C, impedes the flow of alternating-current (AC) charge carriers by temporarily storing the energy as an electric field. A *capacitor* is an electronic component deliberately manufactured to have a specific amount of capacitance. The standard unit of capacitance is the *farad,* abbreviated F. This is a large unit. More common units are the *microfarad* (10^{-6} F), abbreviated μF, and the *picofarad* (10^{-12} F), abbreviated pF.

Principles and Uses

Imagine two flat, parallel sheets of metal such as copper or aluminum, both of which are excellent electrical conductors. If these plates are connected to the terminals of a battery, they will become charged electrically, one positively and the other negatively. If the plates are small, they will become charged almost instantly, attaining a relative voltage equal to the voltage of the battery. If the plates are gigantic, it will take a while for the negative one to "fill up" with electrons, and it will take an equal amount of time for the other one to get electrons "drained out." But however large the plates might be, the voltage between them will eventually become equal to the battery voltage, and an electric field will exist in the space between them. Energy will be stored in this electric field. The amount of energy that can be stored depends on the surface area of the plates, the separation between them, and the type of material that is placed between them. The greater the amount of energy that can be stored, the larger is the value of the capacitance.

Blocking   When a capacitor is used for the purpose of preventing the flow of DC but allowing the passage of AC, the capacitor is called a *blocking capacitor.* Blocking capacitors facilitate the application of different DC bias voltages to two points in a circuit. Such a situation is common in multistage amplifiers (Fig. 2-2). Blocking capacitors are usually of fixed value, and should be selected so that attenuation does not occur at any point in the operating-frequency range. Generally, the blocking capacitor should have a value that allows ample transfer of signal at the lowest operating frequency, but the value should not be larger than the minimum to accomplish this. At very high frequencies (VHF), the value of a blocking capacitor in a high-impedance circuit might be less than

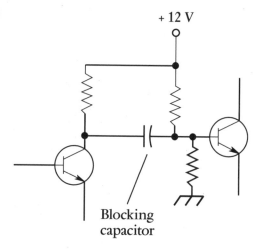

Figure 2-2
Example of a block-
ing capacitor
between two amplifi-
er stages.

+ 12 V

Blocking
capacitor

1 pF. At audio frequencies in low-impedance circuits, values can range up to about 100 μF. Blocking capacitors are used in the feedback circuits of some oscillators. The value of the feedback capacitor in an oscillator should be the smallest that will allow stable operation.

Bypassing   A capacitor that provides a low impedance for a signal, while allowing DC bias, is called a *bypass capacitor.* Bypass capacitors are frequently used in the emitter circuits of transistor amplifiers to provide stabilization. They are also used in the source circuits of field-effect-transistor amplifiers, and in the cathode circuits of tube-type power amplifiers. The bypass capacitor places a circuit element at signal ground potential, although the DC potential can be several hundred volts. An example of the use of a bypass capacitor is shown in Fig. 2-3. A bypass capacitor is so named because the AC signal is provided with a low-impedance path around a high-impedance element. A bypass capacitor is different from a blocking capacitor; a bypass capacitor is usually connected in such a manner that it shorts the signal to ground, while a blocking capacitor is intended to conduct the signal from one part of a circuit to another while isolating the DC potentials.

Filtering   A *filter capacitor* is used to smooth out the ripples in the output of a rectifier circuit. Such a capacitor might have a relatively small value, such as in an RF automatic-gain-control, or its value might be very large, as in a low-voltage, high-current power supply. Filter capacitors are often used in conjunction with other components such as inductors and resistors.

Figure 2-3
Example of a bypass
capacitor in a bipolar-
transistor circuit.

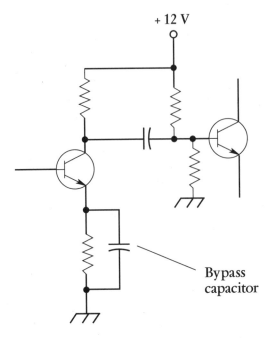

Bypass
capacitor

The filter capacitor operates because it holds the charge from the output-voltage peaks of a rectified signal or AC source. The smaller the load resistance, the greater the amount of capacitance required in order to make this happen to a sufficient extent. Filter capacitors in high-voltage power supplies can hold their charge even after the equipment has been shut off. Resistors of a fairly large value should be placed in parallel with the filter capacitors in such a supply, so that the shock hazard is reduced. Before servicing any high-voltage equipment, the filter capacitors should be discharged to chassis ground with a shorting stick.

Tuning   Capacitors are widely used, in combination with inductors and/or resistors, to control frequency or frequency-response characteristics of a device. Some capacitors have adjustable values that facilitate changing the *resonant frequency* of a tuned circuit.

Types and Characteristics

In a capacitor, the electric-field concentration is multiplied when a *dielectric* of a certain type is placed between the plates. Plastics work well for this purpose. This increases the effective surface area of the plates, so a physically small component can be made to have large capacitance.

In general, capacitance is directly proportional to the surface area of the conducting plates or sheets. Capacitance is inversely proportional to the separation between conducting sheets; in other words, the closer the sheets are to each other, the greater the capacitance. The capacitance also depends on the *dielectric constant* of the material between the plates. A vacuum has a dielectric constant of 1; some substances have dielectric constants that multiply the effective capacitance many times.

Paper Capacitors In the early years of electronics technology, capacitors were commonly made by placing paper, soaked with mineral oil, between two strips of foil. The assembly was rolled up, and wire leads were attached to the two pieces of foil. Finally, the rolled-up foil and paper were enclosed in a cylindrical case. These components are known as *paper capacitors*. They can still sometimes be found in electronic equipment. They have values ranging from about 0.001 to 0.1 μF, and can handle low to moderate voltages, usually up to about 1000 volts.

Mica Capacitors Perhaps you have seen *mica*, a naturally occurring, transparent substance that flakes off in thin sheets. This material makes an excellent dielectric for capacitors. *Mica capacitors* can be made by alternately stacking metal sheets and layers of mica, or by applying silver ink to sheets of mica. The metal sheets are wired together into two meshed sets, forming the two terminals of the capacitor. Mica capacitors have low loss; that is, they waste very little power as heat, provided their voltage rating is not exceeded. Voltage ratings can be up to several thousand volts if thick sheets of mica are used. But mica capacitors are physically bulky in proportion to their capacitance. The main application for mica capacitors is in radio receivers and transmitters. Their capacitances are a little lower than those of paper capacitors, ranging from a few tens of picofarads up to about 0.05 μF.

Ceramic Capacitors Porcelain (ceramic) works well as a dielectric. Sheets of metal are stacked alternately with wafers of ceramic. This material, like mica, has low loss, and therefore allows for high efficiency. For low values of capacitance, only one layer of ceramic is needed, and two metal plates can be glued to a disk of porcelain, one on each side. This type of component is known as a *disk-ceramic capacitor*. Alternatively, a tube or cylinder of ceramic can be employed, and metal ink applied to the inside and outside of the tube. Such units are called *tubular capacitors.* Ceramic capacitors have values ranging from a few picofarads to about 0.5 µF. Their voltage ratings are comparable to those of paper capacitors.

Plastic-Film Capacitors Various synthetic substances make good dielectrics for the manufacture of *plastic-film capacitors.* Polyester, polyethylene, and polystyrene are commonly used. The method of manufacture is similar to that for paper capacitors when the plastic is flexible. Stacking methods can be used if the plastic is more rigid. The geometries can vary, and these capacitors are therefore found in several different shapes. Capacitance values for plastic-film units range from about 50 pF to several tens of microfarads. Most often they are in the range of 0.001 to 10 µF. Plastic-film capacitors are employed in the AF and RF ranges at low to moderate voltages. The efficiency is good, although not as high as that for mica-dielectric units.

Electrolytic Capacitors All the aforementioned capacitors provide relatively small values of capacitance. They are also *nonpolarized capacitors,* meaning that they can be connected in a circuit in either direction. An *electrolytic capacitor* provides considerably greater capacitance than any of the previously described types, but it must be connected in the proper direction in a circuit to function well. Electrolytics are *polarized capacitors.* Electrolytic capacitors are manufactured by rolling up aluminum foil strips separated by paper saturated with an electrolyte liquid. The electrolyte is a conducting solution. When a DC voltage is applied to the component, the aluminum oxidizes because of the electrolyte. The oxide layer is nonconducting, and forms the dielectric for the capacitor. The layer is extremely thin, and this results in a high capacitance per unit volume. Electrolytic capacitors can have values up to thousands of microfarads, and some units can handle thousands of volts. These capacitors are most often seen in AF circuits and in DC power supplies.

Tantalum Capacitors Another type of electrolytic capacitor uses tantalum rather than aluminum. The tantalum can be foil, as is the aluminum in a conventional electrolytic capacitor. It might also take the form of a porous pellet, the irregular surface of which provides a large area in a small volume. An extremely thin oxide layer forms on the tantalum. Tantalum capacitors have high reliability and excellent efficiency. They are used in military applications because they have a low failure rate. They can be used in AF and digital circuits in place of aluminum electrolytics. The main disadvantage of tantalum capacitors is that they are comparatively expensive, although if high reliability is needed, the extra cost can be a good investment.

Semiconductor Capacitors   Semiconducting materials can be employed to make capacitors. A semiconductor diode conducts current in one direction, and refuses to conduct in the other direction. When a voltage source is connected across a diode so that it does not conduct, the diode acts as a capacitor. The capacitance varies depending on how much *reverse voltage* is applied to the diode. The greater the reverse voltage, the smaller the capacitance. This makes the diode a *variable capacitor.* Some diodes are especially manufactured to serve this function. Their capacitances fluctuate rapidly along with pulsating DC. Such a component is called a *varactor diode* or *varactor.* Capacitors can be formed in the semiconductor materials of an IC in much the same way. Sometimes, IC diodes are fabricated to serve as varactors. Another way to make a capacitor in an IC is to sandwich an oxide layer into the semiconductor material, between two layers that conduct well. Semiconductor capacitors usually have small values of capacitance. They are physically tiny, and can handle only low voltages. The advantages are miniaturization, and an ability, in the case of the varactor, to change capacitance at a rapid rate (kilohertz or megahertz).

Air-Variable Capacitors   By connecting two sets of metal plates so that they mesh, and by affixing one set to a rotatable shaft, an *air-variable capacitor* is made. The rotatable set of plates is called the *rotor,* and the fixed set is called the *stator.* Such devices were once common in radio receivers for the purpose of frequency tuning. They are still used in transmitter output tuning networks. Air-variables have maximum capacitance that depends on the number of plates in each set, and also on the spacing between the plates. Common maximum values are 50 to 1000 pF; minimum values are a few picofarads. The voltage-handling capability depends on the spacing between the plates; some air-variables can handle many kilovolts. Air-variables are used primarily in RF circuits and systems. They are highly efficient, and are nonpolarized, although the rotor is usually connected to common ground (the chassis or circuit board).

Trimmer Capacitors   When it is necessary to change the value of a capacitor occasionally but not often, a *trimmer capacitor* can be used. It consists of two plates, mounted on a ceramic base and separated by a sheet of mica, ceramic, or some other dielectric. The plates are flexible, and can be squashed together more or less by means of a screw.

Sometimes two sets of several plates are interleaved to increase the capacitance. A trimmer can be connected in parallel with an air-variable capacitor, so that the range of the air-variable can be adjusted. Some air-variable capacitors have trimmers built in. Typical maximum values for trimmers range from a few picofarads up to about 200 pF. They handle low to moderate voltages, and are highly efficient. They are nonpolarized.

Coaxial Capacitors   A somewhat uncommon, but highly effective, capacitor uses two telescoping sections of tubing. This is called a *coaxial capacitor*. It works because there is a certain effective surface area between the inner and the outer tubing sections. A sleeve of plastic dielectric is placed between the sections of tubing. This allows the capacitance to be adjusted by sliding the inner section in or out of the outer section. Coaxial capacitors are especially useful in antenna systems for tuning and/or impedance matching. Their values are generally from a few pico-farads up to about 100 pF.

Tolerance and Temperature Coefficient

Capacitors are rated according to how nearly their values can be expected to match the rated capacitance. The most common tolerance is 10 percent; some capacitors are rated at 5 or even at 1 percent. In all cases, the tolerance ratings are plus-or-minus. Therefore, a 10 percent capacitor can range from 10 percent less than its assigned value to 10 percent more.

Some capacitors increase in value as the temperature increases. These components have a *positive temperature coefficient*. Some capacitors' values get smaller as the temperature rises; these have a *negative temperature coefficient*. Some capacitors are manufactured so that their values remain constant over a certain temperature range. Within this span of temperatures, such capacitors have a *zero temperature coefficient*. The temperature coefficient is specified in percent per degree Celsius.

Sometimes, a capacitor with a negative temperature coefficient can be connected in series or parallel with a capacitor having a positive temperature coefficient, and the two opposite effects will cancel each other out over a range of temperatures. In other instances, a capacitor with a positive or negative temperature coefficient can be used to cancel out the effect of temperature on other components in a circuit, such as inductors or resistors.

Inductors

An *inductor* is an electronic component that stores energy in the form of a magnetic field. This property is called *inductance* and is symbolized by the letter *L*. The standard unit of inductance is the *henry*, abbreviated H. This is a large unit. More common units are the *millihenry* (10^{-3} H), abbreviated mH, and the *microhenry* (10^{-6} H), abbreviated μH.

In Electronic Systems

Inductors are most often used to pass, or short-circuit, signals of low frequency, while blocking, or choking off, signals of higher frequencies. Inductors can also be used to adjust the resonant frequency of a circuit or system.

RF Chokes   An inductor can be used for the purpose of blocking RF signals while allowing lower-frequency and DC signals to pass. Such chokes are often used in electronic circuits when it is necessary to apply an AF or DC bias to a component without allowing RF to enter or leave. Radio-frequency chokes typically have inductances of several microhenrys to a few millihenrys. The exact value depends on the impedance of the circuit, and on the frequency of the signals to be choked. If the impedance of a circuit is *Z* ohms at the frequency to be choked, an RF choke is usually selected to have an impedance of approximately 10*Z* ohms. Radio-frequency chokes are commercially manufactured in a variety of configurations. Most have powdered-iron or ferrite cores and solenoidal windings. Some have air cores or toroidal windings. Figure 2-4 shows an example of an RF choke installed in the base circuit of a bipolar-transistor amplifier. The choke allows the base to be at DC ground without causing the high-frequency signal to be shorted out.

Line Filters   A *line filter* is a device that can be inserted in the AC cord for an appliance, device, or system. Line filters generally consist of series inductors and/or parallel capacitors. Figure 2-5 shows an example of a simple line filter. Line filters are available from various commercial sources. They typically have several outlets and are rated at 10 to 20 amperes for 117-volt service. A circuit breaker protects the equipment in the event of a severe transient (for example, the induced voltage from a nearby lightning stroke). The breaker also protects the components in the filter from damage in case a piece of equipment shorts out. A line

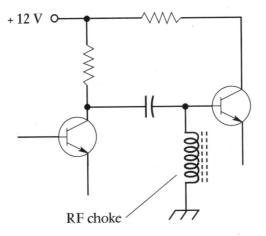

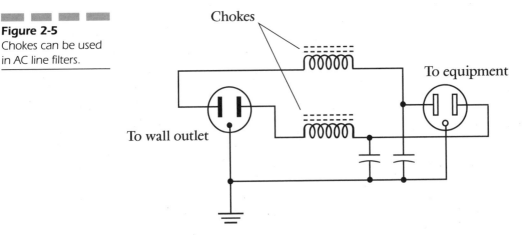

filter is useful for transient suppression in utility lines. Transient voltage "spikes" can cause some electronic equipment, particularly systems containing microcomputers, to malfunction. Line filters are also helpful in reducing *electromagnetic interference* (EMI) that is sometimes conducted along AC power lines. Installed between a radio transmitter and the utility lines, such a filter can choke off RF currents and help keep house wiring from acting as an antenna. Installed in the power cords of home entertainment equipment such as stereo audio amplifiers, line filters can keep RF from entering the apparatus through the power supply.

RF Transformers At RF, transformers are often used for the purpose of impedance matching. RF transformers are also used for coupling between or among amplifiers, mixers, and oscillators. A special form of

RF transformer, known as a *balun,* is sometimes used in antenna systems to match impedances and optimize the balance. All such transformers consist of inductances in various combinations. An RF transformer might consist of *solenoidal windings* with an air core, solenoidal windings with a powdered-iron or ferrite core, or *toroidal windings* with a powdered-iron or ferrite core. Air-core transformers are efficient, but they are rather bulky and are easily affected by nearby metallic objects. Solenoidal transformers with powdered-iron or ferrite cores are less bulky and almost as efficient as the air-core type. The solenoidal core can be moved in and out of the coil to vary the inductances of the windings. Toroidal transformers offer the advantage of being immune to the presence of nearby metallic objects, since all of the magnetic flux is contained within the powdered-iron or ferrite core. This facilitates miniaturization. When no reactance is present, the impedance-transfer ratio of an RF transformer is equal to the square of the turns ratio. Thus, a 2:1 turns ratio results in an impedance-transfer ratio of 4:1; a turns ratio of 1:5 produces a 1:25 impedance-transfer ratio.

AF Transformers   Inductances are used for the purpose of matching impedances at AF. The output of an audio amplifier might have an impedance of 200 ohms, and the speaker or headset an impedance of only 8 ohms. The transformer provides the proper termination for the amplifier. This assures the most efficient possible transfer of power. AF transformers are available in various power ratings and impedance-matching ratios. There are many kinds of AF transformers to meet various frequency-response requirements. Such transformers are physically similar to AC power transformers, except smaller. They are wound on laminated or powdered-iron cores.

Antenna Loading

In RF transmitting installations, it is often necessary to lengthen an antenna electrically, without making it physically longer. This is especially true at very low, low, and medium frequencies (VLF, LF, and MF). The deliberate altering of the resonant frequency of an antenna is known as *loading.* The resonant frequency can be lowered by means of one or more *loading inductors.*

In a quarter-wave resonant vertical antenna working against ground, an inductor can be connected in series with the element. This will reduce the resonant frequency of the antenna and make it possible to

Figure 2-6
An antenna loading inductor, often used with a capacitance hat, lowers the resonant frequency.

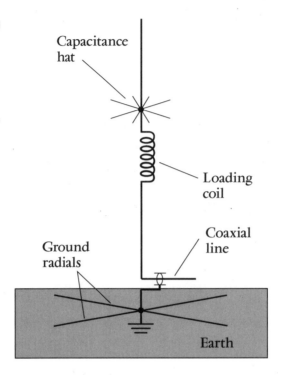

use an element that is physically shorter than 1/4 wavelength. The most common positions for the inductor are at the base (*base loading*) and near the center (*center loading*) of the radiating element. A *capacitance hat* can be added above the coil to lower the resonant frequency further (Fig. 2-6). This will also help to offset the narrowing of antenna bandwidth that always accompanies inductive loading.

For a given operating frequency, larger and larger inductances make it possible to bring shorter and shorter radiating elements to quarter-wave resonance. In theory, there is no limit to how low the resonant frequency of an inductively loaded vertical antenna can be made for a particular physical antenna height. Nor is there any theoretical limit to how short a radiator can be made for a given frequency. But in practice, losses become prohibitive when the physical height of the radiator is less than about 0.05 wavelength, regardless of how serious an attempt is made to optimize the ground system and minimize the coil and conductor losses. Although shorter antennas can work in the sense that they allow some communication, tiny loaded antennas always represent a compromise.

In a balanced antenna system such as a half-wave resonant dipole, identical coils can be placed in each half of the antenna in a symmetrical arrangement with respect to the feed point. For a given physical antenna length, larger and larger inductances result in lower and lower resonant frequencies. As with the quarter-wave vertical antenna, there is no theoretical limit to how low the resonant frequency of an inductively loaded half-wave dipole can be made for a particular physical antenna length. Nor is there any theoretical limit to how short a radiator can be made for a given frequency. But losses become prohibitive when the overall antenna length becomes less than about 0.1 wavelength.

Inductively loaded antennas always have low *radiation resistance*, because the free-space physical length of the radiator is shorter than the length of a full-size antenna. As the length of a radiating element becomes less than the above-mentioned minimum limits, the radiation resistance drops to near zero. Under these conditions it is difficult to get an antenna to adequately perform. There are, however, a few commercially manufactured downsize loop antennas that represent exceptions to this rule. These antennas employ low-loss elements and tuning/loading systems to obtain fairly efficient operation despite very small physical diameters. Further information about antenna systems can be found in Chap. 4.

Types and Characteristics

There are various types of inductors. Most take the form of wire coils. However, transmission lines and integrated circuits can sometimes function as inductors.

Air-Core Coils   The simplest inductors, besides straight or looped lengths of wire, are cylindrical coils. A coil can be wound on a piece of plastic, wood, or other nonferromagnetic material, and it will work well as a low-value inductor as long as it is kept reasonably clear of other electrically conducting objects. An air-core inductor cannot have very much inductance. In practice, the maximum attainable inductance for such coils is about 1 mH. Air-core coils are used mostly in RF transmitters, receivers, and antenna networks. In general, the higher the frequency of AC, the less inductance is needed to produce significant effects. Air-core coils can be made to have almost unlimited current-carrying capacity by using heavy-gauge wire and making the radius of the coil large. Air is almost lossless. For these reasons, air-core coils can be made highly efficient.

Ferromagnetic Cores   Ferromagnetic substances can be crushed into dust and then bound into various shapes, providing core materials that greatly increase the inductance of a coil having a given number of turns. Depending on the mixture used, the increase can range from a factor of a few times up to millions of times. A small coil can thus be made to have a large inductance. Powdered-iron-core inductors, like air-core coils, are common in RF circuits. Besides increasing the inductance compared with air-core coils, ferromagnetic cores tend to confine much of the magnetic flux within themselves. The main trouble with ferromagnetic cores is that, if the coil carries more than a certain amount of current, the core material will reach a state called *core saturation*. This means that the ferromagnetic material is holding as much flux as it can. Any further increase in coil current will not produce a corresponding increase in the magnetic flux in the core. The consequence, in practical terms, is that the inductance depends on the current in the coil. The inductance will decrease when the coil current exceeds the critical value. Core saturation also results in lower inductor efficiency. In extreme cases, ferromagnetic cores can waste considerable power as heat. If a core gets hot enough, it can fracture. This will permanently change the inductance of the coil, and will also reduce its current-handling ability.

Solenoidal Coils   Cylindrical (solenoidal) coils can be made to have variable inductance by sliding ferromagnetic cores in and out of them. This is a common practice in older radio receivers. The resonant frequency of an RF circuit can be adjusted in this way. Because moving the core in and out changes the effective permeability within a coil of wire, this method of tuning is called *permeability tuning*. The in/out motion can be precisely controlled by attaching the core to a screw shaft, and anchoring a nut at one end of the coil. As the screw shaft is rotated clockwise, the core enters the coil, the inductance increases, and the resonant frequency goes down. As the screw shaft is rotated counterclockwise, the core moves out of the coil, the inductance decreases, and the resonant frequency goes up.

Toroidal Cores   Inductor coils do not have to be wound on cylindrical forms or on cylindrical ferromagnetic cores. In recent years, a different coil geometry has become common. This is the *toroid-core inductor*, often called simply a *toroid*. It gets its name from the doughnut shape of the ferromagnetic core. Toroidal coils have advantages over their solenoidal counterparts. First, fewer turns of wire are needed to get a certain inductance with a toroid, as compared with a solenoid. Second, a toroid can be physically smaller for a given inductance and current-carrying capacity. Third, and perhaps most important, is the fact that, for practical purposes, all of the magnetic flux in a toroidal inductor is confined to the core material. This virtually prevents unwanted magnetic coupling to components near the toroid. There are some limitations inherent in toroidal coils. It is more difficult to permeability-tune a toroidal coil than a solenoidal one. Toroidal coils having many turns are difficult to wind compared with solenoidal coils. When *mutual inductance* is desired between coils, the respective coils must be wound on the same toroidal form. Toroids do not always work well in high-power RF applications. In antenna tuning networks, for example, the old-fashioned, air-core, solenoidal coils often give the best performance.

Pot Cores   There is another way to confine the magnetic flux in a coil so that unwanted mutual inductance does not occur. This is to wrap a ferromagnetic core completely around the outside of a solenoidal coil, making the core into a shell. This is known as a *pot-core inductor*. The core is manufactured in two halves, inside one of which the coil is wound. Then the parts are assembled and held together by a bolt and nut. The entire assembly looks like a miniature oil tank. The wires come out of the core through small holes. Pot cores have the same advantages as toroids. The core prevents the magnetic flux from extending outside the physical assembly. Inductance is greatly increased compared to solenoidal windings having a comparable number of turns. In fact, pot cores are even better than toroids if the main objective is to get an extremely large inductance within a small volume of space.

Transmission-Line Inductors   At radio frequencies of more than about 100 MHz, yet another type of inductor becomes practical. This is the *transmission-line inductor*. A transmission line is generally used to get energy from one place to another. In radio communications, transmission lines get energy from a transmitter to an antenna, and from an antenna to a receiver. A *parallel-wire transmission line* consists of two wires running alongside each other. The spacing between the wires is held constant by polyethylene rods molded at regular intervals to the wires, or by a web of polyethylene. A *coaxial transmission line* has a wire conductor surrounded by a tubular braid or pipe. The wire is kept at the center of this tubular shield by means of polyethylene beads or, more often, by solid or foamed polyethylene dielectric. Short lengths of any transmission line behave as inductors, as long as the line is less than 90 electrical degrees in length. At 100 megahertz (MHz), 90 electrical degrees, or a quarter wavelength, in free space is 75 centimeters (cm), or a little more than 2 feet (ft). In general, if f is the frequency in megahertz, then a quarter wavelength (s) in free space, in centimeters, is given by

$$s=7500/f$$

The length of a quarter-wavelength section of transmission line is shortened from the free-space quarter wavelength by the effects of the dielectric. In practice, a quarter wavelength along the line can be anywhere from about 66 percent of the free-space length (for coaxial lines with solid polyethylene dielectric) to 95 percent of the free-space length (for parallel-wire line with spacers molded at intervals of several inches). The factor by which the wavelength is shortened is called the *velocity factor* of the line. The shortening of the wavelength is a result of a reduction in the speed with which electromagnetic (EM) fields travel in the line, as compared with their speed in free space (the speed of light). If the velocity factor of a line is given by v, then the above formula for the length of a quarter-wave line, in centimeters, becomes

$$s=7500\,v/f$$

Very short lengths of line (a few electrical degrees) produce small values of inductance. As the length approaches a quarter wavelength, the inductance increases. The inductance of a transmission-line section changes as the frequency changes. At first, as the frequency rises from zero, the inductance becomes larger as the frequency increases. At a certain limiting frequency, the inductance becomes theoretically infinite. Above that frequency, the line becomes capacitive rather than inductive. As the frequency increases further still, the transmission line becomes inductive again, and then alternately capacitive and inductive, with resonance occurring at regular frequency intervals.

Impedance

Impedance is the opposition that a component or circuit offers to AC. Impedance is a two-dimensional quantity. That is, it consists of two independent components: *resistance* and *reactance*.

Opposition to Current

In a DC circuit, resistance is the opposition to current. It is measured in ohms, and is always zero or more. The larger the value of resistance, the greater the opposition to DC. In an AC circuit, resistance opposes the flow of current in the same way as for DC. But in AC circuits, there is another phenomenon, reactance, that also puts up opposition to the flow of current. Reactance can be either positive (inductive) or negative (capacitive). Resistance does not change with the frequency of the AC signal. But reactance does, if the circuit contains any reactive components (inductances and/or capacitances).

Engineers use a special notation to indicate reactance, and to differentiate it from resistance. While resistance can be delineated along a half line corresponding to the nonnegative real numbers, reactance needs an entire number line, including negative values as well as positive values and zero, to be depicted.

Reactance and resistance can vary independently. The resistive component of impedance is abbreviated R, and the reactance is indicated by jX. While R can be any nonnegative real number, X can be any real number whatsoever. The so-called j *operator* is equal to the positive square root of -1. The j operator is used because it creates a mathematical model that perfectly explains the behavior of complex impedances.

In DC circuits, resistances are indicated by points on the nonnegative number line. In AC circuits, a two-dimensional plane is used. Any impedance, denoted $R + jX$, corresponds to a unique point on this plane. Any point on the plane matches a unique complex impedance value.

Imprecision in Jargon

There will be times when you'll hear that the "impedance" of some device or component is a certain number of ohms, and no mention will be made of resistance or reactance. For example, in acoustic or ultrasonic wireless applications, you might come across "high-impedance" transducers or "low-impedance" amplifier outputs. You might read that a certain acoustic transducer has an impedance of "600 ohms." Figures like this generally refer to devices that have purely resistive impedances. Thus, the "600-ohm" transducer really has a complex impedance of $600 + j0$, and the "low-impedance" amplifier output means that transducers having impedances of, say, $8 + j0$ should be used. If you're not specifically given the complex impedance (that is, both the resistance and reactance values) when a single-number ohmic figure is quoted, it is best to assume that the engineers are talking about a *nonreactive impedance*. That means the impedance is a pure resistance, and that the imaginary, or reactive, factor is zero.

Engineers will sometimes speak of nonreactive impedances, or of complex impedance vectors, as "low-*Z*" or "high-*Z*." These are, of course, relative terms. An 8-ohm speaker would qualify as a low-*Z* device in most applications; a 600-ohm amplifier input would generally be considered high-*Z*. Often, impedance-matching networks are used to transfer low impedances to high ones, or vice versa, so that low-*Z* and high-*Z* circuits or devices will work well together. Impedance-matching networks are generally designed to work with circuits and devices that have no reactance. If there is reactance in a load, it must be eliminated for optimum system performance. Some impedance-matching networks contain circuitry that can eliminate reactance as well as match impedances.

Inductance-Capacitance Circuits

The term *inductance-capacitance (LC) circuit* refers to any of various combinations of inductors and capacitors. The most common LC circuits are *resonant circuits, impedance-matching networks,* and *selective filters.*

Series LC Circuit

When an inductor and capacitor are connected in series, they resonate at a specific frequency. Resonance occurs when the inductive and capacitive reactances are equal in magnitude, canceling each other and leaving a pure resistance. The resistive impedance of a series-resonant circuit is low; if the components had no loss, it would be zero. But because there is always some loss in real-life components, the resistance is usually a few ohms. Figure 2-7A illustrates a simple series-resonant LC combination in an unbalanced configuration.

At RF, a quarter-wave section of transmission line, open at the far end, behaves like a series-resonant circuit. A half-wave section, short-circuited at the far end, also has this property. Such circuits can be used in place of LC combinations, especially at very high frequencies (VHF) and ultra-high frequencies (UHF), where the physical lengths of such sections are reasonable.

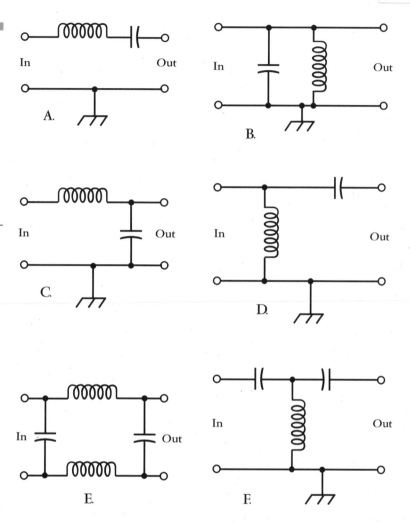

Figure 2-7
Inductance-capacitance (LC) circuits. Unbalanced series-resonant (A), unbalanced parallel-resonant (B), unbalanced low-pass (C), unbalanced high-pass (D), balanced low-pass (E), and unbalanced T-section high-pass (F).

Parallel LC Circuit

When an inductor and capacitor are connected in parallel, they resonate at a specific frequency. This happens when the inductive and capacitive reactances are equal, canceling each other and leaving a pure resistance. The resistive impedance of a parallel-resonant circuit is high. In theory, assuming zero loss in the inductor and capacitor, the resistance at resonance is infinite. In practical circuits it is finite because the components have some loss, but it can be many thousands of ohms. Figure 2-7B shows a parallel-resonant LC combination in an unbalanced configuration.

A section of transmission line can be used as a parallel-resonant circuit. A quarter-wave section, short-circuited at the far end, behaves this way. A half-wave section, open at the far end, also has this property. These devices can be used in place of LC circuits in the same applications, and at the same frequencies, as the series-resonant-equivalent sections described above.

Resonant Frequency

The frequency at which a simple series or parallel LC circuit will resonate depends on the values of the inductor and capacitor. A general formula can be applied when the net inductance L and the net capacitance C are known. If L is given in henrys and C is given in farads, then the resonant frequency f can be calculated as follows:

$$f=1/[6.28(LC)^{1/2}]$$

The above formula also applies for L in microhenrys, C in microfarads, and f in megahertz. Be careful not to mix units. If you plug in a capacitance in microfarads and an inductance in millihenrys, you cannot expect a meaningful result for the frequency.

Other LC Configurations

There are dozens of different ways that inductors and capacitors can be interconnected to obtain various effects. Some examples are shown in Fig. 2-7C through F.

Illustration C shows an LC combination that acts as a *low-pass filter* in an unbalanced circuit. Drawing D shows an LC combination that behaves as a *high-pass filter* in an unbalanced circuit. Drawing E is a low-pass LC filter that can be used in a balanced circuit. Illustration F is an unbalanced "T-section" LC high-pass filter. The configuration at F is often used with variable capacitors to match the load impedances of radio antenna systems to the output impedances of transmitters.

The *cutoff frequency* of a circuit such as those shown in drawings C through F will depend on the values of the inductances and capacitances. In general, the larger the net inductance and capacitance, the lower the cutoff frequency will be; conversely, small net inductances and capacitances result in high cutoff frequencies. When designing impedance-matching networks, optimum component values must be found by trial and error. This is why a *transmatch*, used in some radio antenna systems, almost always has a tapped or variable inductor.

More about Resonance

Resonance is a condition in which the frequency of an applied signal coincides with a natural response frequency of a circuit or object. In RF resonance, there are equal amounts of *capacitive reactance* and *inductive reactance*. Resonance always occurs at one or more specific frequencies.

RF Resonant Devices

In a parallel-tuned or series-tuned LC circuit, the reactances cancel each other at the resonant frequency. The inductive reactance is positive, and the capacitive reactance is negative; when their magnitudes are the same, the net reactance is zero. A parallel-tuned LC circuit has an extremely high, purely resistive impedance at the resonant frequency; the impedance diminishes as the frequency departs from resonance. In a series-tuned LC circuit, neglecting conductor losses, the impedance at resonance is extremely low, and rises as the frequency departs from resonance.

In RF applications, resonance can occur within a metal enclosure known as a *cavity*. Cavities are used as tuned circuits at ultra-high and microwave frequencies. A cavity exhibits resonance on a specific *fundamental frequency* and all its *harmonics* (whole-number multiples of the fundamental). A length of transmission line exhibits resonance on the frequency at which it measures 1/4 electrical wavelength, and on all harmonics.

Piezoelectric crystals exhibit resonant effects at RF. The resonant frequency of a crystal depends on the thickness of the crystal, the manner in which it is cut, and, in some cases, the temperature. The presence of a coil or capacitor in series or parallel with a crystal affects the resonant frequency to some extent.

Acoustic Resonance

In an enclosed chamber with walls that reflect sound, resonance can occur at certain wavelengths because the echoes combine in and out of phase. This causes increased sound volume in some places, and greatly reduced volume in other places. Usually this is an unwanted effect. In the design of auditoriums and music halls, acoustic resonance can be minimized by using baffles and acoustically absorbent materials, and by shaping the chamber in such a way that no two major inside surfaces (walls, ceiling, and floor) are parallel.

Speaker enclosures almost always have resonances at certain frequencies. This effect can be used to advantage when it is necessary to get good bass (low-frequency) response from a relatively small speaker. However, resonances in ill-designed speaker cabinets can cause "peaks" and "valleys" in the speaker output at certain frequencies, resulting in poor fidelity, especially in the reproduction of music.

Virtually all musical instruments, whether or not electronic components are used, make use of acoustic resonance effects. In a guitar, violin, cello, piano, or other string instrument, resonances occur in tightly stretched wires, and also within the enclosure, which is usually made of wood. In wind instruments, the effect occurs within a chamber whose dimensions can be varied at will. A wind instrument treats sound waves in much the same way as an RF resonant cavity treats EM fields.

Resonant Frequency

The fundamental resonant frequency of a tuned LC circuit depends on the values of the components. The larger the product of the inductance and capacitance, the lower the resonant frequency will be. The formula in the preceding section can be used to calculate the fundamental resonant frequency of an LC circuit.

A *resonance curve* is a graph of the response of a circuit or object to variations in the frequency of an applied sound or signal. Resonant curves are almost always plotted in *rectangular coordinates* (also known as the *cartesian plane*) with frequency as the independent variable. The dependent variable can be any characteristic that displays a peak or dip at the resonant frequency or frequencies. In RF circuits, such parameters include current, voltage, attenuation, gain, and impedance.

Examples of RF resonant curves are shown in Fig. 2-8. At A, the impedance of a hypothetical parallel-resonant LC circuit is plotted against the frequency. The resonant frequency is about 7 MHz. The impedance is theoretically infinite at resonance, but in a practical circuit, losses in the components and wiring result in a finite, but very large, impedance at resonance. At B, the impedance is plotted as a function of frequency for a series-resonant circuit. In this case the resonant frequency is also about 7 MHz. The impedance is minimum at resonance. In a lossless series-resonant circuit, the impedance would be zero, but in a practical circuit, losses in the components and wiring result in a very small, but nonzero, impedance.

Figure 2-8
Examples of parallel-resonant response (A) and series-resonant response (B).

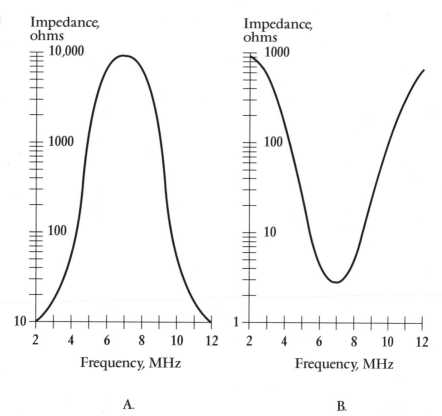

A.

B.

Consonance

When two objects are near each other but not in physical contact, and both have identical or harmonically related resonant frequencies, then one object can be affected by the other. This is called *consonance* and is important in acoustics, AF applications, and RF applications.

An example of *acoustic consonance* is shown by two tuning forks with identical fundamental frequencies. If one fork is struck and then brought near the other, the second fork will begin vibrating. If the second fork has a fundamental frequency that is a harmonic of the frequency of the first fork, the second fork will vibrate at its own resonant frequency.

Electromagnetic consonance takes place when two antenna elements, both having identical or nearly identical resonant frequencies, are in close proximity. If one antenna is fed with an RF signal at its resonant frequency, currents will be induced in the other antenna, and it, too, will radiate. Parasitic arrays, such as the Yagi antenna and the quad antenna, operate on this principle.

Selective Filters

The term *selective filter* refers to circuits designed to tailor the way an electronic circuit or system responds to signals at various frequencies. There are many kinds of selective filters. Some are used at AF; others are used at RF.

Bandpass Filter

Any resonant circuit, or combination of resonant circuits, designed to discriminate against all frequencies except a specific frequency f_0, or a band of frequencies between two limiting frequencies f_0 and f_1, is called a *bandpass filter*. In a parallel LC circuit, a bandpass filter shows a high impedance at the desired frequency or frequencies, and a low impedance at unwanted frequencies. In a series LC configuration, the filter has a low impedance at the desired frequency or frequencies, and a high impedance at unwanted frequencies. Figure 2-9A shows a simple parallel-tuned LC bandpass filter; Fig. 2-9B shows a simple series-tuned LC bandpass filter.

Some bandpass filters are built with components other than actual coils and capacitors, but all such filters operate on the same principle. The *crystal filter* uses piezoelectric materials, usually quartz, to obtain a bandpass response. A *mechanical filter* uses vibration resonances of certain substances, usually ceramics. In optics, a simple color filter, discriminating against all light wavelengths except within a certain range, is a form of bandpass filter.

Bandpass filters are sometimes designed to have very sharp, defined, resonant frequencies. Sometimes the resonance is spread out over a fairly wide range. The *attenuation-versus-frequency characteristic* of a bandpass filter is called the *bandpass response*. A bandpass filter can have a single, well-defined resonant frequency f_0, as shown in Fig. 2-9C, or the response might be more or less rectangular, having two well-defined limit frequencies f_0 and f_1, as shown at D. The bandwidth might be only a few hertz, such as with an audio filter designed for reception of Morse code. Or the bandwidth might be several megahertz, as in a helical filter

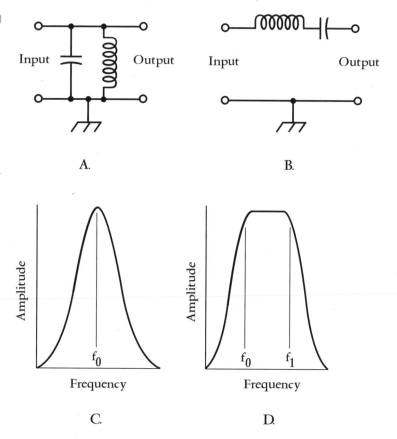

Figure 2-9
At A, elementary par-
allel-resonant band-
pass filter. At B,
elementary series-res-
onant bandpass filter.
At C, sharp bandpass
response. At D,
broad bandpass
response.

designed for the front end of a VHF or UHF radio receiver. A bandpass response is always characterized by high attenuation at all frequencies except within a particular range. The actual attenuation at desired frequencies is called the *insertion loss.*

Band-Rejection Filter

A *band-rejection filter,* also called a *band-stop filter,* is a resonant circuit designed to pass energy at all frequencies, except within a certain range. The attenuation is greatest at the resonant frequency f_0, or between two limiting frequencies f_0 and f_1. Figure 2-9E shows a simple parallel-resonant LC band-rejection filter; drawing F shows a simple series-resonant LC band-rejection filter. Note the similarity between band-rejection and bandpass filters. The fundamental difference is that

Figure 2-9 (cont.)
At E, elementary
parallel-resonant
band-rejection filter.
At F, elementary
series-resonant band-
rejection filter. At G,
sharp band-rejection
response. At H,
broad band-rejection
response.

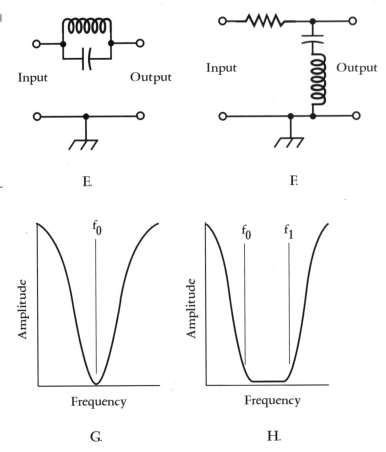

the band-rejection filter consists of parallel LC circuits connected in series with the signal path, or series LC circuits in parallel with the signal path; in bandpass filters, series-resonant circuits are connected in series, and parallel-resonance circuits in parallel.

Band-rejection filters need not necessarily be made up of coils and capacitors, but they often are. Quartz crystals are sometimes used as band-rejection filters. Lengths of transmission line, either short-circuited or open, are useful as band-rejection filters at the higher radio frequencies. A common example of a band-rejection filter is a *parasitic suppressor,* used in high-power RF amplifiers.

All band-rejection filters show an attenuation-versus-frequency characteristic marked by low loss at all frequencies except within a prescribed range. Figure 2-9G and H shows two types of band-rejection response. A sharp response (at G) occurs at or near a single resonant frequency f_0. A rectangular response is characterized by low attenuation below a limit f_0 and above a limit f_1, and high attenuation between these limiting frequencies.

Notch Filter

A *notch filter* is a narrowband-rejection filter. Notch filters are found in many radio communications receivers. The notch filter is extremely convenient for reducing interference caused by strong, unmodulated carriers within the passband of a receiver.

Notch-filter circuits are generally inserted in one of the intermediate-frequency (IF) stages of a superheterodyne receiver, where the bandpass frequency is constant. There are several different kinds of notch-filter circuit. One of the simplest is a trap configuration, inserted in series with the signal path (see Fig. 2-9E). The notch frequency is adjustable, so that the deep null can be tuned to any frequency within the receiver passband.

A properly designed notch filter can produce attenuation in excess of 40 decibels (dB) in the center of the notch. Some sophisticated types, especially AF designs, can provide 60 dB of attenuation at the notch frequency. Audio notch filters generally employ operational amplifiers with resistance-capacitance circuits. The frequency is adjusted by means of a potentiometer. In some AF notch filters, the notch width (sharpness) is adjustable.

High-Pass Filter

A *high-pass filter* is a combination of capacitance, inductance, and/or resistance, intended to produce large amounts of attenuation below a certain frequency and little or no attenuation above that frequency. The frequency at which the transition occurs is called the *cutoff frequency*. At the cutoff frequency, the power attenuation is 3 dB with respect to the minimum attenuation. Above the cutoff frequency, the power attenuation is less than 3 dB. Below the cutoff, the power attenuation is more than 3 dB.

The simplest high-pass filter consists of a parallel inductor or a series capacitor. Generally, high-pass filters have a combination of parallel inductors and series capacitors, such as the simple circuits shown in Fig. 2-9I and J. The filter at I is called an *L-section* high-pass filter;

Figure 2-9 (cont.)
At I, L-section high-
pass filter. At J,
T-section high-pass
filter. At K, high-pass
response.

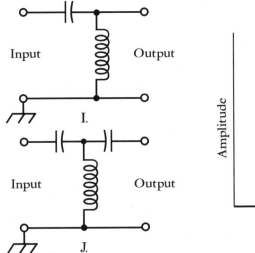

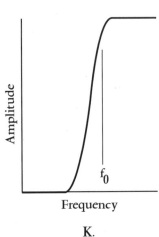

that at J is called a *T-section* high-pass filter. These names are derived from the geometric shapes of the filters as they appear in schematic diagrams.

Resistors are sometimes substituted for the inductors in a high-pass filter. This is especially true if active devices are used, in which case many filter sections can be cascaded.

High-pass filters are used in a wide variety of situations in electronic apparatus. One common use for the high-pass filter is at the input of a television (TV) receiver. The cutoff frequency of such a filter is about 40 MHz. The installation of such a filter reduces the susceptibility of the TV receiver to EMI from sources at lower frequencies.

A *high-pass response* is an attenuation-versus-frequency curve that shows greater attenuation at lower frequencies than at higher frequencies. The sharpness of the response can vary considerably. Usually, a high-pass response is characterized by a high degree of attenuation up to a certain frequency, where the attenuation rapidly decreases. Finally the attenuation levels off at near zero insertion loss. The cutoff frequency of a high-pass response is that frequency at which the insertion power loss is 3 dB with respect to the minimum loss. The *ultimate attenuation* is the level of power attenuation well below the cutoff frequency, where the signal is virtually blocked. A good high-pass response is shown in Fig. 2-9K. The curve is smooth, and the insertion loss is essentially zero everywhere well above the cutoff frequency.

Low-Pass Filter

A low-pass filter is a combination of capacitance, inductance, and/or resistance, intended to produce large amounts of attenuation above a certain frequency and little or no attenuation below that frequency. The frequency at which the transition occurs is called the cutoff frequency. At the cutoff frequency, the power attenuation is 3 dB with respect to the minimum attenuation. Below the cutoff frequency, the power attenuation is less than 3 dB. Above the cutoff, the power attenuation is more than 3 dB.

The simplest low-pass filter consists of a series inductor or a parallel capacitor. More sophisticated low-pass filters have combinations of series inductors and parallel capacitors, such as the examples shown in Fig. 2-9L and M. The filter at L is an L-section low-pass filter; the circuit at M is a pi-section low-pass filter. As above, these names are derived from the geometric arrangement of the components as they appear in diagrams.

Resistors are sometimes substituted for the inductors in a low-pass filter. This is especially true when active devices are used, in which case many filter stages can be cascaded. This substitution reduces the physical bulk of the circuit, and it saves money.

Low-pass filters are used in many different applications in RF electronics. One common use of a low-pass filter is at the output of a high-frequency (HF) transmitter. The cutoff frequency is about 40 MHz. When such a low-pass filter is installed in the transmission line between a transmitter and antenna, VHF harmonics are greatly attenuated. This

Figure 2-9 (cont.)
At L, L-section low-pass filter. At M, pi-section low-pass filter. At N, low-pass response.

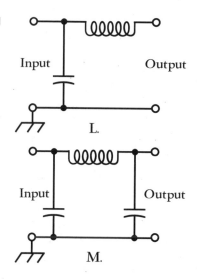

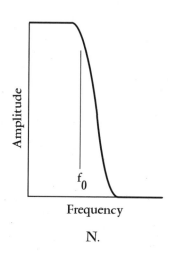

reduces the probability of EMI to TV receivers using outdoor antennas. In a narrowband transmitter, a low-pass filter might be built in for reduction of harmonic output.

A *low-pass response* is an attenuation-versus-frequency curve that shows greater attenuation at higher frequencies than at lower frequencies. The sharpness of the response can vary considerably. Usually, a low-pass response is characterized by a low degree of attenuation up to a certain frequency; above that point, the attenuation rapidly increases. Finally the attenuation levels off at a large value. Below the cutoff frequency, the attenuation is practically zero.

The *cutoff frequency* of a low-pass response is that frequency at which the insertion power loss is 3 dB with respect to the minimum loss. The *ultimate attenuation* is the level of attenuation well above the cutoff frequency, where the signal is virtually blocked. A good low-pass response looks like the attenuation-versus-frequency curve shown in Fig. 2-9N. The curve is smooth, and the insertion loss is essentially zero everywhere well below the cutoff frequency.

Diodes

The term *diode* means "two elements." Almost all diodes are made from silicon or other semiconducting materials.

Theory of Operation

When *P-type semiconductor* and *N-type semiconductor* materials are joined, a *P-N junction* is the result. Such a junction has properties that make semiconductor materials useful as electronic devices. A diode is formed by a single P-N junction. The N-type material comprises the *cathode*, and the P-type material forms the *anode*.

In a diode, electrons flow in the direction opposite the arrow in the schematic symbol. (Physicists consider current to flow from positive to negative, and this is in the same direction as the arrow points.) Current will not normally flow the other way unless the voltage is very high. If you connect a battery and a resistor in series with a P-N junction, current will flow if the negative terminal of the battery is connected to the N-type material (cathode) and the positive terminal is connected to the P-type material (anode). No current will flow if the battery is reversed.

When the cathode is negative with respect to the anode, the N-type material, possessing an excess of electrons to begin with, gets even more. The P-type semiconductor, with an innate shortage of electrons, is made even more deficient. The N-type material constantly feeds electrons to the P-type in an attempt to create an electron balance, and the battery or power supply keeps robbing electrons from the P-type material. This is known as *forward bias*.

It takes a certain minimum voltage for P-N-junction conduction to occur in the forward direction. This is called the *forward breakover voltage* of the junction. It is about 0.3 volt (V) for *germanium diodes* and 0.6 V for *silicon diodes*. If the voltage across the junction is less than the forward breakover voltage, the junction will not conduct. This effect can be of use in amplitude limiters, waveform clippers, and threshold detectors.

When the polarity is switched so the N-type material is positive with respect to the P-type, the situation is called *reverse bias*. Electrons in the N-type material are pulled toward the positive charge, away from the junction. In the P-type material, holes are pulled toward the negative charge, also away from the junction. The electrons (in the N-type material) and holes (in the P-type) are *majority charge carriers*. They become depleted in the vicinity of the P-N junction under reverse-bias conditions. A shortage of majority carriers means that the P-N junction cannot conduct well. The *depletion region* acts like an insulator. When this takes place, the junction resembles a capacitor. A specially manufactured diode called a *varactor* takes advantage of this effect, forming a voltage-variable capacitor when reverse-biased.

Avalanche Effect

When people think of semiconductor diodes, they usually think of rectification, where the device conducts current in one direction but not in the other. But under some conditions, diodes behave much differently than they do in a rectifier circuit.

If the reverse voltage becomes great enough, any diode will conduct. This is known as the *avalanche effect*. The term comes from the way the charge carriers behave in the semiconductor material. If the reverse electric field becomes strong enough, the electrons and holes "break loose" and flow freely across the P-N junction in the opposite direction from normal. The reverse current, which has been near zero up to this avalanche point, rises dramatically. The voltage required to cause the avalanche effect is called the *avalanche voltage*, and it varies among differ-

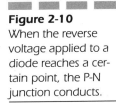

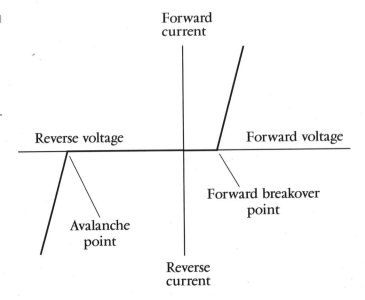

ent kinds of diodes. Figure 2-10 is a graph of the characteristic current-versus-voltage curve for a hypothetical semiconductor diode, showing the *avalanche point*.

Avalanche effect is undesirable in rectifiers, because it degrades the efficiency of a power supply. Avalanche effect can be prevented by choosing diodes with an avalanche voltage higher than the peak inverse voltage produced by the power supply circuit. For extremely high voltage power supplies, it is necessary to use series combinations of diodes to prevent avalanche effect.

A specialized type of diode, called a *zener diode*, takes advantage of avalanche effect. Zener diodes are specially manufactured to have precise avalanche voltages. They form the basis for *voltage regulation* in many power supplies. Zener diodes are available in various voltage and power ratings for a wide variety of voltage-regulator applications.

Types and Uses

Passive diodes are used in many electronic, radio, and wireless applications. The following are some of the most common functions that passive diodes perform.

Rectification The hallmark of a *rectifier diode* is that it passes current in only one direction. This makes it useful for changing AC to DC. In simplified terms, when the cathode (consisting of N-type semiconductor material) is negative with respect to the anode (made of P-type material), current flows; when the cathode is positive relative to the anode, current does not flow. Rectifier diodes are available in various sizes, intended for different purposes. Most rectifier diodes are silicon devices. A few are made from selenium.

Switching   The ability of diodes to conduct with forward bias, and to insulate with reverse bias, makes them useful for switching in some electronic applications. Diodes can switch at extremely high rates, much faster than any mechanical device. One type of diode, made for use as a switch at RF, has a special semiconductor layer sandwiched in between the P-type and N-type material. This layer, made of material called *intrinsic semiconductor,* reduces the capacitance of the diode, so that it can work at higher frequencies than an ordinary diode. The intrinsic material is sometimes called *I-type semiconductor.* A diode with an intrinsic layer is called a *PIN diode.* DC bias, applied to one or more PIN diodes, allows RF currents to be effectively channeled without using bulky, noisy relays and cables. A PIN diode also makes a good RF detector, especially at frequencies above 30 MHz.

Voltage Regulation   Most diodes have avalanche voltages much higher than the reverse bias ever gets. The value of the avalanche voltage depends on how a diode is manufactured. Zener diodes are made to have well-defined, constant avalanche voltages. Suppose a certain zener diode has an avalanche voltage, also called the *zener voltage,* of 50 V. If a reverse bias is applied to the P-N junction, the diode acts as an open circuit below 50 V. When the voltage reaches 50 V, the diode starts to conduct. The more the reverse bias tries to increase, the more current flows through the P-N junction. This effectively prevents the reverse voltage from exceeding 50 V. There are other ways to get voltage regulation besides the use of zener diodes, but zener diodes often provide the simplest and least expensive alternative. Zener diodes are available with a wide variety of voltage and power-handling ratings. Power supplies for solid-state equipment commonly employ zener-diode regulators.

Amplitude Limiting   The forward breakover voltage of a germanium diode is about 0.3 V; for a silicon diode it is about 0.6 V. A diode will not conduct until the forward bias voltage is at least as great as the forward breakover voltage. But the diode will always conduct when the forward bias exceeds the breakover value. In this case, the voltage across the diode will be constant: 0.3 V for germanium and 0.6 V for silicon. This property can be used to advantage when it is necessary to limit the amplitude of a signal. By connecting two identical diodes back to back in parallel with the signal path, the maximum peak amplitude is limited, or clipped, to the forward breakover voltage of the diodes. This scheme is sometimes used in radio receivers to prevent "blasting" when a strong signal comes in. One problem with a diode limiter circuit is that it introduces distortion when clipping occurs. This might not be a problem for reception of encoded digital signals, or for signals that rarely reach the limiting voltage. But for analog signals with amplitude peaks that rise well past the limiting voltage, it can seriously degrade the audio quality, perhaps even rendering the signal unreadable.

Detection Diodes can be used to extract the information from a modulated carrier signal. This process is called *detection* or *demodulation*. Diodes are used for detection of amplitude-modulated (AM) and single-sideband signals. In some cases, diode detection works because the diodes operate as HF signal rectifiers; in other instances, *nonlinearity* exists in the diode P-N junction(s), encouraging the production of *beat frequencies*.

Mixing When two waves having different frequencies are combined in a nonlinear circuit, signals at other frequencies are produced. This is called *mixing*. The new waves are at the sum and difference frequencies of the original waves. Harmonics are also generated, as are so-called *higher-order distortion products*. These signals result from various arithmetic sums and differences of the fundamental frequencies, the beat frequencies, and harmonics. Diodes are often used to obtain the nonlinearity necessary to achieve mixing.

Frequency Multiplication When current passes through a diode, half of the cycle is cut off. This occurs no matter what the frequency, from 60-Hz utility current through RF, as long as the diode capacitance is not too great. The output waveform from the diode looks much different than the input waveform; the device is nonlinear. Whenever there is nonlinearity of any kind in a circuit (that is, whenever the output waveform is shaped differently from the input waveform), there will be harmonic frequencies in the output. Sometimes a circuit is needed that will produce harmonics. Then nonlinearity is introduced deliberately. Diodes are ideal for this. For a diode to work as a *frequency multiplier*, it must be of a type that would also work well as a detector at the same frequencies. This means that the diode should act like a rectifier, not like a capacitor.

LEDs and IREDs   Depending on the mixture of semiconductors used in manufacture, visible light of almost any color can be produced when current passes through a P-N junction. Infrared-emitting devices also exist. One common color for a *light-emitting diode (LED)* is bright red. An *infrared-emitting diode (IRED)* produces emissions at wavelengths too long to see. The intensity of the radiation from an LED or IRED depends to some extent on the forward current. As the current rises, the brightness increases up to a certain point. If the current continues to rise, no further increase in brilliance takes place. The LED or IRED is then said to be in a state of *saturation.*

Digital Displays   Because LEDs can be made in various different shapes and sizes, they are good for use in digital displays. You've probably seen digital clock radios that use them. They are common in cars. They make good indicators for "on/off," "a.m./p.m.," "battery low," and other conditions. In recent years, LED displays have been largely replaced by *liquid-crystal displays (LCDs)*. This technology has advantages over LEDs, including much lower power consumption and better visibility in direct sunlight. LCD technology has exploded in recent years, finding extensive applications in computer systems, especially in notebook and handheld computers.

Communications   LEDs and IREDs are useful in communications because their intensity can be modulated to carry information. When the current through the device is sufficient to produce output, but not enough to cause saturation, the LED or IRED output will follow along with rapid current changes. Special LEDs and IREDs produce coherent radiation; these are called *laser diodes*. A laser LED or IRED generates a cone-shaped beam of low intensity. The beam can be focused; the resulting rays have some of the same advantages found in larger lasers. These rays can carry information through space, clear air, or clear water in optical and infrared wireless systems.

CHAPTER 2

Active
Electronic
Components

The term *active* refers to components that require a source of power (independent of the signal itself) to process a signal in some way. Active devices are used to generate, amplify, display, measure, and store signals or information.

Active Diodes

Under certain conditions, diodes behave as active, rather than passive, electronic components. There are several types of diodes that do this.

Varactor Diode

When a diode is reverse-biased, there is a region at the P-N junction with dielectric properties. This is called the *depletion region*, because it has a shortage of majority charge carriers. The width of this zone depends on several things, including the *reverse bias* voltage. As long as the reverse bias is less than the *avalanche voltage*, the width of the depletion region can be changed by varying the bias voltage. This results in a change in the capacitance of the junction. The capacitance, which is always quite small (on the order of picofarads), varies inversely with the square root of the reverse bias.

Some diodes are manufactured especially for use as variable capacitors. Such a component is called a *varactor* or *varicap*. These devices are usually made from silicon or gallium arsenide (GaAs) semiconductor material. A common use for a varactor is in a circuit called a *voltage-controlled oscillator* (VCO). Varactors are also employed in tuned circuits to control the frequency, or the frequency response, of amplifiers, oscillators, or filters.

Oscillator Diodes

A *Gunn diode* is a semiconductor device that operates as an oscillator in the ultra-high-frequency (UHF) and microwave parts of the radio spectrum. A Gunn diode oscillates because of the *Gunn effect*, named after J. Gunn of International Business Machines (IBM) who first observed and studied it in the 1960s. A Gunn diode does not work like a rectifier, detector, or mixer; instead, the oscillation takes place as a result of a

Figure 3-1
A Gunn-diode oscilla-
tor circuit.

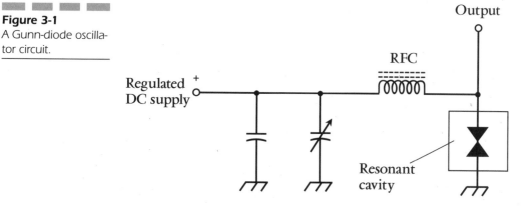

property called *negative resistance*. Gunn-diode oscillators are often tuned using varactors. A Gunn-diode oscillator, connected directly to a microwave horn antenna, is known as a *Gunnplexer.*

In practice, a Gunn diode is mounted inside a resonant enclosure. A direct-current (DC) voltage is applied to the device and, if this voltage is large enough, the diode will oscillate. A schematic diagram of an oscillator using a Gunn diode is shown in Fig. 3-1. Most Gunn diodes require about 12 volts (V) for proper operation. Oscillation can take place at frequencies in excess of 20 gigahertz (GHz). A Gunn diode can produce up to 1 watt (W) of radio-frequency (RF) power output, but more commonly it works at levels of about 0.1 W. Gunn diodes are usu-ally made from gallium arsenide.

Gunn diodes are not particularly efficient. Only a small fraction of the consumed DC power actually results in RF power. Gunn diodes tend to be sensitive to changes in temperature and bias voltage. The fre-quency can vary considerably, even with a small change in the ambient temperature. For this reason, the temperature must be carefully regulat-ed. The change in frequency with voltage can be useful for frequency-modulation (FM) purposes, but voltage regulation of some sort is essential.

An *IMPATT diode* is a microwave oscillating device like a Gunn diode. The acronym IMPATT comes from the words "*imp*act *a*valanche *t*ransit *t*ime." An IMPATT diode can be used as an amplifier for a microwave transmitter that employs a Gunn-diode oscillator. As an oscillator, an IMPATT diode produces about the same amount of output power, at comparable frequencies, as the Gunn diode.

Another type of diode that will oscillate at microwave frequencies is the *tunnel diode,* also known as the *Esaki diode.* It produces only a small amount of power, but it can be used as a local oscillator in a microwave radio receiver. Tunnel diodes work well as amplifiers in microwave receivers, because they generate very little unwanted noise. This is especially true of gallium arsenide devices.

Circuit design using Gunn, IMPATT, and tunnel diodes is a sophisticated topic, and is beyond the scope of this book. College-level electrical-engineering texts are good sources of information on this subject.

Photodiode

A silicon diode, housed in a transparent case and constructed in such a way that visible light can strike the barrier between the P-type and N-type materials, forms a *photodiode*. A reverse bias is applied to the device. When light falls on the junction, current flows. The current is proportional to the intensity of the light, within certain limits.

Silicon photodiodes are more sensitive at some wavelengths than at others. The greatest sensitivity is in the near infrared (IR) part of the spectrum, at wavelengths a little longer than visible red light. When light of variable brightness falls on the P-N junction of a reverse-biased silicon photodiode, the output current follows the light-intensity variations. This makes silicon photodiodes useful for receiving modulated-light signals, of the kind used in fiberoptic systems.

Optoisolator

A *light-emitting diode* (LED) or *infrared-emitting diode* (IRED) and a photodiode can be combined in a single package to get an *optoisolator*. This device creates a modulated-light signal and sends it over a small, clear gap to a receptor. The LED or IRED converts an electrical signal to visible light or IR; the photodiode changes the visible light or IR back into an electrical signal.

A common problem in electronics engineering is the fact that, when a signal is electrically coupled from one circuit to another, the impedances of the two stages interact. This can lead to nonlinearity, unwanted oscillation, loss of efficiency, or other problems. Optoisolators do away with this because there is no electrical coupling between stages. If the input impedance of the second circuit changes, the impedance that the first circuit "sees" is not affected.

Bipolar Transistors

The word *transistor* is a contraction of "current-transferring resistor." This is an excellent description of what a *bipolar transistor* does. A bipolar transistor has two P-N junctions connected together. This is done in either of two ways: a P-type layer sandwiched between two N-type layers, or an N-type layer between two P-type layers. Bipolar transistors, like diodes, can be made from various semiconductor substances.

NPN versus PNP

Figure 3-2A is a simplified drawing of an NPN transistor; Fig. 3-2B shows its schematic symbol. The P-type, or center, layer is called the *base*. The thinner of the N-type semiconductors is the *emitter*, and the thicker is the *collector*. Sometimes these are labeled B, E, and C in schematic diagrams, although the transistor symbol alone is enough to reveal which is which.

A PNP transistor is just the opposite of an NPN device, having two P-type layers, one on either side of a thin, N-type layer. Figure 3-2C and D shows a simplified drawing of, and the schematic symbol for, a PNP device. The emitter layer is thinner, in most units, than the collector layer.

You can always tell whether a bipolar transistor in a diagram is NPN or PNP. With the NPN the arrow points outward; with the PNP it points inward. The arrow is always at the emitter.

Generally, PNP and NPN transistors do the same things in electronic circuits. The only differences are in the polarities of the applied voltages

Figure 3-2

A simplified drawing (A) and schematic symbol (B) for an NPN transistor; a drawing (C) and symbol (D) for a PNP transistor.

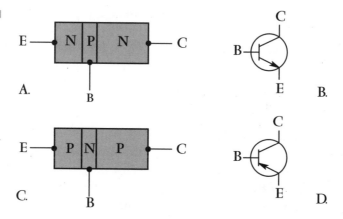

and the directions of the currents. In most applications, an NPN device can be replaced with a PNP device or vice versa, and the power-supply polarity reversed, and the circuit will still work provided the new device has the appropriate specifications.

There are many different kinds of NPN or PNP bipolar transistors. Some are used for RF amplifiers and oscillators; others are intended for audio frequencies (AF). Some can handle high power, and others cannot, being made for weak-signal work. Some bipolar transistors are manufactured for the purpose of switching, rather than signal processing. If you look through a catalog of semiconductor components, you'll find hundreds of different bipolar transistors, each with its own unique set of specifications.

Biasing

The normal method of biasing an NPN transistor is to have the emitter negative and the collector positive. The reverse is true for a PNP transistor. The base current determines in large measure how the transistor will behave. In the following examples, it is assumed that an NPN bipolar transistor is used. For PNP devices, simply reverse the polarities (plus to minus and vice versa).

Forward Bias Suppose a source of positive DC voltage is connected to the base of an NPN bipolar transistor. This is *forward bias*. If this bias is less than the forward breakover voltage of the emitter-base (E-B) junction, no current will flow. But when the applied voltage reaches breakover, the E-B junction will conduct.

Despite the reverse bias of the B-C junction, the emitter-collector current, called *collector current* and denoted I_C, will flow once the E-B junction conducts. Then even a small increase in the base current I_B will cause a large increase in I_C. This is how a bipolar transistor amplifies. Small changes in the signal current at the E-B junction produce large fluctuations in the collector current. The output is taken from the collector when gain (amplification) is needed.

Saturation If the base current continues to increase, a point will eventually be reached where the collector current levels off. The transistor is then said to be in a state of *saturation*. It is conducting as much as it possibly can. A bipolar transistor loses its ability to amplify AC signals when it is at, or near, a condition of saturation. However, when the device is used in switching applications, it might be alternately switched between cutoff (zero collector current) and saturation.

Zero Bias When the base is not connected to anything, or is at the same potential as the emitter, the transistor is operating at *zero bias*. Under no-signal conditions this will result in *cutoff*, or a collector current of zero. When a signal is applied to the base, it must swing to a sufficiently large positive voltage to overcome the forward breakover voltage of the emitter-base junction, in order to produce collector current for any part of the cycle. Zero bias is sometimes used for AF or RF power amplification.

Reverse Bias Suppose that a negative DC voltage is connected to the base of a bipolar NPN transistor, so the base becomes negative with respect to the emitter. This is *reverse bias* of the E-B junction. It is assumed that this voltage is not so high that avalanche breakdown takes place at the junction. A signal can be injected to overcome the combined reverse-bias and forward-breakover voltage of the junction, but such a signal must swing to large positive voltages during the positive peaks of the cycle. Reverse bias is sometimes used for RF power amplification, especially in digital signal modes.

Current Amplification

Because a small change in the base current I_B results in a large collector-current (I_C) variation when the bias is just right, a transistor can operate as a *current amplifier.* It might be more technically accurate to say that it is a "current-fluctuation amplifier," because it is the magnification of current changes, not the absolute level of the current, that is significant.

Static Current Amplification Current amplification is often called *beta* by engineers. It can range from a factor of just a few times up to hundreds of times. One method of expressing the beta of a transistor is as the *static forward current transfer ratio,* abbreviated H_{FE}. Mathematically,

$$H_{FE} = I_C / I_B$$

Thus, if a base current of 1 milliampere (mA) results in a collector current of 35 mA, $H_{FE} = 35/1 = 35$. If $I_B = 0.5$ mA yields $I_C = 35$ mA, then

Figure 3-3
Three different operating points result in varying degrees of amplification (A). Overdrive reduces amplification (B).

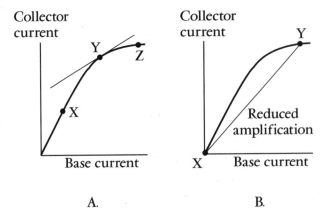

A.

B.

$H_{FE}=35/0.5=70$. The H_{FE} rating is important because it gives engineers an indication of the greatest current amplification that can be obtained with a particular transistor.

Dynamic Current Amplification   Another way of specifying current amplification is as the ratio of the difference in I_C to the difference in I_B. Abbreviate the words "the difference in" by the letter d. Then, according to this second definition,

$$\text{Current amplification}=dI_C / dI_B$$

A graph of collector current versus base current (I_C versus I_B) for a hypothetical transistor is shown in Fig. 3-3A. Three different points are shown, corresponding to various bias values. The ratio dI_C/dI_B is different for each of the points in this graph. Geometrically, dI_C/dI_B at a given point is the slope of a line tangent to the curve at that point. The tangent line for point Y is shown as a thin, straight line; the tangent lines for points X and Z are not shown, because they lie almost exactly along the curve. The steeper the slope of the line at any given point, the greater is dI_C/dI_B.

Point X provides the highest dI_C/dI_B of the three points shown in the graph, as long as the input signal is small. This value is very close to H_{FE}. For small-signal amplification, point X represents a good bias level. Engineers would say that it is a good *operating point* for the transistor.

At point Y, dI_C/dI_B is smaller than at point X. (It might actually be less than 1.) At point Z, dI_C/dI_B is practically zero. Transistors are rarely biased at these points.

Overdrive

Even when a transistor is biased for best operation (near point X in drawing A), a strong input signal can drive it to point Y or beyond during part of the cycle. Then dI_C/dI_B is reduced, as shown in Fig. 3-3B. Points X and Y in the graph at illustration B represent the instantaneous current extremes during the signal cycle. The thin line represents the effective value of dI_C/dI_B under these conditions.

In a situation such as that shown by Fig. 3-3B, there will be distortion in a bipolar-transistor amplifier. The output waveform will not have the same shape as the input waveform. This can be tolerated in certain applications, such as FM radio transmitter power amplifiers. But in other applications it will cause trouble. In a high-fidelity audio system, such a condition would grossly degrade the quality of the sound.

The more serious problem with overdrive is the fact that the transistor is in or near saturation during part of the cycle. When this happens, the efficiency of the circuit is reduced. The transistor is doing futile work for a portion of every wave cycle. This can cause excessive collector current, and can overheat the base-collector (B-C) junction. In the worst case it can destroy the component.

Gain versus Frequency

Another important specification for a transistor is the range of frequencies over which it can be used as an amplifier. All transistors have an *amplification factor*, or *gain*, that decreases as the signal frequency increases. Some devices will work well only up to a few megahertz; others can be used to several gigahertz.

Gain can be expressed in various ways. The preceding discussion involved *current gain*, expressed as a ratio. You will also sometimes hear about *voltage gain* or *power gain* in amplifier circuits. These, too, can be expressed as ratios. For example, if the voltage gain of a circuit is 15, then the output signal voltage (rms, peak, or peak-to-peak) is 15 times the input signal voltage. If the power gain of a circuit is 25, then the output signal power is 25 times the input signal power. Gain is also sometimes expressed in terms of a logarithmic unit called the *decibel*.

There are two expressions commonly used for the gain-versus-frequency behavior of a bipolar transistor.

The *gain-bandwidth product,* symbolized as f_T, is the frequency at which the gain becomes equal to 1 (no amplification or loss) with the emitter connected to ground. If you try to make an amplifier using a transistor at a frequency higher than its f_T, you will inevitably fail. Thus, f_T represents an absolute upper limit for the frequency at which a bipolar transistor will work as an amplifier.

The *alpha cutoff frequency* of a transistor is the frequency at which the gain becomes 0.707 times its value at 1 kilohertz (kHz). A transistor might still have considerable gain at its alpha cutoff. By looking at the alpha cutoff frequency, you can get an idea of how rapidly the transistor loses gain as the frequency goes up. Some devices "die off" faster than others.

Field-Effect Transistors

All transistors are semiconductor devices, but there are numerous different configurations. Transistors can be categorized broadly into two types: the bipolar transistor (discussed above) and the *field-effect transistor* (FET). This section discusses the *junction FET* (JFET) and the *metal-oxide-semiconductor FET* (MOSFET).

Principle of the JFET

A JFET can have any of several different forms. They all work the same way: the current varies because of the effects of an electric field within the device. Some JFETs work well as weak-signal amplifiers and oscillators; others are made for power amplification.

The electrical effects inside a JFET can be likened to the control of water flowing through a garden hose. *Charge carriers* (negative *electrons* or positive *holes*) pass from the *source* (S) electrode to the *drain* (D). This results in a drain current I_D that is generally the same as the source current I_S. This is analogous to the fact that the water comes out of a garden hose at the same rate it goes in, as long as there are no leaks.

The rate of flow of charge carriers—that is, the current—depends on the voltage at a regulating electrode called the *gate* (G). Fluctuations in gate voltage V_G cause changes in the current through the *channel,* or the path between the source and the drain. The channel current is generally the same as I_S and I_D. This is analogous to the fact that the flow rate of water at any point in a hose is the same as the flow rate at either end, again assuming there are no leaks. Small fluctuations in the control voltage V_G can cause large variations in the flow of charge carriers through the JFET. This translates into voltage amplification in electronic circuits.

FETs have some advantages over bipolar devices. Perhaps the most important is that FETs usually generate less internal noise than bipolar transistors. This makes them excellent for use in sensitive radio receivers at very high or ultra-high frequencies. FETs also have high input impedances. The gate controls the flow of charge carriers by means of an electric field, rather than via an electric current. Because of this, FETs are useful in devices that must cause the least possible disturbance in circuits or devices to which they are connected.

N-channel and P-channel

Simplified drawings of an N-channel JFET and its schematic symbol are shown in Fig. 3-4A and B. The N-type material forms the channel, or the path for charge carriers. In the N-channel device, the *majority carriers* are electrons. The source is at one end of the channel, and the drain is at the other. You can think of electrons as being injected into the source and collected from the drain as they pass through the channel. The drain is positive with respect to the source.

In an N-channel device, the gate consists of P-type material. Another, larger section of P-type material, called the *substrate*, forms a boundary on the side of the channel opposite the gate. The JFET is formed in the substrate during manufacture by a process known as *diffusion.*

The voltage on the gate produces an electric field that interferes with the flow of charge carriers through the channel. The more negative V_G becomes, the more the electric field chokes off the current through the channel, and the smaller I_D becomes.

A P-channel JFET (drawings C and D in Fig. 3-4) has a channel of P-type semiconductor. The majority charge carriers are holes. The drain is negative with respect to the source. In a sense, holes are injected into the source and are collected from the drain. The gate and the substrate are of N-type material. In the P-channel JFET, the more positive V_G gets, the more the electric field chokes off the current through the channel, and the smaller I_D becomes.

You can recognize the N-channel device by the arrow pointing inward at the gate, and the P-channel JFET by the arrow pointing

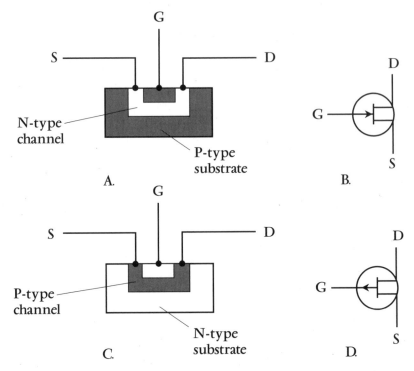

outward. Also, you can tell which is which (sometimes arrows are not included in schematic diagrams) by the power-supply polarity. A positive drain indicates an N-channel JFET, and a negative drain indicates a P-channel type.

In electronic circuits, N-channel and P-channel devices can do the same kinds of things. The main difference is the polarity. An N-channel JFET can usually be replaced with a P-channel JFET, and the power-supply polarity reversed, and the circuit will still work if the new device has the right specifications.

Depletion and Pinchoff

Either the N-channel or the P-channel JFET works because the voltage at the gate causes an electric field that interferes, more or less, with the flow of charge carriers along the channel.

As the drain voltage V_D increases, so does the drain current I_D, up to a certain level-off value. This is true as long as the gate voltage is constant and is not too large. As the gate voltage continues to increase (positively for a P-channel device and negatively for an N-channel device), a *depletion region* begins to form in the channel. Charge carriers cannot flow in this region; they must pass through a narrowed channel. The more V_G increases, the wider the depletion region gets. Ultimately, if the gate voltage becomes high enough, the depletion region will completely obstruct the flow of charge carriers. This is called *pinchoff.*

Again, think of the garden-hose analogy. Increasing gate voltages correspond to stepping down with your foot harder and harder on the hose. When pinchoff takes place, you've cut off the water flow entirely, perhaps by bearing down with all your body mass. Biasing beyond pinchoff is something like loading yourself up with heavy weights as you stand on the hose, thereby shutting off the water flow with extra force. It takes larger and larger input signals to overcome bias as it increases beyond pinchoff.

Amplification

When V_G is fairly large, the JFET is pinched off, and no current flows through the channel. As V_G decreases, the channel opens up, and current begins flowing. As V_G decreases further, the channel gets wider, and the drain current I_D increases. As V_G approaches the point where the source-gate (SG) junction is at forward breakover, the channel conducts as well as it possibly can.

If V_G becomes such that forward breakover occurs at the SG junction, the junction will begin to conduct. Then the JFET will no longer work properly. Some of the current in the channel will be shunted off through the gate, a situation that is never desirable in a JFET. This is analogous to a garden hose springing a leak.

The best amplification for weak signals is obtained when the gate bias V_G is such that the drain current I_D changes greatly when the gate voltage changes only a little. For power amplification, results are often best when the JFET is biased at, or even beyond, pinchoff.

In the output circuit of a JFET amplifier, the drain current passes through a resistor. Small fluctuations in gate voltage cause large changes in drain current, and these variations in turn produce wide swings in the voltage across the drain resistor. The AC component of this voltage can be drawn off through a capacitor or passed through a transformer; it appears at the output as a signal of much greater AC voltage than that of the input signal at the gate.

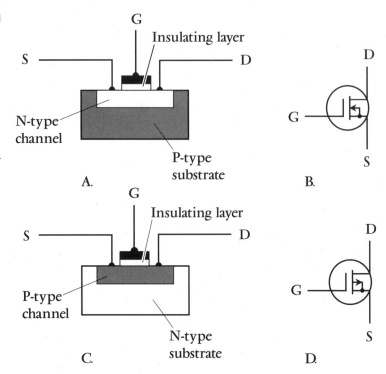

A.

B.

C.

D.

The MOSFET

A simplified cross-sectional drawing of an *N-channel MOSFET* (pronounced "MOSS-fet"), and its schematic symbol, are shown in Fig. 3-5A and B. A simplified drawing and the symbol for a *P-channel MOSFET* are at Fig. 3-5C and D. The N-channel device is diffused into a substrate of P-type semiconductor material. The P-channel device is diffused into a substrate of N-type material.

When the MOSFET was first developed, it was called an *insulated-gate FET* or *IGFET*. This is perhaps more descriptive of the device than the currently accepted name. The gate electrode is actually insulated, by a thin layer of dielectric, from the channel. As a result, the input impedance is even higher than that of a JFET; the gate-to-source resistance of a typical MOSFET is comparable to that of a capacitor. This means that a MOSFET draws essentially zero current, and therefore no power, from the signal source. Some MOSFETs have input resistance exceeding a trillion (10^{12}) ohms.

One trouble with MOSFETs is that they can be easily damaged by static electric discharges. When building or servicing circuits containing

MOS devices, technicians must use special equipment to ensure that their hands do not carry static charges that might ruin the components. If a static discharge occurs through the dielectric of a MOS device, the component will be destroyed permanently. Humid weather does not eliminate this hazard, but dry conditions can exaggerate it.

In practical electronic circuits, an N-channel JFET can sometimes be replaced directly with an N-channel MOSFET; P-channel devices can be similarly interchanged. But the characteristic curves for MOSFETs are not the same as those for JFETs. The main difference is that the SG junction in a MOSFET is not a P-N junction. Therefore, forward breakover cannot occur. A gate voltage (V_G) of more than $+0.6$ V can be applied to an N-channel MOSFET, or a V_G more negative than -0.6 V to a P-channel device, without a current leak taking place.

A family of *characteristic curves* for a hypothetical N-channel MOSFET is shown in the graph of Fig. 3-6. The device will work with positive gate bias voltage (V_G) as well as with negative gate bias. A P-channel MOSFET behaves in a similar way, being usable with either positive or negative V_G. In general, the drain current increases as the drain voltage increases. However, the maximum attainable drain current depends on the gate voltage.

Figure 3-6
A family of characteristic curves for a hypothetical N-channel depletion-mode MOSFET.

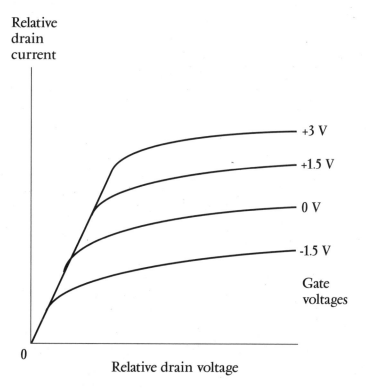

Relative drain current

+3 V

+1.5 V

0 V

-1.5 V

Gate voltages

0

Relative drain voltage

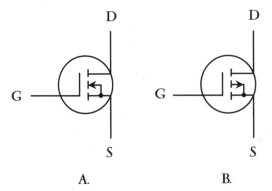

Depletion versus Enhancement

The JFET works by varying the width of the channel. Normally the channel is "wide open." As the depletion region gets wider and wider, choking off the channel, the charge carriers are forced to pass through a narrower and narrower path. This is called the *depletion mode* of operation for a field-effect transistor, whether it is a JFET or a MOSFET. But MOS technology allows an entirely different means of operation, a phenomenon that cannot be made to take place in JFETs. This is known as the *enhancement mode*. An enhancement-mode MOSFET normally has a pinched-off channel. It is necessary to apply a bias voltage V_G to the gate so that a channel will form. If $V_G=0$ in such a MOSFET, that is, if the device is at zero bias, the device is pinched off, the drain current (I_D) will be zero when there is no signal input.

The schematic symbols for enhancement-mode MOSFETs are similar to the symbols for depletion-mode MOSFETs. The only difference is that, in the enhancement-mode symbols, the vertical lines are broken instead of solid. These symbols are shown in Fig. 3-7. The device at A is N-channel; the device at B is P-channel.

Integrated Circuits

An *integrated circuit* (IC) is an electronic device containing many diodes, transistors, resistors, and/or capacitors fabricated onto a wafer, or *chip*, of semiconductor material. The chip is enclosed in a small package with pins for connection to external components. ICs have probably stimulated more evolution than any other single development in electronics.

Boxes and Cans

Most ICs look like gray or black plastic boxes with protruding pins. Common configurations are the *single inline package* (SIP), the *dual inline package* (DIP), and the *flatpack*. Another package looks like a transistor with too many leads. This is a *metal-can package*, sometimes also called a *TO package*.

The schematic symbols for most ICs are simple geometric shapes such as triangles or rectangles. A component designator is usually written inside the figure, and the wires emerging from it are labeled according to their specific functions. The internal circuitry is usually too complicated to be rendered in all its detail. An IC is, in effect, a "black box." The engineer or technician can forget about its inner workings, just as you don't think about what happens inside your stereo receiver or computer as you use it. Even with the simplicity of the IC symbol, apparatus using many ICs can have nightmarish schematic diagrams, with dozens of parallel lines so close together that you must run a pencil along them to keep track of them.

Assets of IC Technology

ICs have several advantages over discrete components. Some of these are as follows.

Compactness   ICs are far more compact than equivalent circuits made from individual transistors, diodes, capacitors, and resistors. A corollary to this is the fact that more complex circuits can be built, and kept down to a reasonable size, using ICs as compared with discrete components. Thus, for example, there are laptop (notebook) computers with capabilities more advanced than early computers that took up entire rooms or buildings.

Speed   The interconnections among components in an IC are physically tiny, making high switching speeds possible. Electric currents travel fast, but not instantaneously. The smaller the separation among components, the less time the charge carriers take to cover the distances. This translates into a larger number of computations that can be done in a given period of time; it also translates into a shorter length of time required to carry out a complex operation.

Low Power Requirement   Another advantage of ICs is that they use less power than equivalent discrete-component circuits. This is especially important if batteries are to be used for operation. Because ICs use so little current, they produce less heat than their discrete-component equivalents. This translates into better efficiency. It also minimizes the problems that plague equipment that gets hot with use, such as frequency drift and generation of internal noise.

Reliability ICs fail less often, per component-hour of use, than appliances that make use of discrete components. This is mainly a result of the fact that all interconnections are sealed within the IC case, preventing corrosion or the intrusion of dust. The reduced failure rate translates into less downtime, or time during which the equipment is out of service for repairs.

Ease of Maintenance IC technology lowers maintenance costs, not only because failures are comparatively rare, but also because repair procedures are simplified when failures do occur. Many appliances use sockets for ICs, and replacement is simply a matter of finding the faulty IC, unplugging it, and plugging in a new one. Special desoldering equipment is used with appliances having ICs soldered directly to the circuit boards.

Modularization of Equipment Modern IC appliances employ *modular construction*. In this scheme, individual ICs perform defined functions within a circuit board; the circuit board or card, in turn, fits into a socket and has a specific purpose. A computer, programmed with customized software, is used by repair technicians to locate the faulty card in an appliance. The whole card can be pulled and replaced, getting the appliance back to the consumer in the shortest possible time. Then the computer can be used to troubleshoot the faulty card, getting the card ready for use in the next appliance that happens to come along with a failure in the same card.

Limitations of IC Technology

No technological advancement ever comes without some compromise. Here are some of the problems and limitations inherent in ICs.

Inductors While some components are easy to fabricate onto chips, other components defy the IC manufacturing process. Inductances, except for extremely low values, are one example. Devices using ICs must generally be designed to work without inductors. Fortunately, resistance-capacitance (RC) circuits are capable of doing most of the things that inductance-capacitance (LC) circuits can do. Some active IC circuits can introduce a 90-degree current phase lag, equivalent to an inductive reactance.

Power   High-power amplifiers cannot, in general, be fabricated onto semiconductor chips. High power necessitates a certain minimum physical bulk, because such amplifiers always generate large amounts of heat. This, however, is not generally a problem. Power transistors and vacuum tubes are available to perform high-power tasks.

Linear Integrated Circuits

A *linear IC* is used to process analog signals such as voices, music, and radio transmissions. The term *linear* arises from the fact that, in general, the amplification factor is constant as the input amplitude varies. If the output voltage is graphed as a function of the input voltage for a linear IC, the result is a straight line. Linear ICs include amplifiers, regulators, timers, multiplexers, and comparators.

Operational Amplifier   An operational amplifier, or *op amp*, is a specialized form of linear IC. The op amp consists of several transistors, resistors, diodes, and capacitors, interconnected so high gain is achievable over a wide range of frequencies. A single op amp might comprise an entire IC, or an IC might consist of two or more op amps. You will sometimes see a *dual op amp* or a *quad op amp*. Some ICs contain one or more op amps in addition to other circuits. Op amps can be used with RC combinations to build *active filters*. These circuits are commonly used at audio frequencies. They are *selective filters* that also provide signal amplification.

Regulator   A *voltage-regulator IC* acts to control the output voltage of a power supply. This is important with precision electronic equipment. These ICs are available in various voltage and current ratings. Typical voltage-regulator ICs have three terminals. They look somewhat like power transistors.

Timer   A *timer IC* is actually a form of oscillator. It produces a delayed output, with the delay being adjustable to suit the needs of a particular device. The delay is generated by counting the number of oscillator pulses; the length of the delay can be adjusted by means of external resistors and capacitors.

Analog Multiplexer   This IC allows several different signals to be combined in a single channel. An analog multiplexer can also be used in reverse; then it works as a *demultiplexer.* Thus, you will sometimes hear about *multiplexer/demultiplexer ICs.*

Comparator   Like an op amp, a comparator IC has two inputs. It compares the voltages at the two inputs (called A and B). If the input at A is significantly greater than the input at B, the output will be about +5 V. If the input at A is not greater than the input at B, the output voltage will be about +2 V. Voltage comparators are available for a variety of applications. They are generally used to actuate, or trigger, other devices such as relays and electronic switching circuits.

Digital Integrated Circuits

A *digital IC* operates using just two states, called *high* and *low.* These are sometimes called *logic 1* and *logic 0,* respectively. Digital ICs consist of many logic gates that perform operations in binary (boolean) algebra. Several different digital IC technologies exist.

Transistor-Transistor Logic (TTL)   In this scheme, arrays of bipolar transistors, some with multiple emitters, operate on DC pulses. This technology has several variants, some of which date back to around 1970. The hallmark of TTL is immunity to noise pulses. The transistors are either completely cut off, or else completely saturated. Thus, TTL is not seriously affected by external noise.

Emitter-Coupled Logic (ECL)   In ECL, the transistors are not operated at saturation, as they are with TTL. This increases the speed of operation of ECL compared with TTL. But noise pulses have a greater effect in ECL, because unsaturated transistors amplify signals as well as switch them.

Complementary-Metal-Oxide-Semiconductor (CMOS)   This type of device employs both N-type and P-type silicon on a single chip. This is analogous to using N-channel and P-channel MOSFETs in a circuit. The main advantages of CMOS technology are extremely low current drain, high operating speed, and relative immunity to noise.

N-channel MOS (NMOS) and P-channel MOS (PMOS)   These technologies offer simplicity of design, along with high operating speed. P-channel MOS is similar to NMOS, but the speed is slower. An NMOS or PMOS digital IC is like a circuit that uses only N-channel MOSFETs, or only P-channel MOSFETs.

Microprocessors and Microcomputers

The most significant application of IC technology, many engineers would argue, is in the computing field. Digital computers are revolutionizing the way people live their lives. There are two primary types of "computing" IC: the *microprocessor* and the *microcomputer.* These two terms are often used interchangeably, but there are subtle technical differences.

Microprocessors

A microprocessor is the IC, or chip, that coordinates the actions of a computer and does the calculations. It is located on the *motherboard* (sometimes called the *logic board*). The microprocessor, together with various other ICs, comprise the *central processing unit* (CPU). The peripheral circuits can be integrated onto the same chip as the microprocessor, but they are usually separate. The external chips contain memory and programming instructions.

The CPU forms the complete "brain" of a computer. You might think of the microprocessor as the computer's "cerebral cortex," which directs the behavior of the machine by deliberate control. The CPU, dominated by the microprocessor, represents the computer's "whole brain." All the memory and the buses, added to this, creates the computer's "central nervous system." Peripherals such as printers, disk drives, pointing/control devices, speech recognition/synthesis apparatus, modems, and monitors are the "hands," "ears," "eyes," and "mouth" of the machine. In the future, you can expect to see robots, vision systems, various home appliances, surveillance apparatus, medical devices, and other exotic equipment under the control of personal computers. At the helm of every such system, there will be a microprocessor.

Microprocessors get more powerful every year. Physically, this translates to an increasing number of digital switching transistors per chip. The number of digital switches that can be fabricated onto a semiconductor chip of a particular size is ultimately limited by the structure of matter. In the extreme, each binary digit (bit) might be represented by the presence or absence of a single electron, or by its energy level (shell), within an atom.

Microcomputers

A *microcomputer* is a small computer with the microprocessor enclosed in a single IC package. Today, microprocessor chips are available in sizes less than $1/4$ inch on a side for specialized control purposes. Microcomputers vary in sophistication and memory storage capacity, depending on the intended use. Simple microcomputers are available for less than $100. They have liquid-crystal displays and small keypads that allow only encoded data entry. A good example of such a microcomputer is a *programmable calculator.*

Microcomputers are often used for the purpose of regulating the operation of electrical and electromechanical devices. This is known as *microcomputer control,* and makes it possible to perform complex tasks with a minimum of difficulty. Microcomputer control is widely used in such devices as radio receivers and transmitters, television sets, automobiles, aircraft, and robots. For example, a microcomputer can be programmed to switch on an oven, heat the food to a prescribed temperature for a certain length of time, and then switch the oven off. Microcomputers are widely used to control automobile engines, optimizing the efficiency and gasoline mileage. Microcomputers can navigate and fly airplanes. It has been said that a modern jet aircraft is really a giant robot, because it can (in theory at least) complete a flight all by itself, without a single human being on board.

One of the most recent, and exciting, applications of microcomputer control is in the field of medical electronics. Microcomputers can be programmed to provide electrical impulses to control erratically functioning body organs, to move the muscles of paralyzed persons, and for various other purposes.

Memory and Storage Devices

Memory refers to electronic media that allow fast storage and retrieval of digital data. There are several forms of memory. Usually, when you hear someone talking about computer memory, they are referring to the *random-access memory* (RAM). However, there are other kinds of memory.

RAM

Within RAM, data is stored in *memory arrays*. An array resembles a grid or matrix. The data can be addressed (selected) from anywhere in the matrix. Data is easily modified and then stored back in RAM, in whole or in part. An example of RAM data is a word-processing file as it is being actively worked on. This chapter, for example, was written in RAM. It was periodically saved on the computer's hard disk, and also on backup diskettes and cartridges.

There are two basic types of RAM: *dynamic RAM* (DRAM) and *static RAM* (SRAM). Dynamic RAM employs IC transistors and capacitors; data is stored as charges on the capacitors. The charge must be replenished frequently, or it will dissipate. Replenishing is done several hundred times per second. Static RAM uses a circuit called a *flip-flop* to store data. This gets rid of the need to replenish the electric charge, because flip-flops hold their state indefinitely until they receive a change-of-state pulse. But SRAM ICs require more elements than DRAM to store a given amount of data.

Volatility

With any RAM, the data is erased when the appliance is switched off, unless some provision is made for *memory backup*. The most common means of memory backup is the use of an electrochemical cell or battery. Modern IC memories need so little current to store their data that a backup battery lasts almost as long in the circuit as it would on the shelf.

A memory from which data disappears when power is removed is called a *volatile memory.* If memory is retained when power is removed, it is a *nonvolatile memory.* The RAM in a typical computer is volatile. You must save data on a hard disk or diskette, or in some other nonvolatile storage medium, before you switch off the computer. Otherwise you will lose everything in RAM.

Read-Only Memory

In contrast to RAM, *read-only memory* (ROM) can be accessed, in whole or in part, but not easily overwritten. A standard ROM chip is programmed at the factory. This permanent programming is known as *firmware.* There are also ROM devices that you can program and reprogram yourself. This scheme is called *programmable ROM* (PROM).

An *erasable programmable ROM* (EPROM) chip is an IC whose memory is of the read-only type, but that can be reprogrammed by a certain procedure. It is more difficult to rewrite data in EPROM than in RAM; the usual process for erasure involves exposure to ultraviolet (UV) radiation. An EPROM IC can be recognized by the presence of a transparent window with a removable cover, through which the UV is focused to erase the data. The IC must be taken from the circuit in which it is used, exposed to the UV for several minutes, and then reprogrammed via a special process. There are EPROMs that can be erased by electrical means. Such an IC is called an *electrically erasable programmable read-only memory* (EEPROM). This type of device does not have to be removed from the circuit for reprogramming.

Bubble Memory

A *bubble memory* system uses magnetic fields within ICs. The scheme allows a fairly large amount of data to be stored in a small physical volume. A so-called *bubble* is a magnetic field a tiny fraction of a millimeter across. Logic highs and lows correspond to the existence or absence, respectively, of a bubble. The IC contains a ferromagnetic film that acts as a reprogrammable permanent magnet on which bubbles are stored.

Magnetic bubbles do not disappear when power is removed from the IC. Bubbles are easily moved by electrical signals. An advantage of bubble memory is that it is nonvolatile RAM that does not need a backup battery. Another asset is that data can be moved from place to place in large chunks. This process is called *block memory transfer.*

Disks and Tapes

Personal and commercial computers almost always use *magnetic disks.* They come in two forms: the *hard disk* and the *diskette.* This is technically known as *storage;* it is sometimes inaccurately called "memory."

Another form of magnetic medium, convenient for mass data storage, is *magnetic tape.* Several tens of gigabytes can be stored on a small cassette via a *tape drive.*

Data can be stored and retrieved optically on *CD-ROM,* an abbreviation for *compact disk read-only memory.* This medium offers advantages over magnetic disks in some applications. Optical and magnetic technologies are combined in a *magneto-optical diskette* and drive.

Vacuum Tubes

When you think of a *vacuum tube* (often called an *electron tube* or simply a *tube*), you probably envision old radios in racks with strangely shaped, glowing glass globes. Maybe you have heard that these devices are obsolete. But some tubes are still in use. One common example is the *cathode-ray tube* (CRT) in a typical television (TV) set. The final power amplifier in a TV broadcast transmitter is usually a tube or set of tubes.

A vacuum tube allows electrons to be accelerated to high speed, resulting in large electric current. This current can be made more or less intense, or focused into a beam and guided in a particular direction. The intensity and/or beam direction can be changed with extreme rapidity, making possible a variety of different useful effects. In any vacuum tube, the charge carriers are *free electrons.* This means that the electrons are not bound to atoms, but instead they fly through space in a barrage, somewhat like photons of visible light, or like the atomic nuclei in a particle accelerator.

Diode Tube

Before the start of the twentieth century, scientists knew that electrons could carry a current through a vacuum. They also knew that hot electrodes would emit electrons more easily than cool ones. These phenomena were exploited in the first electron tubes, known as *diode tubes,* for the purpose of *rectification,* or converting AC to DC.

In any tube, the electron-emitting electrode is called the *cathode.* The cathode is usually heated by means of a wire *filament,* similar to the glowing element in an incandescent bulb. The electron-collecting electrode is known as the *anode* or *plate.* In some tubes, the filament also serves as the cathode. This is called a *directly heated cathode.* The negative supply voltage is applied directly to the filament. The filament voltage for most tubes is 6 or 12 V. The schematic symbol for a diode tube with a directly heated cathode is shown in Fig. 3-8A.

In many tubes, the filament is enclosed within a cylindrical cathode, and the cathode gets hot from infrared (IR) radiation. This is known as an *indirectly heated cathode.* The cathode itself is grounded. The filament normally receives 6 to 12 V AC. The symbol for a diode tube with an indirectly heated cathode is shown in Fig. 3-8B.

In either the directly heated or indirectly heated cathode, electrons are driven from the element by the heat of the filament. The cathode of a tube is analogous to the source of a FET, or to the emitter of a bipolar transistor. Because the electron emission in a tube depends on the filament (sometimes called the *heater*), tubes need from 30 seconds to a few minutes to warm up before they are ready for operation. This waiting period can be an annoyance, and it seems bizarre at first to people who have never dealt with tubes before. But in effect it is no different from waiting for a personal computer to boot up.

The plate of a tube is a cylinder that is concentric with the cathode and filament (Fig. 3-8C). The plate is connected to the positive DC supply voltage. Tubes operate at plate potentials ranging from a few volts to several kilovolts. Because the plate readily attracts electrons but is not a good emitter of them, and because the opposite is true of the cathode, a diode tube works well as a rectifier for AC. Diode tubes can also work as *envelope detectors* for amplitude-modulated signals, although they are no longer used for that purpose.

In a diode tube, the flow of electrons from cathode to plate depends mainly on the DC power supply voltage. The greater this voltage, the greater the current through the tube.

Figure 3-8
At A, schematic symbol for diode tube with directly heated cathode. At B, symbol for diode tube with indirectly heated cathode. At C, construction of a diode tube with indirectly heated cathode. For clarity, the electrodes are shown transparent (in reality they are opaque).

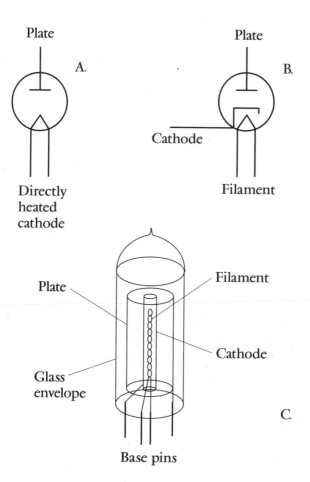

Triode Tube

The flow of current in a vacuum tube can be controlled via an electrode between the cathode and the plate. This electrode, called the *control grid,* is a wire mesh or screen that lets electrons physically pass through. But the control grid (also called simply the *grid*) interferes with the electrons if it is provided with a voltage that is negative with respect to the cathode. The greater this negative *grid bias,* the more the grid impedes the flow of electrons through the tube.

A tube with a control grid, in addition to the cathode and plate, is known as a *triode.* This is illustrated schematically in Fig. 3-9A. In this case the cathode is indirectly heated, so the filament is not shown. This omission is common in schematic diagrams showing vacuum

Figure 3-9
Symbols for triode
(A), tetrode (B), pen-
tode (C), hexode (D),
and heptode (E)
vacuum tubes.

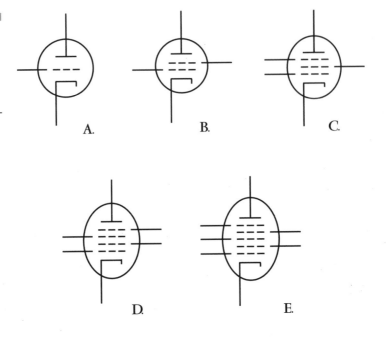

tubes with indirectly heated cathodes. When the cathode is directly heated, the filament symbol serves as the cathode symbol.

A triode tube is, in some respects, the analog of an N-channel FET. The cathode of the tube is analogous to the source of the FET; the grid of the tube is analogous to the gate of the FET; the plate of the tube corresponds to the drain of the FET. The major differences between the triode tube and the N-channel FET are that the tube is constructed differently, and the tube works with much higher voltages.

In a vacuum-tube-type RF power amplifier, the control grid is where the signal input is often applied, although the input can be applied to the cathode. A triode amplifier or oscillator circuit diagram looks like an FET circuit diagram, except that the tube plate and grid bias voltages are far greater than the FET drain and gate voltages.

Multigrid Tubes

Many vacuum tubes have more than one grid. The extra elements allow for improved gain and stability in tube-type amplifiers. A grid can be added between the control grid and the plate. This is a spiral of wire or a coarse screen, and is called the *screen grid* (or simply the *screen*). This grid normally carries a positive DC voltage, roughly one-fourth to one-third that of the plate voltage.

The screen grid reduces the capacitance between the control grid and plate, minimizing the tendency of a tube amplifier to oscillate. The screen grid can also serve as a second control grid, allowing two signals to be injected into a tube. This tube has four elements, and is known as a *tetrode*. Its schematic symbol is shown in Fig. 3-9B.

The electrons in a tetrode can strike the plate with extreme speed, especially if the plate voltage is high. The electrons might bombard the plate with such force that some of them bounce back or knock other electrons from the plate. Many of these electrons end up "leaking out" through the screen grid rather than going through the plate circuit of the tube. The result is diminished plate current and increased screen current. Because the plate current produces the useful output from the tube, this *secondary emission* hinders performance. If secondary emission is extensive, it can cause screen current so high that the screen grid can be physically damaged.

The problem of secondary emission can be dealt with by placing yet another grid, called the *suppressor grid* (or *suppressor*), between the screen and the plate. The suppressor reduces the capacitance between the control grid and the plate still more than is the case with the tetrode. Greater gain and stability are possible with a *pentode*, or tube with five elements, than with a tetrode or triode. The schematic symbol for a pentode is shown in Fig. 3-9C. The suppressor is usually grounded, or connected directly to the cathode. It carries a negative charge with respect to the screen and plate.

In some older radio and TV receivers, tubes with four or five grids were sometimes used. These tubes had six and seven elements, respectively, and were called *hexode* and *heptode*. The usual function of such tubes was signal mixing. The schematic symbol for a hexode is shown in Fig. 3-9D; the symbol for a heptode is illustrated in Fig. 3-9E. You will probably never hear about these devices in modern electronics, because solid-state components are used for signal mixing nowadays. You might elicit a raised eyebrow if you talk about a "pentagrid converter" or a "heptode mixer."

Applications

The most common application of vacuum tubes in modern technology is in RF amplifiers, especially at very high frequencies and/or power levels of more than 1 kilowatt (kW). Two configurations are employed: *grounded-cathode* and *grounded-grid*.

The input impedance of a grounded-cathode power amplifier is moderate; the plate impedance is high. Impedance matching between the amplifier and the load (usually an antenna) is obtained by tapping the coil of the output tuned circuit, or by using a transformer.

Grounded-cathode RF power amplifiers sometimes oscillate unless they are *neutralized,* or provided with some negative feedback. The oscillation usually occurs at some frequency far removed from the operating frequency. Known as *parasitic oscillation,* it can rob the amplifier of output power at the desired frequency, as well as causing interference to communications on the frequency or frequencies of oscillation. Neutralization can be a tricky business. When the tube is replaced, the neutralizing circuit must be readjusted. The adjustment is rather critical. Improved stability, without the need for neutralization, can be obtained by grounding the grid, rather than the cathode, of the tube in an RF power amplifier.

The grounded-grid configuration requires more driving power than the grounded-cathode scheme. While a grounded-cathode amplifier might produce 1 kW of RF output for 10 W of input, a grounded-grid amplifier needs about 50 to 100 W of driving power to produce 1 kW of RF output. The cathode input impedance is low, and the plate output impedance is high. The output impedance is matched by the same means as with the grounded-cathode arrangement.

Aside from RF power amplification, vacuum tubes are used in TV receivers and computer monitors. In any CRT, an *electron gun* emits a high-intensity stream of electrons. This beam is focused and accelerated as it passes through anodes that carry positive charge. The anodes of a CRT work differently than the anode of an RF vacuum tube. Rather than hitting the anode, the electrons pass on until they strike a screen whose inner surface is coated with phosphor. The phosphor glows visibly as seen from the face of the CRT.

Video cameras use another form of vacuum tube. A camera tube converts visible light into varying electric currents. The two most common types of camera tube are the *vidicon* and the *image orthicon.*

A *traveling-wave tube* is a form of electron-beam tube that is useful at ultra-high and microwave frequencies. There are several variations on this theme; the two most common are the *magnetron* and the *Klystron.*

Antiques

Vacuum tubes aren't used in receivers anymore, except as picture tubes in TV sets. This is because, at low signal levels, solid-state components (bipolar transistors, FETs, and integrated circuits) can do everything that tubes ever could, with greater efficiency and using lower voltages and currents.

Some electronics hobbyists like to work with antique radios. There is a certain charm in a broadcast receiver that takes up as much space as, and that weighs as much as, a small refrigerator. It brings back sentimental memories of a time when drama was broadcast on local AM radio, complete with whining heterodynes from other stations and sferics from distant storms. The action was envisioned in listeners' minds as they sat on the floor around the radio. The high (and potentially lethal) voltages attracted dust to interior components and wiring, giving the radio's innards a film of grit.

Perhaps you'd like to collect and operate radio antiques, just as some people collect and drive vintage cars. But you should be aware that replacement receiving tubes are hard to find, and they can also be rather costly. When your relic breaks, you'll have to become a spare-parts sleuth.

Displays

A *display* is a visual indication of the operating status of a piece of electronic equipment. A display might be as simple as the frequency readout in a communications receiver or transmitter. Or it might be as complicated as the video monitors used with computers. The physical layout of a display is important from the standpoint of operating efficiency and convenience.

Cathode-Ray Tubes

A *CRT* is a display device used in many video applications. Common examples are TV picture tubes, oscilloscope displays, and some computer displays.

In any CRT, an *electron gun* emits a high-intensity stream of electrons. This beam is focused and accelerated as it passes through electrodes that

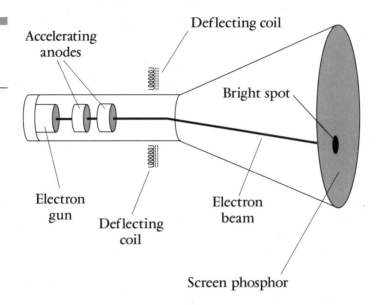

Figure 3-10
Simplified cross-sec-
tional drawing of a
cathode-ray tube.

carry positive charge. The electrons continue until they strike a screen whose inner surface is coated with phosphor. The phosphor glows visibly as seen from the face of the CRT. Unless something moves the beam around the screen of the CRT, you see nothing but a spot in the center of the screen. Beam deflection makes displays possible. This is done with electrostatic or magnetic fields.

A cross-sectional rendition of a monochrome, or single-color, CRT is shown in Fig. 3-10. There are two sets of *deflecting coils:* one for horizontal beam motion, and the other for vertical motion. (In the drawing, only one set of coils is shown for clarity.) These coils generate magnetic fields because they carry electric currents. The greater the electric currents in the coils, the stronger the magnetic fields will be, and the more the electron beam will be deflected.

The horizontal coils receive a certain current waveform, which causes the beam to *sweep* across the screen. After each sweep, the beam jumps instantly back for the next sweep. The vertical coils get another waveform. It makes the beam move down the screen. All this time, the electron beam is modulated, so the moving spot changes brightness in an intricate, complicated way. The end result is what you see on the screen.

In a color picture tube or computer monitor, there are three electron beams, one each for red, green, and blue colors. Each beam works independently of the other two. There are really three images superimposed on the screen: a red (R) image, a green (G) image, and a blue (B) image. They combine to form the color pictures you see. This color scheme is known as the *RGB color model.*

Liquid-Crystal Displays

A *liquid-crystal display* (LCD) is a solid-state, flat-screen device that can show geometric shapes. The simplest LCDs are used as alphanumeric displays in calculators, meters, wristwatches, and radios. More sophisticated LCDs are used in computers, particularly laptops and handheld units, and in portable television receivers.

All LCDs contain a fluid that changes its light-transmitting and light-reflecting properties in regions. The fluid is confined between transparent, electrically conductive plates. When a voltage is applied to the plates, the electric field causes a change in the molecules of the liquid. This alters the way light passes through the display, within the region containing the electric field.

Modern LCDs not only can change from light to dark (so-called *black on gray*), but can also exhibit colors. There is a limit to how fast the liquid can change state. In recent years, LCDs have been developed that are fast enough, at room temperature, to function as video displays in miniature fast-scan television sets. Such LCDs are also suitable for computer graphics applications. The speed of the LCD is affected by the temperature; extreme cold causes the LCD to change state more slowly.

One of the most significant advantages of the LCD, compared with a CRT display, is that the LCD needs much less power to operate. This makes it ideal for all kinds of portable electronic devices, where the batteries must last as long as possible. The LCD is an electrostatic device, consuming minimal power to change state, and practically no power to maintain a given state.

Another asset of the LCD is image resolution comparable to that of a CRT display. This has not always been the case; early LCD displays were notoriously bad in this respect. The newer LCDs allow graphics work to be done on laptop computers, where only text-based applications could be done before. The improvement in image resolution, along with the speed enhancement that has occurred since the mid-1980s, can be expected to continue in the future. Eventually, variations of the LCD will probably supplant the CRT in most, if not all, video applications.

Some types of LCD can be difficult to read from certain viewing angles. This situation has improved in recent years with the advent of *active-matrix*, also known as *thin-film-transistor* (TFT), displays. But in older LCD-equipped equipment, the display can force the user to physically change position in order to read or observe it with ease.

RF Antennas

An *antenna* is a radio-frequency (RF) transducer. Antennas can be broadly categorized into two major classes: receiving types and transmitting types. A *receiving antenna* converts an electromagnetic (EM) field to an alternating current (AC). A transmitting antenna converts AC to an EM field. Most transmitting antennas can function effectively for reception. Some receiving antennas can efficiently transmit EM signals; others cannot.

Radiation Resistance

When RF AC flows in an electrical conductor, some EM energy is radiated into space. If a resistor is substituted for the antenna, in combination with a capacitor or inductor to mimic any inherent reactance in the antenna, the transmitter behaves in the same manner as when connected to the actual antenna. For any antenna operating at a specific frequency, there exists a specific resistance, in ohms, for which this can be done. This is known as the *radiation resistance* (R_R) of the antenna at the frequency in question. Radiation resistance, like ordinary resistance, is specified in ohms.

Determining Factors

The radiation resistance of an antenna depends on several things. The main consideration is the length of the radiating element, as measured in free-space wavelengths. Objects near the radiating element, such as trees, buildings, utility wires, or other antenna elements, also affect the radiation resistance.

Suppose an extremely thin, straight, lossless vertical antenna is placed over perfectly conducting ground. Further suppose that there are no objects nearby that affect the radiation resistance. Then R_R will be a simple, predictable function of the vertical-antenna height in wavelengths. This function is approximately graphed in Fig. 4-1A. For a quarter-wavelength ideal vertical antenna, the radiation resistance is about 37 ohms. As the conductor length decreases, the radiation resistance decreases. As the conductor becomes longer than a quarter wavelength, the radiation resistance increases, and becomes very high as the height approaches a half wavelength.

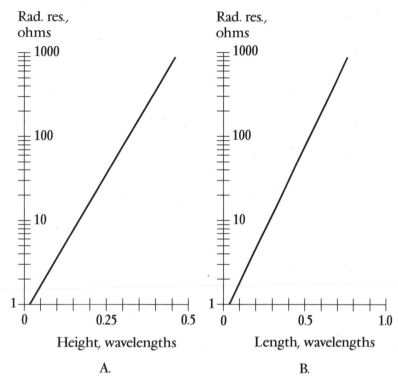

Figure 4-1
Approximate values
of radiation resistance
for vertical antennas
over perfectly con-
ducting ground (A)
and for center-fed
antennas in free
space (B).

Suppose a conductor is placed in free space and is fed at the exact center. Again, assume that the conductor is straight, thin, lossless, and far away from objects that might affect it. Then the value of the radiation resistance, as a function of the antenna length in wavelengths, is approximately shown by Fig. 4-1B. When the conductor measures a half wavelength (or a quarter wavelength on either side of the feed line), the radiation resistance is about 73 ohms. As the conductor becomes shorter, the radiation resistance gets smaller, approaching zero. As the conductor length approaches a full wavelength (or a half wavelength on either side of the feed line), the feed-point radiation resistance becomes very high.

Significance

Ideally, an antenna should be designed so it has a large value of radiation resistance. This is because the efficiency of an antenna depends on the ratio of the radiation resistance to the total antenna system resistance R. The greater the ratio R_R/R, the better the antenna will work, all other things being equal. In effect, the radiation resistance appears in series with any *loss resistance* (R_L) that exists in the antenna system. When the radiation resistance is much higher than the loss resistance, most of the power appears in the radiation resistance, and is radiated into space. Conversely, if the radiation resistance is much lower than the loss resistance, most of the power appears in the loss resistance, and is dissipated as heat in the earth, the antenna loading apparatus, and surrounding objects such as trees and electrical wiring.

The total antenna system resistance R is equal to R_R+R_L. The antenna efficiency is given by

$$\text{Eff}=R_R/R=R_R/(R_R+R_L)$$

which is the ratio of the radiation resistance to the total antenna system resistance. As a percentage,

$$\text{Eff} = 100\,[R_R/(R_R+R_L)]$$

Antenna efficiency is always less than 1 (or 100 percent). No antenna is perfect; there is always some loss resistance. The loss resistance R_L in a practical antenna system is generally at least a few ohms, even when every effort is made to minimize it. In cases where a vertical antenna is operated against poorly conducting earth, the loss resistance can be 100 ohms or more. If R_R is extremely low (say, 0.5 ohm), even the most carefully engineered antenna will be inefficient in most practical situations. This is the main trouble with short radiators operating against earth ground at low frequencies. In contrast, if R_R is very high (say, 2000 ohms), the antenna will radiate most of the power supplied to it even if R_L is considerable.

Some short radiators can function efficiently if the earth is kept out of the antenna circuit and if the antenna is set up in a favorable location. Small loop antennas, for example, if designed with large-diameter tubing, low-loss tuning networks, and excellent electrical connections, will radiate most of the power they receive even when they are considerably smaller than a quarter wavelength in circumference. But in practice, there is a limit to how far this can be carried. No one has yet constructed a loop antenna 10 millimeters (mm) in diameter that will efficiently radiate EM energy at 500 kilohertz (kHz), for example. Such an antenna might be made to work fairly well if cooled to temperatures near absolute zero, so that the wires become superconductive. The problems with this scheme, from the standpoint of the everyday radio communications user, are high expense and a complicated maintenance requirement.

Half-Wave Antennas

A *half-wave antenna* is a radiating element that measures an electrical half wavelength. In practice, such an antenna can have a length anywhere from practically zero to about 95 percent of the physical dimensions of a half wavelength in free space. In theory, a half wavelength in free space is given in feet according to the equation:

$$L=492/f$$

where L is the linear distance and f is the frequency in megahertz. A half wavelength in meters is given by

$$L=150/f$$

In practice, an additional factor must be incorporated into the above equations, because EM fields travel somewhat more slowly along the conductors of an antenna than they do in free space. For ordinary wire, the results as obtained above should be multiplied by a *velocity factor v* of 0.95 (95 percent). For tubing or large-diameter conductors, v is somewhat smaller, and can range down to about 0.90 (90 percent).

A half-wave antenna can be made shorter than a physical half wavelength. This is accomplished by inserting inductance in series with the radiator. The antenna can be made longer than the physical half wavelength by inserting capacitance in series with the radiator.

Basic Dipole

The terms *dipole, dipole antenna,* and *doublet* are used to describe a half-wavelength radiator fed at the center with a two-wire or coaxial transmission line. Each "leg" of the antenna is a quarter wavelength long (Fig. 4-2A). Such an antenna can be oriented horizontally or vertically, or at a slant. The radiating element is usually straight.

Figure 4-2
Basic half-wave
antennas. At A,
dipole antenna. At B,
folded-dipole anten-
na. At C, zeppelin
antenna.

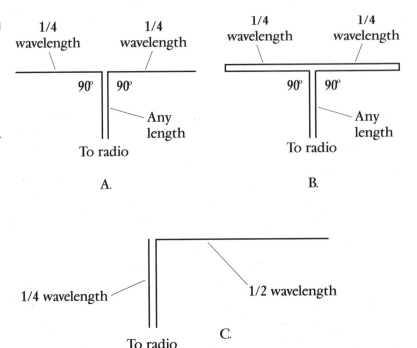

For a straight-wire radiator, properly insulated at the ends and placed well away from obstructions, the length L, in feet (ft), at a design frequency f, in megahertz (MHz), for a practical half-wave dipole is close to

$$L=467/f$$

The length in meters is close to

$$L=143/f$$

These values assume $v=0.95$.

A half-wavelength conductor displays resonant properties for EM energy. In free space—that is, when there are no objects near the radiator—the impedance at the center of a dipole is about 73 ohms, purely resistive. The impedance is also a pure resistance at all *harmonic frequencies*. At *odd harmonics*, the impedance is about the same as, or slightly higher than, the impedance at the *fundamental frequency*, which is the frequency at which the antenna is half-wave resonant. At *even harmonics*, the impedance is very high, on the order of 1000 ohms or more.

Because of its relative simplicity, the dipole antenna is popular among shortwave listeners and radio amateurs. This is especially true at frequencies below 10 MHz, where more complicated antennas are often impractical. At higher frequencies, *parasitic elements* are often added to the dipole, creating *power gain* and enhanced *directivity*. Dipoles can also be fed in various multiple configurations to obtain power gain and directivity.

A simple dipole, having a straight element fed at the center, exhibits maximum radiation and response in the plane perpendicular to the line containing the element. The radiation and response diminish as the angle relative to the element decreases. Directly off the ends of the element, the radiation and response are theoretically zero.

Folded Dipole

A *folded-dipole antenna* is a half-wavelength, center-fed antenna constructed of two parallel wires with their ends connected together (Fig. 4-2B).

The folded dipole has exactly the same gain and radiation pattern, in free space, as a simple dipole antenna. However, the feed-point impedance of the folded dipole is four times that of the ordinary dipole. Instead of approximately 73 ohms, the folded dipole presents a resistive impedance of about 300 ohms. This makes the folded dipole desirable for use with high-impedance, parallel-wire transmission lines. It also can be used to obtain a good match with 75-ohm coaxial cable when four antennas are connected in phase, or with 50-ohm coaxial cable when six antennas are connected in phase.

Folded dipoles are often found in vertical *collinear arrays*, such as those used in repeaters at very high frequencies (VHF) and ultra-high frequencies (UHF). Folded dipoles have somewhat greater bandwidth than ordinary dipoles, and this makes them useful for reception in the frequency-modulation (FM) broadcast band, between 88 and 108 MHz. A folded dipole cut for 98 MHz will provide decent reception over the entire band. The dimensions of a half-wave folded dipole for a given frequency can be calculated using the formulas for an ordinary dipole.

Zeppelin Antenna

A *zeppelin antenna*, also called a *zepp*, is a half-wavelength radiator, fed at one end with a quarter-wavelength section of open-wire line

(Fig. 4-2C). The antenna gets its name from the fact that it was originally used for radio communications aboard zeppelins and dirigibles. The entire antenna system, feed line and all, was simply dangled in flight.

The zeppelin antenna can be thought of as a current-fed, full-wavelength radiator, part of which has been folded up to form the transmission line. The impedance at the feed point is extremely high, and at the transmitter end of the line, the impedance is low. A zeppelin antenna will operate well at all harmonics of the design frequency. If a transmatch is available, a half-wavelength radiator can be end-fed with an open-wire line of any length, not just a quarter wavelength. This arrangement is popular among radio amateurs.

The primary advantage of the zeppelin antenna is ease of installation. The feed line does not have to come away from the antenna element at any particular angle, although the larger the angle, the better (180 degrees is ideal). A zepp can be "flown" with a kite or helium balloon, provided a separate tether is used to keep the antenna from breaking loose in a gust of wind, and provided all possible precautions are taken to ensure that the antenna will not attract lightning, come down on utility wires, or present any other sort of hazard to people or property.

The main difficulty with the zepp is that the feed line radiates to some extent, because the system is not perfectly balanced. Feed-line radiation can be kept to a minimum by carefully cutting the radiator to a half wavelength at the design frequency, and by using the antenna only at this frequency or one of its harmonics. The half-wave dipole formulas can be used for calculating the length of a zeppelin antenna. Some "pruning" will probably be necessary to minimize radiation from the line. A field-strength meter can be used to observe the line radiation while adjusting the length of the antenna for optimum performance at the design frequency.

The radiation pattern from a zepp antenna is the same as that from a dipole at the design frequency, provided the radiation from the line is not excessive. At harmonic frequencies, the radiation pattern becomes more complex. At the frequencies at which a zepp is typically used (10 MHz and below), this complexity is not normally a concern. Radiation patterns from most horizontal wire antennas below 10 MHz are modified by nearby objects such as utility wires and steel-frame buildings anyhow, and are therefore almost impossible to predict based on theoretical models.

Quarter-Wave Antennas

A *quarter wavelength* is the distance that corresponds to 90 degrees of phase as an EM field is propagated. In free space, it is related to the frequency by a simple equation:

$$L = 246/f$$

where L represents a quarter wavelength in feet, and f represents the frequency in megahertz. If L is expressed in meters, then the formula is

$$L = 75/f$$

In media other than free space, such as in an electrical conductor or transmission line, EM waves propagate at speeds less than their speed in free space. The wavelength of a disturbance having a given frequency depends on the speed of propagation. In general, if v is the velocity factor in a given medium, then

$$L = 246v/f$$

if L is expressed in feet, and

$$L = 75v/f$$

if L is expressed in meters. For a typical wire conductor, $v = 0.95$ (95 percent); for metal tubing, v can range down to about 0.90 (90 percent). For a vertical antenna constructed from aluminum tubing such that $v = 0.92$, for example, the formulas are

$$L = 226v/f$$

if L is expressed in feet, and

$$L = 69v/f$$

if L is expressed in meters. There is some margin for error in these determinations, because the exact value of v depends on the diameter of the tubing relative to the wavelength of the EM field.

A quarter-wave antenna must be operated against a good RF ground. The RF ground causes a quarter-wavelength radiator to behave, in some

respects, like a half-wave, center-fed antenna. Quarter-wavelength antennas are almost always vertically oriented, so they radiate and receive
vertically polarized EM fields.

Ground-Mounted Quarter-Wave Vertical

The simplest form of vertical antenna is a quarter-wave radiator mounted at ground level. The radiator is fed with a coaxial cable. The center
conductor is connected to the base of the radiator, and the shield is connected to a ground system. The feed-point impedance of a ground-
mounted vertical is a pure resistance, and is equal to 37 ohms plus the
ground-loss resistance. The ground-loss resistance can dramatically
affect the efficiency of this type of antenna system. A set of *radials*
helps to minimize the loss. These radials can be buried just below the
ground surface, and should be at least a quarter wavelength long for
best results.

At some frequencies, the height of a quarter-wave vertical is unman-
ageable unless *inductive loading* is used to reduce the physical length of
the radiator. This technique is used mostly at very low, low, and medium frequencies, and to some extent at high frequencies. A quarter-wave
vertical antenna can be made resonant on several frequencies by the
use of multiple loading coils, or by inserting *traps* at specific points
along the radiator. Traps are *band-rejection filters* that cause multiple resonances to appear in a single antenna element. The problem with
inductive loading and trap tuning is that these schemes reduce the
radiation resistance, thereby reducing the efficiency of a ground-
mounted antenna. These designs also tend to reduce the antenna bandwidth, or the range of frequencies over which it can be operated
effectively.

The chief advantage of a ground-mounted vertical antenna is its
convenience. It takes up practically no real estate (unless it is so tall that
guy wires are needed). Vertical antennas are omnidirectional in the
horizontal plane, neglecting the effects of obstructions such as large
buildings and hills. Vertical antennas are preferred at very low frequencies (VLF) and low frequencies (LF), because they offer superior *surface
wave* reception and radiation, an important mode at long wavelengths.

A major problem with ground-mounted vertical antennas is the fact
that, unless an extensive radial system is installed, the system will usually be inefficient. Another potential problem is that vertically polarized antennas in general receive more human-made noise than

horizontal antennas. Ground-mounted antennas are closer to such noise sources than antennas mounted at a height, such as on top of a tower, and the noise from human-made sources tends to be vertically polarized. The EM fields from ground-mounted transmitting antennas are more likely to cause interference to electronic devices than are the EM fields from antennas installed at a height.

Ground-Plane Antenna

A *ground-plane antenna* is a vertical radiator operated against a system of quarter-wave radials and elevated at least a quarter wavelength above the earth's surface. The radiator itself can be any length, but should be tuned to resonance at the desired operating frequency.

When a ground plane is elevated at least 90 electrical degrees above the surface, only three or four radials are necessary in order to obtain an almost lossless system. The radials are run outward from the base of the antenna at an angle between 0 and 45 degrees with respect to the horizon. Figure 4-3A illustrates a typical ground-plane antenna.

A ground-plane antenna is an unbalanced system, and should be fed with coaxial cable. The feed-point impedance of a ground-plane antenna having a quarter-wave radiator is about 37 ohms if the radials are horizontal; the impedance increases as the radials are "drooped," reaching about 50 ohms at an angle of 45 degrees. The radials can be run directly downward, in the form of a quarter-wave cylinder concentric with the feed line. Then the feed-point impedance is approximately 73 ohms. This configuration is known as a *coaxial antenna* (Fig. 4-3B) and is sometimes used at frequencies above 10 MHz.

Figure 4-3
At A, basic ground-plane antenna. At B, coaxial antenna.

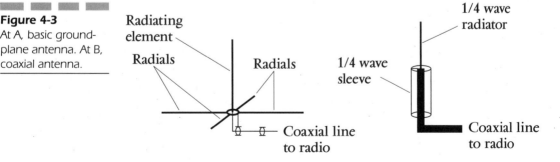

Loop Antennas

Any receiving or transmitting antenna, consisting of one or more turns of wire forming a DC short circuit, is called a *loop antenna*. Loop antennas can be categorized as either small or large.

Small Loops

A *small loop antenna* has a circumference of less than 0.1 wavelength at the highest operating frequency. Such antennas are suitable for receiving, but generally not for transmitting, because the radiation resistance is extremely low. A small loop is the least responsive along its axis, and is most responsive in the plane perpendicular to its axis. In fact, the response pattern is almost exactly the same shape as that of a half-wave dipole antenna, if the loop's axis coincides with the line containing the dipole element.

A small loop might have only one turn of wire, or it might contain many turns. A small loop is nonresonant; that is, it does not discriminate against or in favor of any particular frequency. However, a capacitor can be connected either in series or in parallel with the loop, in order to provide a resonant response. An example of such an arrangement is shown in Fig. 4-4.

Small loops are useful for direction finding, and also for reducing interference caused by human-made noise or strong local signals. The null along the axis is sharp, covering a narrow angle; it is also deep, causing a dramatic decrease in the strength of signals or noise that fall into it.

Loopsticks

For receiving applications at low, medium, and high frequencies up to approximately 20 MHz, a *loopstick antenna* is sometimes used. This antenna, a variant of the small loop, consists of a coil wound on a rod-shaped, powdered-iron core. A series or parallel capacitor, in conjunction with the coil, forms a tuned circuit. The operating frequency is determined by the resonant frequency of the inductance-capacitance (LC) combination.

Loopsticks display directional characteristics similar to those of the dipole antenna. The sensitivity is maximum off the sides of the coil, and

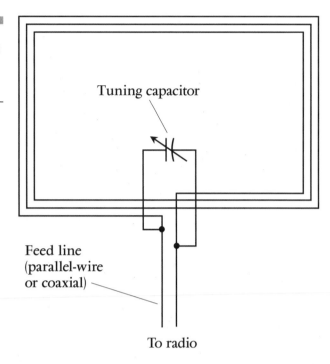

Figure 4-4
A small loop antenna with a capacitor for adjusting the resonant frequency.

Tuning capacitor

Feed line
(parallel-wire
or coaxial)

To radio

a sharp null occurs off the ends. This has little effect with sky-wave signals, which tend to arrive from varying directions. However, the null can be used to advantage to eliminate interference from local signals and from human-made sources of noise. The antenna is physically tiny; it might be less than 5 centimeters (cm) in length and 1 cm in diameter. This makes it easy to orient the antenna in any direction.

Loopstick antennas are not generally used at VHF and UHF, because the required inductance is so small that the coil would have very few turns, and capacitive effects would predominate. This would offset the directional advantages of the antenna.

Large Loops

If a loop has a circumference greater than 0.1 wavelength at the operating frequency, it is generally classified as a *large loop antenna*. Such an antenna normally has a circumference of either 0.5 or 1 wavelength, and is self-resonant without the need for a tuning capacitor. The half-wavelength loop presents a very high impedance at the feed point, and the maximum radiation occurs in the plane of the loop (perpendicular to the axis). The full-wavelength loop presents an impedance of about 50 ohms at the feed point, and the maximum radiation occurs along the axis.

A large loop can be used for either transmitting or receiving. The half-wavelength loop exhibits a slight power loss relative to a dipole, but the full-wavelength loop shows a little bit of gain over a dipole in its favored directions. The full-wavelength loop forms the driven element for the well-known *quad antenna*, which is popular at high frequencies, especially among amateur radio operators.

Ground Systems

It has been said that an antenna can be only as good as its ground system. This is true for some types of antennas, but not for all. In general, unbalanced antenna systems such as a quarter-wave vertical require a low-loss RF ground to operate "against," but most balanced systems, such as the half-wave dipole, do not. A low-loss DC ground, however, is advisable for any antenna system and, in fact, for all electronic installations.

Electrical Ground versus RF Ground

All electronic equipment should, if possible, be connected to a good *electrical ground* system. Grounding is important for personal safety. It can help protect equipment from damage if lightning strikes nearby. Grounding minimizes the chances for *electromagnetic interference* (EMI) to and from the equipment.

There's an unwritten commandment in electrical and electronics engineering: "Never touch two grounds at the same time." This refers to the tendency for a voltage to exist between different system points when you don't suspect it, and even when you don't think it is possible. Many people have received severe electrical shocks because they forgot this simple but important rule.

In recent years, appliances have been equipped with three-wire cords. One wire is connected to the "common" or "ground" part of the hardware, and leads to a D-shaped or U-shaped prong in the plug. This ground prong should never be cut off or otherwise defeated, because such modification can result in dangerous voltages appearing on exposed metal surfaces. You can recognize three-wire electrical systems by the appearance of the wall outlets. If each outlet has a sideways-D-shaped hole below the two rectangular holes, then you have a three-wire system. But for the system to be effective, the contact in this D-shaped hole must actually be connected to an earth ground. If you are in doubt about this, have an electrician check it out.

Many consumer electronic devices, such as computers, television (TV) receivers, high-fidelity stereos, and telephone sets, can be affected by strong EM fields. Computers and TV sets also generate radio waves of their own. Some amateur radio operators find that they cannot use their computers and their "ham rigs" at the same time. The computer can interfere with radio reception, and a high-powered transmitter occasionally disrupts the operation of the computer. A good *RF ground* system will help minimize the chances of your experiencing EMI.

Figure 4-5 shows a good RF ground scheme (A) and a bad one (B). In a good RF ground system, each device is connected to a common *ground bus*, which in turn runs to the earth ground via a single conductor. This conductor should be as short as possible. A bad ground system contains *ground loops* that increase the susceptibility of equipment to EMI.

A good ground system minimizes the possibility of "RF in the shack" problems when high-power RF transmitting equipment is used. As previously stated, a good RF ground is generally more important with an

Figure 4-5
At A, the correct method for grounding multiple units. At B, an incorrect method creates RF ground loops.

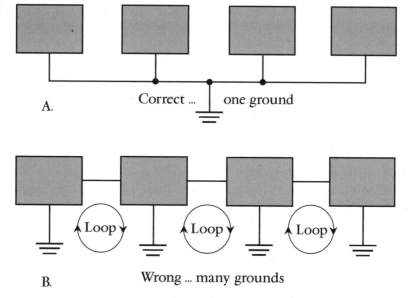

A.   Correct ... one ground

B.   Wrong ... many grounds

unbalanced antenna system than with a balanced antenna system. But in any EM communications or broadcast installation, it is good engineering practice to maintain the best RF ground possible, irrespective of other considerations.

Radial and Counterpoise Systems

A *radial* is a conductor used to enhance the ground system of an unbalanced, vertical antenna. Radials can be constructed from wire or metal tubing. They generally measure a quarter wavelength or more.

When a vertical antenna is mounted at ground level, the earth alone will rarely serve as an adequate RF ground. The ground conductivity is improved by the installation of radials. The radials are run outward from the base of the antenna, and are connected to the shield or "braid" of the coaxial feed line. The radials can be installed just above the ground, or buried under the surface. The greater the number of radials of a given length, the better the antenna will work. Also, the longer the radials for a given number, the better. Some amplitude-modulation (AM) broadcast stations, operating in the band between 535 and 1605 kHz, have antennas with 360 radials, one for every azimuth degree, each measuring a half wavelength at the broadcast frequency.

If a vertical antenna is placed well above the ground level, forming a ground-plane antenna, there need only be three or four radials of exactly one-fourth wavelength. The radials in this type of antenna sometimes run down at a slant, rather than horizontally outward; such radials are called *drooping radials*.

Radials can be used to obtain a good RF ground for any type of EM communications or broadcast installation. A large number of radials measuring at least a quarter wavelength, attached to a central ground rod, can improve transmitting efficiency and reduce received noise. The ground rod should be located as close to the station as possible, and the wire from the rod to the station should be as straight, and as large in diameter, as possible.

A *counterpoise* is a means of obtaining an RF ground or ground plane without a direct earth-ground connection. A grid of wires, a screen, or a metal sheet is placed above the ground surface, and oriented horizontally, to provide capacitive coupling to the earth. This greatly reduces ground loss at radio frequencies. Ideally, the radius of a counterpoise should be at least a quarter wavelength at the lowest operating frequency for a given system. The counterpoise is especially useful at locations where the soil conductivity is poor, rendering a direct ground connection ineffective.

Gain and Directivity

The *power gain* of a transmitting antenna is the ratio of the maximum *effective radiated power* (ERP) to the actual RF power applied at the feed point. Power gain is usually expressed in *decibels* (dB). If the ERP is P_{ERP} W and the applied power is P W, then

$$\text{Power gain (dB)}=10 \log_{10} (P_{ERP}/P)$$

Power gain is always measured in the favored direction of an antenna. The favored direction is the azimuth direction in which the antenna performs the best.

For power gain to be defined, a *reference antenna* must be chosen with a gain assumed to be unity, or 0 dB. This reference antenna is usually a half-wave dipole in free space. Power gain figures taken with respect to a dipole are expressed in units called dBd. The reference antenna for power-gain measurements can also be an *isotropic radiator*, which theoretically radiates and receives equally well in all directions in three dimensions. In this case, units of power gain are called dBi. For any given antenna, the power gains in dBd and dBi are different by approximately 2.15 dB:

$$\text{Power gain (dBi)}=2.15+\text{power gain (dBd)}$$

For efficient transmitting antennas, the gain is usually the same for reception as for transmission of signals. Therefore, when directional antennas are used at both ends of a communications circuit, the effective gain over a pair of dipoles is the sum of the individual antenna power gains in dBd.

Directivity Plots

The directional characteristics of any transmitting or receiving antenna, when graphed on a polar coordinate system, are called the *directivity plot*.

Figure 4-6
Directivity plots for a
dipole antenna. At A,
H-plane; at B,
E-plane.

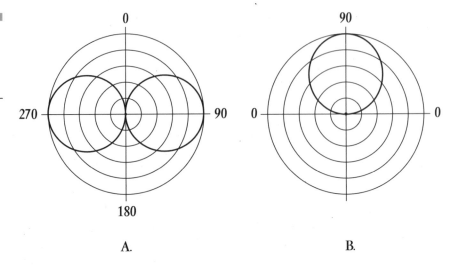

A.

B.

The simplest possible directivity plot is that of an isotropic antenna, although this is a theoretical ideal.

Antenna radiation and response patterns are represented by plots such as those shown in Fig. 4-6. The location of the antenna is assumed to be at the center (origin) of the coordinate system. The greater the radiation or reception capability of the antenna in a certain direction, the farther from the center the points on the chart are plotted.

A dipole antenna, oriented horizontally so that its conductor runs in a north-south direction, has a horizontal-plane (H-plane) pattern similar to that in Fig. 4-6A. Azimuth directions are given in degrees, where 0 represents geographic north, 90 represents east, 180 represents south, and 270 represents west. The elevation-plane (E-plane) pattern depends on the height of the antenna above effective ground at the viewing angle. With the dipole oriented so that its conductor runs perpendicular to the page, and the antenna a quarter wavelength above effective ground, the E-plane antenna pattern will resemble the graph shown at B. Elevation angles are given in degrees. Only the top half of the coordinate system is used, because the bottom half represents points that are beneath the earth's surface.

The patterns in Fig. 4-6 are simple. Many antennas have complicated directivity plots. For all antenna pattern graphs, the power gain relative to a dipole is plotted on the radial axis. The values thus range from 0 to 1 on a linear scale. Sometimes a logarithmic scale, showing gain in decibels, is used.

Forward Gain

The *forward gain* of a directional antenna is an indicator of how well it concentrates transmitted RF energy in its favored direction. This specification also tells how well a directional receiving antenna focuses its response in a favored direction. Forward gain is sometimes called *antenna power gain* in transmitting applications.

Forward gain is expressed as the level, in decibels, of the ERP in the antenna's favored direction (also called the *main lobe*) to the ERP from a reference antenna in its favored direction. The reference antenna is usually a half-wave dipole. A typical two-element *phased array* has a forward gain of 3 dBd. As more in-phase elements are added, the gain increases. A typical two-element *parasitic array* such as a *Yagi antenna* has a gain of about 5 dBd when optimized.

Some antennas have forward gain figures exceeding 25 dBd. At microwave frequencies, large *dish antennas*, such as those used in radio telescopes, can be built to have forward gain upward of 35 dBd. The limit is determined only by practical constraints, and by the wavelength. In general, as the wavelength decreases (the frequency gets higher), it becomes easier to obtain high forward gain figures.

Front-to-Back Ratio

The *front-to-back ratio* of a directive antenna, abbreviated f/b, is an expression of the ability of an antenna to concentrate its radiation or response in its favored direction. The term applies to *unidirectional antennas*, that is, antennas having one favored direction. Examples of such systems are Yagi antennas, *quad antennas*, some phased arrays, *helical antennas*, *horn antennas*, dish antennas, and certain types of *longwire antennas*.

The f/b ratio is nearly always specified in decibels. The EM field strength in the favored direction is compared with the field strength exactly opposite the favored direction at the same distance from the antenna. Measurements can be made with a calibrated field-strength meter.

Figure 4-7 shows a simple, hypothetical directivity plot for a unidirectional antenna pointed north. The circles in this polar graph, typical of antenna directivity plots, represent relative field strength from the antenna. The outermost circle depicts the field strength in the direction of the center of the main lobe, and represents 0 dB. The next smaller

Figure 4-7
Directivity plot for a
hypothetical anten-
na. Front-to-back and
front-to-side ratios
can be determined
from such a graph
(see text).

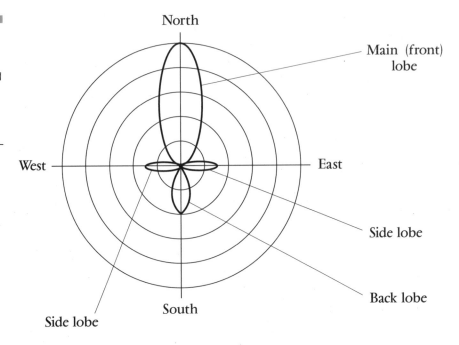

circle represents a field strength 5 dB down with respect to the main
lobe. Continuing inward, circles represent 10 dB down, 15 dB down, and
20 dB down. The center, or origin, of the graph represents 25 dB down,
and also shows the position of the antenna. From this illustration, it
can be determined that the f/b ratio of this system is 15 dB; the *back
lobe* is oriented toward the south.

Front-to-Side Ratio

The *front-to-side ratio* of a directive antenna, abbreviated f/s, is another
expression of the directivity of an antenna system. The term applies to
unidirectional antennas (having one favored direction) and to bidirectional antennas (having two favored directions opposite each other). There are
numerous types of directional arrays that fall into these categories.

The f/s ratio, like the f/b ratio, is expressed in decibels. The EM field
strength in the favored direction is compared with the field strength at
right angles to the favored direction at the same distance from the
antenna. An example is shown in Fig. 4-7. From this plot, it can be determined that the f/s ratio for the antenna in question is about 17 or 18 dB.
The side lobes are oriented toward the east and west.

Phased Arrays

A *phased array* is an antenna having two or more *driven elements*—that is, two or more elements directly connected to the feed line. The elements are fed with a certain relative phase, and they are spaced at a certain distance, resulting in a directivity pattern that exhibits gain in some directions and little or no radiation in other directions.

Concept

Phased arrays can be simple, consisting of only two elements. Two examples of simple pairs of phased dipoles are shown in Fig. 4-8. At A, the two dipoles are spaced one-quarter wavelength apart in free space, and they are fed 90 degrees out of phase. The result is that the signals from the two antennas add in phase in one direction, and cancel in the opposite direction, as shown by the arrows. In this particular case, the radiation pattern is unidirectional. However, phased arrays can have directivity patterns with two, three, or even four different optimum directions. A bidirectional pattern can be obtained, for example, by spacing the dipoles one wavelength apart and feeding them in phase, as shown at B.

More complicated phased arrays are sometimes used by radio transmitting stations. Several vertical radiators, arranged in a specified pattern and fed with signals of specified phase, produce a designated directional pattern. This is done to avoid interference with other broadcast stations on the same channel, and/or to optimize coverage for the intended listening audience.

Phased arrays can have fixed directional patterns, or they might have rotatable or steerable patterns. The pair of phased dipoles shown in Fig. 4-8A can, if the wavelength is short enough to allow construction from metal tubing, be mounted on a rotator for 360-degree directional adjustability. With phased vertical antennas, the relative signal phase can be varied, and the directional pattern thereby adjusted to a certain extent.

Longwire Antennas

A wire antenna measuring a full wavelength or more, and fed at a high-current point or at one end, comprises a *longwire antenna*. Such antennas

Figure 4-8
At A, a unidirectional
phased system. At B,
a bidirectional
phased system. These
are both examples of
end-fire arrays.

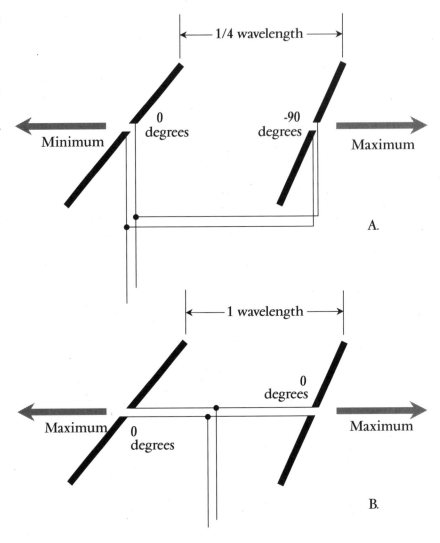

Figure 4-8
At A, a unidirectional phased system. At B, a bidirectional phased system. These are both examples of end-fire arrays.

are sometimes used for receiving and transmitting at medium and high frequencies.

Longwire antennas offer some power gain over a half-wave dipole. As the wire is made longer, the main lobes get more nearly in line with the antenna, and their amplitudes increase. As the wire is made shorter, the main lobes get farther from the axis of the antenna, and their amplitudes decrease.

Longwire antennas have certain advantages. They offer considerable gain and low-angle radiation, provided they are made long enough.

The gain is a function of the length of the antenna: the longer the wire, the greater the gain. A longwire antenna must be as straight as possible for proper operation.

There are two main disadvantages to the longwire antenna. First, it cannot conveniently be rotated to change the direction in which maximum gain occurs. Second, a great deal of real estate is needed to install such an antenna, especially in the medium-frequency (MF) spectrum and the longer high-frequency (HF) wavelengths. For example, a 10-wavelength longwire antenna at 7 MHz measures 1340 ft from end to end. This is approximately $\frac{1}{4}$ mi or 0.4 km. It is not the sort of antenna that most people can install in their backyards!

Broadside Array

A *broadside array* is a phased group of antennas, arranged in such a way that the maximum radiation occurs in directions perpendicular to the plane containing the driven elements. This requires that all of the antennas be fed in phase. A *phasing harness* is required to accomplish this. Figure 4-9 shows the geometric arrangement of a broadside array.

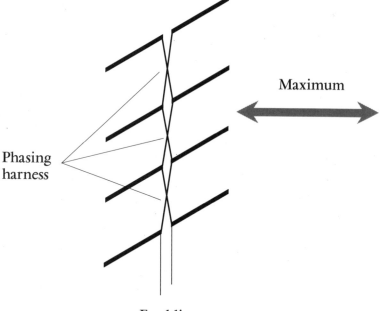

Figure 4-9
A broadside array. The elements are all fed in phase. Maximum radiation and response occur perpendicular to the plane containing the elements.

Phasing harness

Maximum

Feed line

The elements can be either half-wave dipoles, or they can be full-wave, center-fed straight conductors. Full-wave elements have a slight inherent gain over half-wave elements. In this example, there are four driven antennas in the array. However, the array could have as few as two or as many as a dozen or more elements.

At HF, broadside arrays are often constructed from two driven antennas. At VHF and UHF, there are usually several driven antennas. The antennas themselves can each consist of a single element, as shown in the figure, or they can consist of Yagi antennas, loops, or other systems with individual directive properties. If a reflecting screen is placed behind the array of dipoles in Fig. 4-9, the system becomes a *billboard antenna*.

The directional properties of a broadside array depend on the number of elements, whether or not the elements have gain themselves, and on the spacing among the elements. In general, the larger the number of elements, the greater the gain.

End-Fire Array

An *end-fire array* is a bidirectional or unidirectional antenna in which the greatest amount of radiation takes place off the ends. An end-fire antenna consists of two or more parallel driven elements if it is a phased array; all of the elements lie in a single plane. A parasitic array is sometimes considered an end-fire antenna, although the term is used primarily for phased systems.

A typical end-fire array might consist of two parallel half-wave dipole antennas, fed 90 degrees out of phase and spaced a quarter wavelength apart in free space. This will produce a unidirectional radiation pattern. Or, the two elements might be driven in phase and spaced at a separation of one wavelength. This results in a bidirectional radiation pattern. In the phasing system, the two branches of transmission line must be cut to precisely the correct lengths, and the velocity factor of the line must be known and be taken into account. The antenna systems shown in Fig. 4-8 are both examples of end-fire arrays.

End-fire systems show some power gain, in their favored directions, compared to a single half-wave dipole antenna. The larger the number of elements, with optimum phasing and spacing, the greater the realizable power gain.

Parasitic Arrays

Parasitic arrays are commonly used at HF, VHF, and UHF for obtaining directivity and forward gain. Common examples include the *Yagi antenna* and the *quad antenna.*

Concept

A *parasitic element* is an electrical conductor that comprises an important part of an antenna system, but that is not directly connected to the feed line. Parasitic elements are used for the purpose of obtaining directivity and power gain. They operate via EM coupling to the driven element.

The principle of parasitic-element operation was first discovered by Japanese engineers named Yagi and Uda. Antennas using such elements were originally called *Yagi-Uda arrays,* but the name was later shortened to Yagi. These two engineers observed that antenna elements parallel to the driven element but not connected to anything, at a specific distance from the driven element, and having a certain length, cause the radiation pattern to show gain in one direction and loss in the opposite direction. When gain is produced in the direction of the parasitic element, the parasitic element is called a *director.* When gain is produced in the direction opposite the parasitic element, the parasitic element is known as a *reflector.*

Yagi Antenna

The Yagi, sometimes called a *beam antenna,* is an array of parallel, straight elements, usually made of aluminum or stainless-steel tubing or rods. One or more of the elements is driven, and one or more is parasitic. The elements all lie in the same plane. This plane can be oriented horizontally, vertically, or slantwise.

Yagi antennas are popular among amateur-radio operators at frequencies from 7 MHz through the UHF bands. Yagi antennas are also used in commercial communications and FM broadcast reception. An *azimuth rotator* allows the antenna to be pointed in any horizontal direction. A combination *azimuth/elevation (az/el) rotator* allows the antenna to be aimed toward any point in space.

The driven element(s) of a Yagi array is/are connected to the feed line. In amateur-radio Yagi antennas, there is usually only one driven element. This element is sometimes physically shortened by *inductive loading*. It can contain traps so it resonates within more than one frequency band. The driven element is half-wave resonant and center-fed. By itself, it would comprise a dipole antenna.

When a Yagi is optimized for forward gain, there is some compromise in f/b ratio. Conversely, when it is optimized for maximum f/b ratio, there is some compromise in forward gain. The highest forward gain for a typical Yagi antenna is exhibited at a slightly different frequency than the highest f/b ratio.

A two-element Yagi can be formed by adding either a director or a reflector alongside the driven element. The optimum spacing for a driven-element/director Yagi is between 0.1 and 0.2 wavelength, with the director tuned to a frequency 5 to 10 percent higher than the resonant frequency of the driven element. The optimum spacing for a driven-element/reflector Yagi is 0.15 to 0.2 wavelength, with the reflector tuned to a frequency 5 to 10 percent lower than the resonant frequency of the driven element. The gain of an optimally built, full-size two-element Yagi is about 5 dBd using either a director or a reflector. If the elements are shortened by inductive loading or with traps, the gain is compromised.

A Yagi with one director and one reflector, along with the driven element, increases the gain and f/b ratio compared with a two-element Yagi. An optimally designed three-element Yagi has approximately 7 dBd gain. An example is shown in Fig. 4-10. This is a basic design scheme, and should not be used as an engineering blueprint.

The gain and f/b ratio of a Yagi both increase as elements are added. This is usually done by placing extra directors in front of a three-element Yagi. Some Yagis have 10, 15, or even 20 elements. An ideal 20-element Yagi can theoretically provide about 19 dBd of forward gain. If the gain needed in an antenna requires more than 20 elements on a single boom, two or more Yagis can be fed together in phase and placed end-to-end (collinear) or stacked, or both.

A pair of Yagis can be oriented at right angles on a single *boom* (main support), and fed 90 degrees out of phase to get *circular polarization*. This scheme is commonly used for satellite communications at VHF and UHF.

Quad Antenna

A quad is a directional antenna commonly employed in amateur radio communication, and also by some users of the Citizens Radio Service.

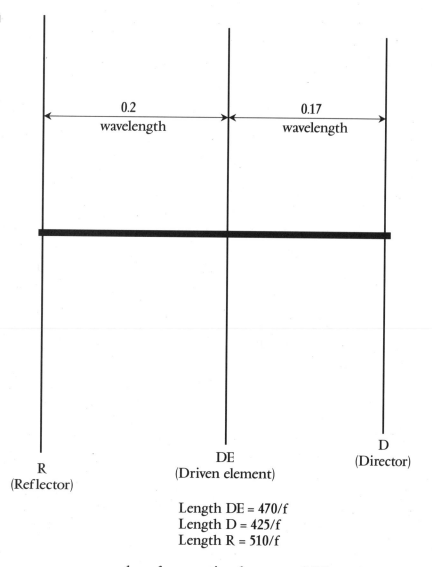

Length DE = 470/f
Length D = 425/f
Length R = 510/f

where f = operating frequency, MHz

This antenna operates according to the same principles as the Yagi antenna, except that full-wavelength loops are used instead of straight elements.

A full-wave loop has approximately 2 dB gain, in terms of ERP, compared to a half-wavelength dipole. This effect, when it was first observed, suggested to the inventors of the quad antenna that it ought to exhibit about 2 dB forward gain over a Yagi with the same number of elements. Experiments demonstrated this to be true.

A two-element quad antenna can consist of a driven element and a reflector, or it can have a driven element and a director. A three-element quad has one driven element, one director, and one reflector. The director has a perimeter of about 0.97 electrical wavelength, the driven element measures exactly one electrical wavelength around, and the reflector has a perimeter of about 1.03 electrical wavelength. These are approximate figures because the optimum lengths of the director and reflector are affected by the distance(s) between the elements. Additional director elements can be added to the basic three-element quad design to form quads having any desired numbers of elements. Provided optimum spacing is used, the gain increases as the number of elements increases. Each succeeding director should be slightly shorter than its predecessor. Long quad antennas are practical at frequencies above about 100 MHz, but they tend to be mechanically unwieldy at lower frequencies.

Geometrically, a two-element quad antenna is shaped more or less like a cube. For this reason, this antenna is sometimes called a *cubical quad*.

UHF and Microwave Antennas

At UHF and microwave frequencies, high-gain antennas are reasonable in size because the wavelengths are short. This section describes several types of high-gain antennas commonly used at these frequencies.

Dish Antenna

A *dish antenna* is used for UHF and microwave radio and TV transmission and reception. Satellite TV receivers employ dish antennas ranging in diameter from 2 ft in digital systems to 6, 8, or even 12 ft in analog systems. Dish antennas are also used by amateur radio operators, particularly for *earth-moon-earth* (EME) communication, also called *moonbounce*, in the UHF and microwave bands.

When a dish is used for receiving, distant signals arrive in parallel wavefronts that reflect off the dish and come together at the *focal point*. When the antenna is used for transmitting, radiated energy is reflected by the dish and sent out as parallel waves. The principle is the same as that of a telescope or searchlight, except that radio waves are involved instead of visible light.

A dish antenna must be correctly shaped and precisely aligned. The most efficient shape is a *paraboloidal reflector*. Such a dish forms the base of a large paraboloid (a parabola rotated around its axis). Less common, but adequate in most applications, is the *spherical reflector,* so named because it is a shallow section of a sphere.

The feed system for a dish antenna usually consists of a coaxial line or waveguide from the receiver and/or transmitter, and a horn or helical driven element at the focal point of the dish. A *preamplifier* might be located at the focal point, along with the driven element, if the antenna is used for receiving. A preamplifier at the antenna boosts the signal before it travels down the transmission line; this optimizes the *signal-to-noise ratio* (sensitivity). Figure 4-11A shows conventional dish feed as it is typically employed in a satellite TV receiving system.

An alternative method of feeding a dish antenna is shown in Fig. 4-11B. The driven element and preamplifier are located at the center

Figure 4-11
Dish antennas with conventional feed (A) and Cassegrain feed (B).

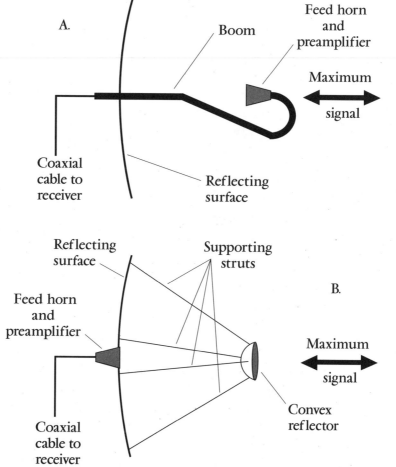

of the dish base. A small, spherical, convex reflector is positioned at the focal point of the main dish. This scheme is called *Cassegrain feed*. Signals coming in are focused at the small convex reflector; it directs the energy in a narrow beam to the driven element. When the antenna is used for transmitting, the driven element beams energy at the small convex reflector; it spreads the wavefronts out over the full surface of the main dish. The main dish then collimates the energy to be transmitted to the destination.

When properly aligned, dish antennas have high power gain and a sharp main radiation/response lobe. The larger the diameter of the dish with respect to a wavelength, the greater the gain, and the narrower the main lobe. For a dish antenna of fixed physical diameter (in feet or meters), the gain increases, and the main lobe narrows, as the frequency increases. This is because higher frequencies translate into shorter wavelengths, rendering the dish a greater number of wavelengths in diameter.

A dish antenna must be at least several wavelengths in diameter for proper operation. Otherwise, the waves diffract around the dish rather than reflecting from it. For most consumers and amateur radio operators, the dish is an impractical choice of antenna at frequencies below about 300 MHz, because the waves are too long below that frequency to allow the use of a dish of manageable size.

Some radio-telescope observatories have gigantic dish antennas that can function at frequencies as low as approximately 30 MHz. The largest dish antenna in the world, located in Arecibo, Puerto Rico, is literally carved in a valley among mountains. Some radio telescopes employ numerous dish antennas spaced miles apart, increasing the *aperture* (effective diameter in wavelengths) far beyond that of any single dish of practical construction.

The main reflecting element of a dish antenna can be made of sheet metal, or it can be fabricated from a screen or a wire mesh. If a screen or mesh is used, the spacing between the wires must be a small fraction of a wavelength.

Helical Antenna

A *helical antenna* is a circularly polarized, high-gain, unidirectional antenna, used mostly in the UHF and microwave parts of the radio spectrum. Circular polarization offers advantages at these frequencies, especially in space communications.

A typical helical antenna is shown in Fig. 4-12. The reflecting surface can be sheet metal or a screen. It can be disk-shaped, square, cone-shaped, or horn-shaped. In the drawing, the reflector is a disk. The

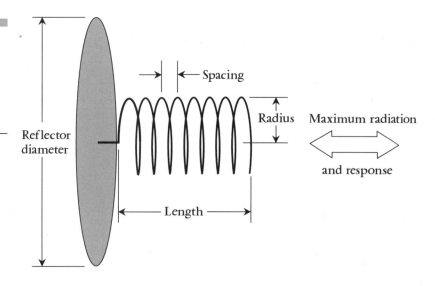

Figure 4-12
A helical antenna consists of a spring-shaped radiator. Usually there is a reflecting element as well.

reflector diameter should be at least 0.8 wavelength at the lowest antici-pated operating frequency. The radius of the helix should be approxi-mately 0.17 wavelength at the center of the intended operating frequency range. The longitudinal spacing between turns of the helix should be approximately 0.25 wavelength in the center of the operating frequency range. The overall length of the helix can vary, but should be at least one full wavelength at the lowest operating frequency. The longer the helix, the greater the forward gain of the antenna.

A single, moderate-sized helical antenna can provide about 15 dBd forward gain. When several helical antennas are fed in phase and placed side by side with a common reflector, the gain increases. Bays (parallel combinations) of helical antennas are common in space communica-tions systems.

Helical antennas are ideally suited for satellite communications because the circular polarization of the transmitted and received sig-nals reduces the amount of fading as the satellite orientation changes. Circular polarization is also an asset for moonbounce. The sense of the circular polarization can be made either clockwise or counterclock-wise, depending on the sense of the helix.

Corner-Reflector Antenna

A *corner reflector*, employed with a half-wave driven element, is illustrat-ed in Fig. 4-13. This provides some gain over the dipole alone. The

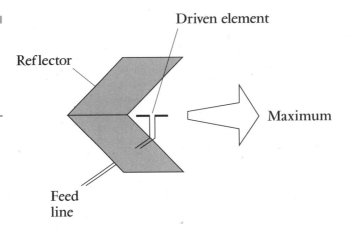

reflector is made of wire mesh, screen, or sheet metal. The *flare angle* of the reflecting element is usually about 90 degrees, although it can be somewhat greater or smaller. In the example shown here, the structure is oriented horizontally, but it could be oriented vertically. Corner reflectors are used mostly at UHF and microwave frequencies when the available space is limited. They work well for TV reception and for satellite communications. Sometimes several half-wave dipoles are fed in phase and arranged along a common line with a single, elongated corner reflector.

Another form of corner reflector, technically not an antenna but mentioned here because the same term is used, consists of three flat metal surfaces or screens, attached together in a manner identical to the way two walls meet the floor or ceiling in a room. Such a device, if it is at least several wavelengths across, will always return EM energy in exactly the same direction from which it arrives. Because of this effect, corner reflectors of this type make excellent radar dummy targets. In optical and infrared (IR) wireless ranging systems, three mirrors can be oriented in this fashion. The distance to an object can thus be accurately determined by measuring the time it takes for a beam of IR or visible light to travel from a transmitting station to the reflector and back. In this application the device is more often called a *tricorner reflector*.

Horn Antenna

A *horn antenna* is a device used for transmission and reception of signals at microwave frequencies. There are several different configura-

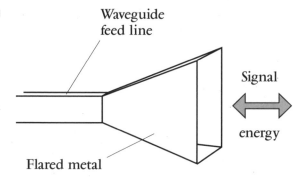

Figure 4-14
A horn antenna, used with a waveguide transmission line at microwave frequencies.

tions of the horn antenna, but they all look similar. Figure 4-14 is a drawing of a commonly used horn antenna. It provides a undirectional radiation and response pattern, with the favored direction coincident with the opening of the horn. The feed line generally consists of a *waveguide*, a hollow metal tube with a circular or rectangular cross section. The waveguide joins the antenna at the narrowest point (throat) of the horn.

Horn antennas are often used in feed systems of large dish antennas. The horn is pointed toward the center of the dish, in the opposite direction from the favored direction of the dish. When the horn is positioned at the focal point of the dish, extremely high gain and narrow-beam radiation are realized.

Transmission Lines

A *transmission line* is a medium by which energy is transferred from one place to another. In wireless communications and broadcasting, this refers to an antenna feed line, or to a cable used in TV or fiberoptic networks.

Wire Lines

Electromagnetic transmission lines can be categorized as either *unbalanced lines* or *balanced lines*. Unbalanced lines include *single-wire line*, *coaxial cable*, and *waveguides*. Coaxial cable, also called "coax" (pronounced CO-ax), is the most common type of unbalanced RF transmission line. Balanced lines are generally of parallel-wire construction, and include *open wire, ladder line, tubular line, ribbon line,* and *four-wire line.*

All wire transmission lines exhibit a property called *characteristic impedance.* This is the ratio of the voltage to the current in a line, when that line is terminated with a load that results in currents and voltages that maintain the same ratio at all points along the length of the line. Characteristic impedance is a constant that depends on the physical construction of the line. Coaxial lines typically have a characteristic impedance, or Z_0, between 50 and 100 ohms. Twinlead is available in 75-ohm and 300-ohm Z_0 values. Open-wire line has a characteristic impedance between 300 and 600 ohms, depending on the spacing between the conductors, and also on the type of dielectric (insulating material) employed to keep the spacing constant between the conductors.

Waveguides

A *waveguide* is an RF transmission line used at UHF and microwave frequencies. It is a hollow metal pipe, usually having a rectangular or circular cross section. The EM field travels down the pipe, provided that the wavelength is short enough. In order to efficiently propagate an EM field, a rectangular waveguide must have sides measuring at least 0.5 wavelength, and preferably more than 0.7 wavelength. A circular waveguide should be at least 0.6 wavelength in diameter, and preferably 0.7 wavelength or more.

An EM field can travel down a waveguide in various ways. If all of the electric lines of flux are perpendicular to the axis of the waveguide, the waveguide is said to operate in the *transverse-electric (TE) mode.* If all of the magnetic lines of flux are perpendicular to the axis, the waveguide is operating in the *transverse-magnetic (TM) mode.* The EM field can be coupled into the waveguide via either the electric or the magnetic components.

The characteristic impedance of a waveguide varies with frequency. In this sense, it differs from coaxial or parallel-wire lines, whose Z_0 values are generally independent of the frequency.

It is important that the interior of a waveguide be kept clean and free from condensation. Even a small obstruction can seriously degrade the performance of a waveguide.

Standing Waves

All transmission lines have some *loss;* they do not transfer energy with perfect efficiency. The loss occurs because of the ohmic resistance of the conductors, because of *skin effect* (a tendency for RF current to flow mostly near the surface of a conductor), and because of losses in the dielectric. This loss is measured in decibels per unit length, such as dB/100 ft. In most lines, this loss increases with increasing frequency. It also increases if the load impedance does not match the characteristic impedance of the line. *Standing waves* are voltage and current variations that exist on an RF transmission line when the load impedance differs from the characteristic impedance of the line.

In a transmission line terminated in a pure resistance having a value equal to the characteristic impedance of the line, no standing waves occur. This is an optimal state of affairs. When standing waves are present, a nonuniform distribution of current and voltage exists. The greater the impedance mismatch, the greater the nonuniformity. The ratio of the maximum voltage to the minimum voltage, or the maximum current to the minimum current, is called *the standing-wave ratio* (SWR) on the line. The SWR is 1:1 only when the current and voltage are in the same proportions everywhere along the line. Otherwise, the SWR is larger than 1:1.

In theory, there is no limit to how large the SWR on a transmission line can be. In the extreme cases of a short circuit, open circuit, pure inductance, or pure capacitance at the load end of the line, the SWR is theoretically infinite, because the current and the voltage fall to zero at certain points, and rise to high values at other points. In practice, line losses prevent the SWR from becoming infinite, but it can reach values of 20:1 or even 40:1 in practical transmission lines when a short or open circuit occurs.

The SWR is important as an indicator of the performance of an antenna system, because a high SWR indicates a severe mismatch between the antenna and the transmission line. This can have an adverse effect on the performance of a transmitter or receiver connected to the antenna system. An extremely large SWR can cause significant signal loss in a transmission line. If a high-power transmitter is used, the currents and voltages on a severely mismatched line can be large enough in some regions of the line to cause physical damage. The current can heat the conductors to the point that polyethylene dielectric material melts; the voltage can cause arcing that melts or burns the dielectric.

Safety

There are risks involved in the installation, operation, and maintenance of antennas. For outdoor antennas, the chief hazards come from electric utility lines, lightning, and large antenna structures and towers. For indoor transmitting antennas, EM fields are believed by some scientists to present a biological hazard. A few safety tips follow. *These suggestions do not constitute a complete set of safety guidelines. For information concerning antenna safety, consult a comprehensive text on antenna design and construction.*

Antennas should never be placed in such a way that they can fall or blow down on power lines. Also, it should not be possible for power lines to fall or blow down on an antenna. When erecting an antenna, sufficient distance from utility wiring should always be maintained. People have been electrocuted because antennas or feed lines came into contact with utility wires.

Radio equipment should not be used during thundershowers, or when lightning is anywhere in the vicinity. Antenna construction and maintenance should never be undertaken when lightning is visible, even if the storm appears to be several miles away. A direct hit is not necessary to cause electrocution. Sometimes, a lightning stroke will occur several miles away from a thunderstorm; this is called a *superbolt.* Under some conditions lightning can occur even when the weather appears fair and there are no showers anywhere in sight. Lightning can take place in snowstorms as well as in rainstorms. Ideally, antennas should be disconnected from electronic equipment, and connected to a good earth ground, whenever the equipment is not actually in use.

Tower and antenna climbing is a job for professionals. Under no circumstances should an inexperienced person attempt to climb such a structure. Sometimes, radio amateurs climb towers in order to install, maintain, and repair antennas. Installation of large antenna towers is a potentially dangerous job. It is important that such structures be guyed if necessary, and that they be able to withstand the weight and wind load of the antenna systems attached to them. The *ARRL Handbook for Radio Amateurs* (American Radio Relay League, Inc.) contains a chapter on safety, including substantial information on antenna and tower precautions. The *ARRL Antenna Book* also has detailed safety information. For further data, consult a professional antenna engineer.

Indoor transmitting antennas expose operating personnel to EM field energy. The extent of the risk (if any) of such exposure has not been firmly established. Radio amateurs using high-power transmitters (more than about 100 W output) should use outdoor antennas if possible. Outdoor antennas are generally more effective for communications than indoor antennas of the same size and gain characteristics. This is especially true if the communications equipment is located in concrete-and-steel structures such as large apartment complexes and office buildings. Locating a transmitting antenna outdoors, and as high as possible, minimizes the chances for EMI to home entertainment equipment and computers. Receiving antennas pick up less noise, and more signal, when they are located outdoors.

CHAPTER 5

Power Sources

Electronic systems require certain voltages and currents for proper operation. The power from a typical wall outlet consists of *alternating current* (AC) at 117 or 234 volts (V), and 50 or 60 hertz (Hz). But most electronic equipment requires *direct current* (DC) at much lower or higher voltages. This chapter discusses various forms of, sources of, and methods of obtaining power for use in electronic equipment.

Alternating Current

AC is an electric current in which the *polarity* reverses at regular intervals. The charge carriers, usually *electrons*, periodically reverse their direction, and the intensity of the current changes from instant to instant.

Period and Frequency

In a *periodic AC wave*, the function of magnitude versus time repeats, and the same pattern recurs countless times. The length of time between one repetition of the pattern, or one *cycle*, and the next is called the *period* of the wave and is denoted by the uppercase letter T.

The *frequency*, denoted f, is the reciprocal of the period. Thus, $f=1/T$ and $T=1/f$. Originally, frequency was specified in *cycles per second*, abbreviated cps. High frequencies were given in kilocycles, megacycles, or gigacycles, representing thousands, millions, or billions of cycles per second. But nowadays the standard unit of frequency is known as the *hertz*. Thus 1 Hz=1 cps, 10 Hz=10 cps, and so on. Higher frequencies are expressed in *kilohertz* (kHz), *megahertz* (MHz), or *gigahertz* (GHz). The relationships are

$$1 \text{ kHz}=1000 \text{ Hz}=10^3 \text{ Hz}$$

$$1 \text{ MHz}=1000 \text{ kHz}=1{,}000{,}000 \text{ Hz}=10^6 \text{ Hz}$$

$$1 \text{ GHz}=1000 \text{ MHz}=1{,}000{,}000{,}000 \text{ Hz}=10^9 \text{ Hz}$$

Sometimes an even bigger unit, the *terahertz* (THz), is used. This is 1,000,000,000,000 (10^{12}) Hz.

Some AC waves have all their energy concentrated at one specific frequency. These waves are called *pure sine-wave AC*. But often there are components at multiples of the main, or fundamental, frequency. There might also be components at other frequencies. Some AC waves are extremely complex, consisting of hundreds, thousands, or millions of component frequencies.

Waveforms

Although people usually envision a *sine wave*, also called a *sinusoid*, when they think of AC, there are infinitely many possible shapes an AC wave can have. Generally, the simpler waveforms are more familiar.

Sine Wave   In a sinusoid, the direction of the current reverses at regular intervals, and the current-versus-time curve is shaped like the rectangular-coordinate graph of the trigonometric sine function. A sine wave looks something like the waveform shown in Fig. 5-1A. Any AC wave that consists of a single frequency will have a perfect sine waveshape; also, any perfect sine-wave current contains only one component frequency. The current at utility outlets in American households has an almost perfect sine waveshape, with a frequency of 60 Hz.

Square Wave   On an oscilloscope, a perfect square wave looks like a pair of parallel, dotted lines, one with positive polarity and the other with negative polarity. The oscilloscope shows a graph of current on the vertical scale, versus time on the horizontal scale. The transitions between negative and positive for a perfect square wave are instantaneous, and if they show up on an oscilloscope at all, they are very thin vertical lines (Fig. 5-1B).

Sawtooth Waves   Some AC waves rise and/or fall in straight, but diagonal, lines as seen on an oscilloscope screen. The slope of the line indicates how fast the amplitude is changing. Such waves are called sawtooth waves because of their appearance. Sawtooth waves are generated by certain electronic test devices. These waves provide ideal signals for control purposes. Integrated circuits can be wired so that they produce sawtooth waves having an exact desired shape.

Complex Waves   The shape of an AC wave can get exceedingly complicated, but as long as it has a definite period, and as long as the polarity keeps switching back and forth between positive and negative, it is true AC. With some waves, it can be difficult, or almost impossible, to determine the period. This is because the wave has two or more components

Figure 5-1
A sine wave (A) and a
square wave (B).

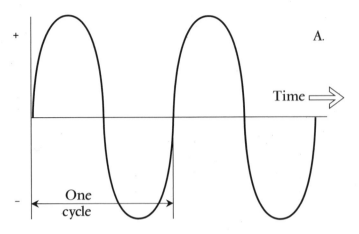

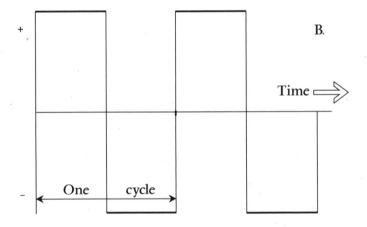

that are nearly the same magnitude. When this happens, the energy is
split up among two or more frequencies.

Amplitude

The *amplitude* of an AC wave is sometimes called *magnitude, level,* or
intensity. Depending on the quantity being measured, it might be specified in *amperes* (for current), in *volts* (for voltage), or in *watts* (for power).

Instantaneous Amplitude   The instantaneous amplitude of a wave
is the amplitude at some precise point in time. This constantly changes.

The manner in which it varies depends on the waveform. Instantaneous amplitudes are represented by individual points on wave curves.

Peak Amplitude   The peak amplitude of an AC wave is the maximum extent, either positive or negative, that the instantaneous amplitude attains. In many waves, the positive and negative peak amplitudes are the same. But sometimes they differ.

Peak-to-Peak Amplitude   The peak-to-peak (pk-pk) amplitude of a wave is the net difference between the positive peak amplitude and the negative peak amplitude. Another way of saying this is that the pk-pk amplitude is equal to the positive peak amplitude plus the negative peak amplitude. Peak-to-peak is a way of expressing how much the wave level "swings" during the cycle. In many waves, the pk-pk amplitude is twice the peak amplitude. This is the case when the positive and negative peak amplitudes are the same.

Root-Mean-Square Amplitude   Often, it is necessary to express the effective level of an AC wave. This is the voltage, current, or power that a DC source would produce to have the same general effect. When you say a wall outlet supplies 117 V, you mean 117 effective volts. The most common figure for effective AC levels is called the root-mean-square, or rms, value. For a perfect sine wave, the rms value is equal to 0.707 times the peak value, or 0.354 times the pk-pk value. Conversely, the peak value is 1.414 times the rms value, and the pk-pk value is 2.828 times the rms value. For a perfect square wave, the rms value is the same as the peak value; the pk-pk value is twice the rms value or the peak value. For sawtooth and irregular waves, the relationship between the rms value and the peak value depends on the shape of the wave. The rms value is never more than the peak value for any waveshape.

Why Is AC Used?

AC is easy to generate from turbines. It can be transformed to higher and lower voltages with comparative ease. These two facts make AC more convenient than DC for utility power transmission.

Electrochemical cells and batteries produce DC directly, but they are impractical for the needs of large populations. To serve millions of consumers, the immense power of falling or flowing water, ocean tides, wind, fossil fuels, or geothermal heat is needed. Any of these energy sources can be used to drive turbines that turn AC generators.

Technology is advancing in the realm of *solar-electric energy*; someday a significant part of our electricity might come from photovoltaic power plants. These would generate DC. If solar-electric energy ever becomes the cheapest method to get electricity, it is possible that DC power transmission will become common.

Thomas Edison is said to have favored DC over AC for electrical power transmission when electric utilities were first being planned and developed. His colleagues argued that AC would work better and, as we know, AC prevailed. But there is at least one advantage to DC in electric power transmission. This becomes apparent in long-distance power lines. Direct currents, at extremely high voltages, are transported more efficiently than alternating currents. The wire has less effective resistance with DC than with AC, and there is less energy lost in the magnetic fields around the wires.

Direct Current

DC is an electrical current in which the charge carriers always flow in the same direction. The intensity might change from instant to instant, but the polarity never changes. Figure 5-2 illustrates four signal graphs.

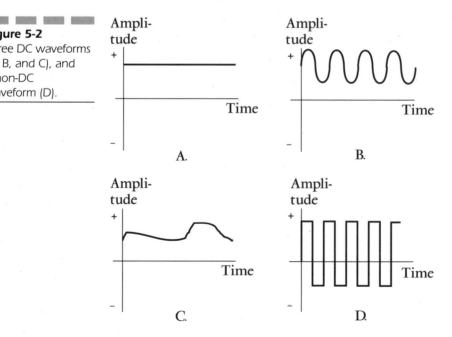

Figure 5-2
Three DC waveforms (A, B, and C), and a non-DC waveform (D).

The graphs at A, B, and C depict DC; the polarity never reverses, even though the amplitude (intensity) might change with time. The rendition at D is not DC because the polarity does not remain the same at all times.

Physicists consider the current in a circuit to flow from the positive pole to the negative pole. The movement of electrons in a DC circuit is contrary to the theoretical direction of the current. In a semiconductor material, the motion of positive charge carriers (*holes*) is the same as the theoretical direction of the current.

Common Sources

Typical sources of DC include power supplies, electrochemical cells and batteries, and photovoltaic cells and panels. The intensity, or amplitude, of a direct current might fluctuate with time, and this fluctuation might be periodic. In some such cases the DC has an AC component superimposed on it (as shown in Fig. 5-2B). An example of this is the output of a photovoltaic cell that receives a modulated-light communications signal. A source of DC is sometimes called a *DC generator*. This definition especially applies to a conventional *AC generator* (*dynamo* or *alternator*) followed by a *rectifier, filter,* and *voltage regulator.*

Batteries and other sources of DC produce a constant voltage. This is called *pure DC* and can be represented by a straight horizontal line on a graph of voltage versus time (as in Fig. 5-2A). The peak and effective values are the same. The pk-pk value is zero because the instantaneous amplitude never changes. In some instances the value of a DC voltage pulsates or oscillates rapidly with time, in a manner similar to the changes in an AC wave. The unfiltered output of a half-wave or a full-wave rectifier, for example, is *pulsating DC.*

Rectification is the process in which AC is changed to DC in a typical power supply. The most common method of rectification employs one or more semiconductor diodes. Part of the AC wave is either cut off or turned upside-down, so that the polarity is always the same, either positive or negative. In pulsating DC, the peak and average (effective) voltages are different from each other. The average value, compared with the peak value, for pulsating DC depends on whether *half-wave rectification* or *full-wave rectification* is used. For example, the peak voltage of household AC is about 165 V; the effective value is 117 V. If this voltage is subjected to full-wave rectification, the effective value remains at 117 V. If half-wave rectification is used, the effective value drops to 58.5 V. The peak value, however, remains at 165 V.

Paths for DC versus AC

A *DC short circuit* is a path that offers little or no resistance to DC. However, AC reactance might nevertheless be present.

The simplest example of a DC short circuit is a length of electrical conductor. This also provides a direct short circuit for AC. An inductor provides a low-resistance path for DC but, unlike a plain length of conductor, offers reactance to AC. Coils are often used in electronic circuits to provide a DC short circuit while offering a high reactance to AC signals. Such a component is called a *choke*. When it is necessary to place a circuit point at a certain DC potential without draining off the AC signal, a choke is used. Chokes also can be employed to eliminate the AC component of a DC power-supply output, and to prevent radio-frequency (RF) signals from traveling along electrical power lines.

A capacitor can act in the opposite sense from a choke in a circuit containing both AC and DC. The capacitor, if it is sufficiently large, will offer practically no reactance to AC signals but will appear as an open circuit to DC signals. When a capacitor is used in this way, it is called a *blocking capacitor*. Blocking capacitors are commonly used for interstage coupling in amplifier systems. A capacitor connected between a circuit point and ground, with the intention of providing an open circuit for DC while shorting out AC, is called a *bypass capacitor*.

It is often desirable to place a component lead at DC ground potential but to maintain an AC signal at that point. An example of this is an amplifier that uses a field-effect transistor (FET) in which the gate is at DC ground. Another example is the grounding of an antenna system through large inductors. This prevents the buildup of hazardous DC voltages on the antenna system but does not interfere with the radiation or reception of RF signals.

Utility Power Transmission

Electrical energy undergoes various transformations from the point of origin to the end users. The initial source consists of potential or kinetic energy in some nonelectrical form, such as falling or flowing water (*hydroelectric energy*), coal, oil, or natural gas (*fossil-fuel energy*), radioactive substances (*nuclear energy*), moving air (*wind energy*), light or heat from the sun (*solar energy*), or heat from the earth's interior (*geothermal energy*).

Generating Plants

In fossil-fuel, nuclear, geothermal, and some solar electric power generating systems, heat is used to boil water, producing steam under pressure to drive turbines. These turbines produce the rotational force (torque) necessary to drive large electric generators. Considerable torque is needed; the greater the power demand, the more force it takes to turn a generator shaft. This is why large amounts of oil, coal, or natural gas are required to stoke fossil-fuel power plants. Nuclear energy systems solve this problem, but they worry many people because of the radioactive byproducts from fission reactors. Coal, oil, gas, and nuclear-fission power plants all produce waste products, but geothermal and solar-heat systems do not. A nuclear-fusion (as opposed to fission) power plant, when it is finally developed, will presumably have no potentially harmful by products.

In a photovoltaic energy-generating system, sunlight is converted into DC electricity by semiconductor devices. This DC must be changed to AC that is used by most household appliances. This is done by a *power inverter.* Energy cannot be collected by photovoltaic cells during the hours of darkness, so *storage batteries* are necessary if a stand-alone photovoltaic system is to provide useful energy at night. Photovoltaic systems cannot generate much power and are therefore used mainly in homes and small businesses. These systems produce no waste products other than chemicals that must be discarded when storage batteries wear out. Photovoltaics can be used in conjunction with existing utilities, without storage batteries, to supplement the total available energy supply to all consumers in a power grid.

In a hydroelectric power plant, the movement of water (waterfalls, tides, river currents) drives turbines that turn the generator shafts. In a wind-driven system, moving air operates devices similar to windmills, producing torque that turns the generator shafts. These power plants do not cause direct pollution, but they are not without problems. The construction of a large hydroelectric dam can disrupt ecosystems, can have an adverse effect on agricultural and economic interests downriver, and can displace people upriver by flooding their land. Many people regard arrays of windmill-like structures as an eyesore, but in order to generate significant electrical power, hundreds or even thousands of the devices must be connected together and operated simultaneously.

Whatever source is used to generate the electricity at a large power plant, the output is AC electricity, with a voltage on the order of 100,000 to 1,000,000 V.

High-Tension Lines

When electric power is transmitted over wires for long distances, power is lost because of *ohmic loss* (resistance) in the conductors. This loss always costs money, but it can be minimized in two ways. First, efforts can be made to keep the wire resistance to a minimum. This involves using large-diameter wires made from metal having excellent conductivity, and by routing the power lines in such a way as to keep their lengths as short as possible. Second, the highest possible voltage can be used. Long-distance power lines are often called *high-tension lines*; in this context the word *tension* refers to voltage, not to physical stress or strain.

The reason that high voltage minimizes power-transmission line loss is apparent upon analysis of equations denoting the relationship among power, current, voltage, and resistance. One equation can be stated as

$$P_{\text{loss}} = I^2 R$$

where P_{loss} is the power (in watts) lost in the line as heat, I is the line current (in amperes), and R is the line resistance (in ohms). For a given span of line, the value of R is a constant. The value of P_{loss} therefore depends on the current in the line. This current depends on the load, that is, on the collective demand of the end users. But power-transmission line current also depends on the line voltage. Current is inversely proportional to voltage at a given fixed power load. That is,

$$I = P_{\text{load}} / E$$

where I is the line current (in amperes), P_{load} is the total power demanded by the end users (in watts), and E is the line voltage (in volts). By making a substitution from this equation into the previous one, the following formula can be obtained:

$$P_{\text{loss}} = (P_{\text{load}} / E)^2 R = P_{\text{load}}{}^2 R / E^2$$

For any given fixed values of P_{load} and R, doubling E will cut P_{loss} to one-fourth of its previous value. Multiply the line voltage by 10, and (in theory at least) the line loss will drop to 1/100 (1 percent) of its former value. Thus it makes sense to generate as high a voltage as possible for efficient long-distance electric power transmission.

The loss in a power line can be reduced even further by using DC rather than AC for long-distance transmission. Direct currents do not produce *electromagnetic (EM) fields* as alternating currents do, so a secondary source of power-line loss—EM radiation—is eliminated by the use of DC. But DC power transmission is difficult and costly, because it necessitates high-power, high-voltage rectifiers at the generating plant, and power inverters at distribution stations where high-tension lines branch out into lower-voltage, local lines.

Power-Line Transformers

High-tension electricity is impractical for end-point use. It is dangerous to people, and it would destroy common appliances. The standard utility voltages are 117 and 234 V rms. These lower voltages are obtained from higher voltages via *power-line transformers.*

Step-down transformers reduce the voltage of high-tension lines (100,000 V or more) down to a few thousand volts for distribution within municipalities. These transformers are physically large (about the size of a small room) because they must carry significant power. Several of them might be placed in a building or a fenced-off area. The outputs of these transformers are fed to power lines that run along city streets.

Smaller transformers, usually mounted on utility poles or underground, step the municipal voltage down to 234 V for distribution to individual homes and businesses. These transformers are about the size of a refrigerator. The 234-V electricity is provided in three phases to the distribution boxes in each house, apartment, or business. Some utility outlets are supplied directly with three-phase electricity. Large appliances such as electric stoves, ovens, and laundry machines employ this power. Smaller wall outlets and light fixtures are provided with single-phase, 117-V electricity.

Brownouts and Blackouts

Normally, the electric outlets in a typical home supply about 117 V for small appliances like lamps, radios, and computers. For large appliances like ovens and clothes dryers, 234 V is common. During times of peak demand, such as during a hot day when many air conditioners are in use, these voltages can drop by a few percent. This is called a *brownout*.

Most brownouts pass without causing problems for ordinary energy consumers. Light bulbs burn a little dimmer than normal; foods take a bit longer than usual to cook in electric ovens or frying pans. But the difference is so slight that, if you notice anything, you might think it's only your imagination.

If a brownout is severe, attended by a drop in voltage of 10 percent or more, there is no doubt that something is out of order with the electric utility. You might hear about it on the local news. The television raster (screen image) will shrink; you'll see black borders around the picture. You will also notice the shrinkage on cathode-ray-tube (CRT) computer displays. Under such conditions, *transients*, or sudden voltage spikes, on the power line are more common than usual. There is also an increased risk of a power *blackout*, or total loss of electricity. Computers, and some computer-controlled appliances, are especially sensitive to transients. If a blackout occurs, a *surge*, or moment of excessive voltage lasting up to one second, is likely to occur when power returns. Transients and surges occasionally cause permanent damage to precision electronic equipment.

You should always use *transient suppressors* with susceptible equipment, especially personal computers. Transients can occur at any time, whether there's a brownout or not. Devices for this purpose, also (imprecisely) called *surge suppressors* or *surge protectors*, cost only a few dollars. If you live in an area where brownouts are common, you should consider getting an *uninterruptible power supply* if you make frequent use of computers. This will keep the machine operating at the voltage it is designed to handle, and will allow you to save your data in the event of a complete blackout.

Utility blackouts can occur over regions as small as a city block, or as large as several states. The most common reasons for small-scale power blackouts include transformer failure, lines blown or knocked down, and short circuits caused by fallen television or radio antennas. These situations don't normally last long. If you experience a blackout that appears to be of small scale, you can call the electric utility and inform them of the fact. Usually the problem will be corrected within a few minutes or hours.

Larger blackouts occur when there are widespread severe thunderstorms, tornadoes, ice storms, massive blizzards, earthquakes, floods, or hurricanes. In August 1992, Hurricane Andrew completely destroyed the utility network over a large part of Dade County, Florida. The entire system in the affected area had to be rebuilt, and some residences and businesses did not see power restored for weeks.

Widespread blackouts can occur when demand exceeds the capacity of generating plants over a large geographic region. In such cases the utility might conduct *rotating blackouts* to force a reduction in the demand. Extremely hot weather in normally temperate regions is probably the most common cause of these blackouts. People use air conditioners at high duty cycles and in large numbers, overloading the utility system.

The most devastating and prolonged blackouts take place in war zones. A single hydrogen bomb detonated in the upper stratosphere would create an *EM pulse* resulting in an immediate blackout covering tens of thousands of square miles. This has actually been suggested as a means of throwing an enemy's general population into disarray without causing major property damage or loss of life.

Power Generators

A *power generator* is a machine designed to produce electricity using a coil and magnetic field. Sometimes it is called a *dynamo*. In an automobile or truck, it is called an *alternator*. In some aircraft it is called a *magneto*.

Principle

Whenever an electrical conductor, such as a wire, moves relative to the flux lines of a magnetic field, current is induced in the conductor. A generator can consist of a rotating magnet inside a coil of wire, or a rotating coil inside a magnet or pair of magnets. The rotating shaft is driven by mechanical energy from moving water, nuclear reactions, the heat from the earth's interior, or burning fuels such as coal, gasoline, natural gas, or hydrogen gas. The output is AC unless it is rectified by some means.

Small portable gasoline-powered generators, capable of delivering approximately 1000 to 5000 watts (W) of useful power output, can be purchased in department stores. Somewhat larger generators can produce enough electricity to supply a home or business. The most massive electric generators, found in power plants, are as large as houses, produce millions of watts, and can provide sufficient electricity for a community.

A generator, like a motor, is an *electromechanical transducer.* The construction of an electric generator is, in fact, almost identical to that of a motor. Some motors can operate as generators when their shafts are turned by external force. A machine that can function in either capacity is called a *motor/generator.*

When AC is generated by a rotating magnet inside a coil of wire, or by a rotating coil of wire inside a powerful magnet, the voltage appears between the ends of the length of wire. The voltage depends on the strength of the magnet, the number of turns in the wire coil, and the speed at which the magnet or coil rotates. The AC frequency depends only on the speed of rotation. Normally, for utility AC, this speed is 3600 revolutions per minute, or 60 revolutions per second. This produces a frequency of 60 Hz.

Efficiency

A generator requires very little mechanical power when there is no load connected to its output. When a load, such as a light bulb or heater, is connected to a generator, it becomes more difficult to turn the shaft. The greater the electrical wattage that is demanded from a generator, the more mechanical power is required to drive it. This is why, for example, you cannot connect a generator to a stationary bicycle and pedal a city into electrification. At first thought this might seem to be a brilliant idea, but a timeless rule of physics comes into play to thwart the scheme: "There's no such thing as a free lunch." The output power from the generator cannot be greater than the power supplied by your legs; in fact it will be less, because no real-world electromechanical system is 100 percent efficient.

There is always some energy lost in a generator, mainly as heat in the generator hardware. If you connect a generator to a stationary bicycle, your legs might provide enough power to run a small radio, but nowhere near enough to supply a household. *Generator efficiency* is defined as the ratio of the power output to the driving power, both measured in the same units (such as watts or kilowatts), multiplied by 100 to get a percentage.

Transformers in Electronic Equipment

A transformer is a device used to change the voltage of AC, or the impedance of an AC circuit path. Transformers operate by means of *inductive coupling.* The windings of a transformer can have a ferromagnetic core, an air core, or a core made of some dielectric material such as porcelain or certain plastics.

Windings

The *primary winding* of a transformer is the winding to which a voltage or signal is applied from an external source. In a *step-down transformer,* the primary winding has more turns than the *secondary winding,* from which power is drawn. In a *step-up transformer,* the primary has fewer turns than the secondary.

The size of the wire used for a primary winding depends on the amount of current it must carry. For a given power level, a step-up transformer must use larger primary-winding wire than a step-down transformer requires. The secondary winding can have taps for obtaining various voltages or impedances. A center-tapped secondary winding is used in some rectifier circuits. The size of the wire in the secondary winding of a transformer depends on whether the transformer is used to step up the voltage or to step it down. A step-up transformer can have smaller secondary-winding wire than a step-down transformer for a given amount of power.

Operation

Small transformers are used in power supplies for electronic devices such as radio receivers and low-power amplifiers. Larger transformers are employed in high-current and/or high-voltage power supplies.

In the following discussions, assume that voltages are always expressed in volts, currents are always expressed in amperes, impedances are always expressed in ohms, and power levels are always expressed in watts.

If a transformer has a primary winding with n_{pri} turns and a secondary winding with n_{sec} turns, then the *secondary-to-primary turns ratio* T is defined as

$$T = n_{sec} / n_{pri}$$

This ratio determines the impedance and voltage ratios.

For a transformer with secondary-to-primary turns ratio T and an efficiency of 100 percent (that is, no power loss), the output (secondary) voltage E_{out} is related to the input (primary) voltage E_{in} by the equation

$$E_{out} = TE_{in}$$

The output, or load, impedance Z_{out} in such a transformer is related to the input impedance Z_{in} according to the equation

$$Z_{out} = T^2 Z_{in}$$

Autotransformer

An *autotransformer* is a special type of step-up or step-down transformer with a single tapped winding. For step-down purposes, the input to the autotransformer is connected across the entire winding, and the output is taken from only part of the winding. For step-up purposes, the situation is reversed: the input is applied across part of the winding, and the output appears across the entire coil.

An autotransformer usually contains less wire than an equivalent transformer with two separate windings. But an autotransformer does not provide *DC isolation* between the input and output. If such isolation is required, a transformer with separate primary and secondary windings must be used.

The step-up or step-down voltage ratio of an autotransformer is determined according to the same formula as for an ordinary transformer. The same holds true for the impedance-transfer ratio.

Efficiency

In any transformer, some power is lost in the coil windings. If the transformer has a ferromagnetic core, some power is lost in the core. *Conductor loss* takes place because of ohmic resistance, and sometimes also because of *skin effect*, especially at high frequencies. In skin effect, most of the current flows on the outside of the transformer wire, effectively reducing the cross-sectional area of the wire and increasing its resistance.

Core loss occurs because of *eddy currents* (circulating currents in the material) and *hysteresis* ("sluggishness" of the material's response to changing magnetic fields). In these cases, the lost power appears as heat. A small amount of power might also be radiated as EM fields from the coil windings. All power dissipated or radiated by a transformer represents loss. This power cannot appear at the transformer output, and therefore it cannot be put to useful work.

Let E_{pri} and I_{pri} represent the primary-winding voltage and current in a hypothetical transformer, and let E_{sec} and I_{sec} be the secondary-winding voltage and current. In a perfect transformer, the product $E_{pri}I_{pri}$ would be equal to the product $E_{sec}I_{sec}$. However, in a real transformer, $E_{pri}I_{pri}$ is always greater than $E_{sec}I_{sec}$. The efficiency *Eff* of a transformer, expressed as a percentage, is

$$Eff = 100 \, E_{sec}I_{sec} / (E_{pri}I_{pri})$$

The power P_{loss} dissipated in the transformer windings and core is equal to

$$P_{loss} = E_{pri}I_{pri} - E_{sec}I_{sec}$$

The efficiency of a transformer varies depending on the load connected to the secondary winding. If the current drain is excessive, the efficiency is reduced. Transformers are generally rated according to the maximum amount of power they can deliver without serious degradation in efficiency.

Rectification

Rectification is a process in which AC is converted to DC. The most common rectifier circuits employ *semiconductor diodes*. Part of the AC wave is either cut off or inverted, so the polarity is always the same (either positive or negative) rather than periodically changing (between positive and negative).

Schemes

Figure 5-3 illustrates the output waveforms of two different rectifier circuits commonly used with utility AC. The input waveform in both cases is assumed to be a sine wave. In the output waveform at A, the negative

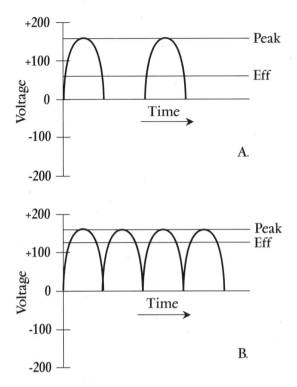

Figure 5-3
At A, half-wave rectifi-
cation of an AC sine
wave. At B, full-wave
rectification of the
same sine wave.

(bottom) part of the cycle has simply been eliminated. At B, the negative portion of the sine wave has been turned upside-down and made positive, a mirror image of its former self. The situation at A is known as *half-wave rectification,* because it involves only half of each wave cycle. At B, the input wave has been subjected to *full-wave rectification;* all of the original current still flows, even though the alternating nature has been changed so that the current never reverses its direction. Both of the resulting waveforms are *pulsating DC.*

The effective value, compared with the peak value, for pulsating DC depends on whether half-wave or full-wave rectification has been used. In the drawings, the peak and effective (eff) voltages are shown as thin horizontal lines. The instantaneous voltages are shown as heavy curves. The instantaneous voltage changes all the time (from instant to instant, hence the term). The peak voltage is the maximum instantaneous voltage. The instantaneous voltage is never greater than the peak. In the half-wave circuit (Fig. 5-3A), the effective voltage is 0.354 times the peak voltage. In the full-wave circuit (Fig. 5-3B), the effective voltage is 0.707 times the peak voltage. Using household AC as an example, the peak value is generally about 165 V; the effective value is 117 V. If full-wave rectification is used, the effective value is 117 V. If half-wave rectification is used, the effective value is about 58.5 V.

Circuits

The half-wave rectifier circuit uses one diode to "chop off" half of the AC input cycle. A half-wave rectifier circuit diagram is shown in Fig. 5-4A. This circuit is useful in supplies that don't have to deliver much current, or that need not be especially well regulated. For high-current equipment, a full-wave rectifier works better. The full-wave scheme is also better when good voltage regulation is needed.

Figure 5-4
At A, a half-wave rectifier circuit. At B, a full-wave center-tap rectifier circuit. At C, a full-wave bridge rectifier circuit.

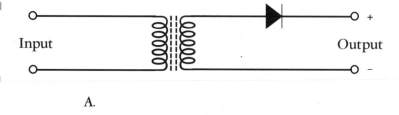

A.

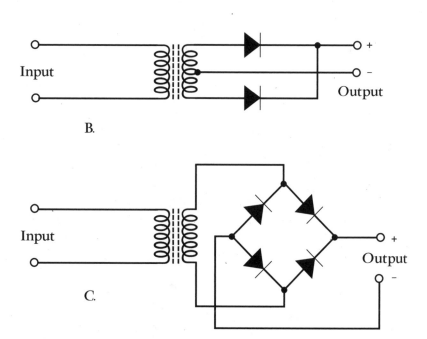

B.

C.

A full-wave rectifier circuit makes use of both halves of the AC cycle to derive its DC output. Suppose you want to convert an AC wave to DC with positive polarity. Then you design the rectifier to allow the positive half of the AC cycle to pass unchanged, and flip the negative portion of the wave upside-down, making it positive instead.

There are two basic circuits for the full-wave supply. One version, shown in Fig. 5-4B, uses a center tap in the transformer, and needs two diodes. This is called *a full-wave center-tap rectifier circuit.* The other circuit, shown in Fig. 5-4C, uses four diodes, and is known as a *full-wave bridge rectifier circuit.* The circuit best suited to a particular application depends on the power requirement, the voltage output needed, and the availability of center-tapped-secondary transformers. Special "bridge rectifiers" are available; these consist of four diodes in a single package, with input and output terminals clearly identified.

Filtering and Regulation

Electronic equipment generally does not function well with the pulsating DC that comes straight from a rectifier. The *ripple* in the output must be smoothed out, so that pure, battery-like DC is supplied. A *filter* circuit does this.

The simplest possible filter is one or more large-value capacitors, connected in parallel with the rectifier output. Electrolytic capacitors are almost always used. They must be hooked up in the correct direction because they are polarized components. Sometimes a large-value coil, called a *filter choke,* is connected in series, in addition to the capacitor in parallel. This provides a smoother DC output than the capacitor by itself. To obtain even better smoothing, two parallel capacitors can be used with a series choke, as shown in Fig. 5-5. This is called a *pi-section* filter.

Figure 5-5
A pi-section power-supply filter circuit.

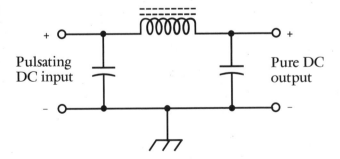

Pulsating
DC input

Pure DC
output

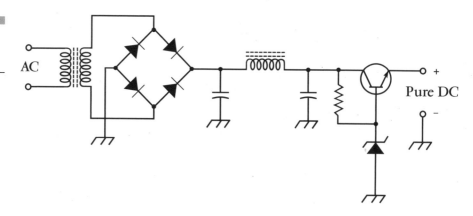

Figure 5-6
A regulated DC
power-supply circuit.

In the circuit shown, the negative side of the power-supply output is connected to chassis ground. A common arrangement in electronic equipment, this is called *negative ground*. Some equipment requires a *positive ground*. Sophisticated equipment often makes use of two separate power-supply outputs, one with a positive ground and the other with a negative ground. Such supplies require separate filters for the positive and negative outputs.

If a special kind of diode, called a *zener diode*, is connected in parallel with the output of a power supply, the diode will limit the output voltage of the supply as long as the diode has a high enough power rating. The limiting voltage depends on the particular zener diode used. There are zener diodes to fit any reasonable power-supply voltage. A zener-diode voltage regulator is inefficient when the supply is used with equipment that draws high current. When a supply must deliver a high level of current, a *power transistor* is used along with the zener diode to obtain regulation.

In recent years, voltage regulators have become available in integrated-circuit (IC) form. Such an IC, sometimes along with some external components, is installed in the power-supply circuit at the output of the filter. This provides excellent regulation at low and moderate voltages.

Figure 5-6 illustrates a complete, full-wave bridge power supply, including a pi-section filter and a voltage-regulation circuit.

Transient Suppression

Many electronic devices are designed to work with utility AC at 60 Hz and 117 V. The figure of 117 V is the rms value, the engineer's way of stating average or effective AC voltage. The instantaneous peaks actually reach about +165 and −165 V.

Figure 5-7
Transients on a utility
line can reach peaks
far higher than the
peak voltage of the
AC waveform.

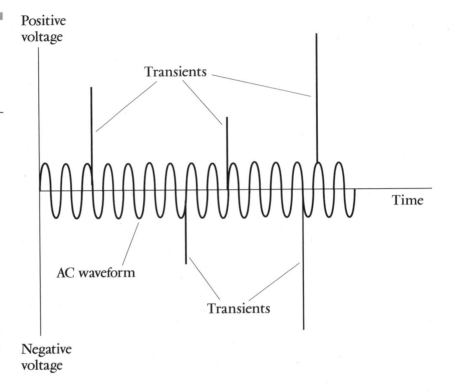

If the voltage fluctuates significantly, the operation of some devices, especially personal computers, can be upset. The most common irregularity is a sudden "spike" known as a *transient*. A transient lasts for only a few millionths of a second, but it can rise to several hundred volts, either positively or negatively (Fig. 5-7). Transients are caused by nearby thundershowers, transformer arcing (sparking), and certain large appliances. Transients cannot be detected without special equipment connected to the power line to monitor the AC waveform.

Somewhat less common is a voltage *surge* or *dip*. These are increases or decreases in voltage that can last for many AC cycles, and up to 1 or 2 seconds. You notice surges and dips because they cause a visible, momentary brightening or dimming of the lights.

Some transients, surges, and dips have no effect on sensitive equipment. But they can occasionally cause computers to crash (lock up). If that happens, the computer will have to be rebooted (reset), and you will lose the data in memory unless it has been stored on a hard disk, diskette, or other permanent medium. It is impossible to predict when a severe transient, surge, or dip will occur. Therefore, it makes sense to protect your equipment as much as possible at all times.

Most transients can be kept from reaching electronic equipment by means of a *transient suppressor.* Such a device is sometimes called *surge suppressor,* although technically it works only with transients, not surges or dips. (Some sophisticated *uninterruptible power supplies* can eliminate surges and dips.) You might occasionally hear somebody talk about a *surge protector,* but this is a real misnomer. The electronic equipment, not the surges or transients, need the protection!

A typical transient suppressor is a box with several three-prong outlets and a power switch. It plugs into the wall outlet. You plug all of the electronic equipment you want to protect into the outlets in the box. You can use the suppressor-box power switch to control power to an entire electronic workstation. Transient suppressors come in a variety of designs, ranging in price from a few dollars to several hundred dollars.

Suppressor boxes almost always have three-wire power cords. It is important that a three-wire outlet be used, and that the third-prong receptacle in the outlet be connected to a substantial electrical ground. It is a good idea to unplug the suppressor box, and thus all the devices connected to it, from the wall socket when you are not using the equipment. This will ensure that the equipment is not damaged in the event of a heavy thunderstorm. If lightning strikes near your home, the resulting transient can reach more than 10,000 V at the wall outlets. That can cause serious, permanent damage to many kinds of electronic devices, including transient suppressors.

Fuses and Circuit Breakers

A *fuse* or *circuit breaker* is a protective device used in power supplies and power sources for appliances of many kinds. Fuses and circuit breakers protect the components of the power supply, as well as the components of the equipment, against possible damage in the event of a malfunction of either the supply or the equipment. Fuses and circuit breakers also reduce the chance of fire in the event excessive current is drawn from a source of electrical power.

The Fuse

A fuse is simple, consisting of a short strip or wire in an enclosure. The thickness of the wire or strip determines the current level at which the fuse will open when it is connected in series with the power-supply line. Some fuses are smaller than pencil erasers, weighing less than 1 gram; others are large and hefty, weighing several pounds. There are hundreds of sizes and shapes in between these extremes. In general, the higher the power (that is, the greater the wattage) of the equipment or circuit, the larger and heavier is its fuse.

In the electrical system of an old house, you might find a *fuse box* containing several fuses, one for each utility circuit in the house. (New houses usually have circuit breakers instead.) Electronic devices such as communications receivers, transmitters, some high-fidelity stereo amplifiers, and personal computers have their own independent fuses, usually in the AC circuit leading into the equipment.

The *time constant* is the length of time, following a sudden short circuit, required for a fuse to blow. Fuses can be obtained with a variety of time constants. Some fuses blow almost instantly when the current exceeds the rated value; such a device is a *quick-break fuse*. Others have a built-in delay that allows a brief period of excessive current without blowing the fuse. This type of fuse is known as a *slow-blow fuse*.

The main problem with fuses is the fact that they must be removed and replaced when they blow out. This can be time-consuming and inconvenient. Replacement fuses must be available for use in case of an accidental short circuit or other temporary malfunction that causes a fuse to blow. A complex system, such as a sophisticated amateur radio station, might have a dozen fuses, each with its own unique specifications. That means the operator must keep several different kinds of spares. *Circuit breakers*, which can be reset without replacement, are generally more convenient.

The Breaker

A circuit breaker, also called simply a *breaker*, is a current-sensitive switch. It is placed in series with the power-supply line to a circuit. If the current in the line reaches a certain value, the breaker opens and removes power from the circuit. Breakers are easily reset when they open, although it is usually necessary to switch the equipment off and wait a certain length of time before resetting. Circuit breakers are used to protect electronic circuits against damage when a malfunction causes them to draw excessive current. Circuit breakers are also used in utility wiring to minimize the danger of fire in the event of a short circuit.

There are many kinds of circuit breakers, for various voltages and limiting currents. Some breakers will open with just a few milliamperes of current, while others require hundreds of amperes. Some breakers open almost immediately if their limiting currents are reached or exceeded. Some breakers have a built-in delay.

Cells and Batteries

One of the most common, and most versatile, sources of DC is the *cell*. The term means "self-contained compartment" and can refer to any of various different things in (and out of) science. In electricity and electronics, a cell is a unit source of DC energy. There are dozens of different types of electrical cells. When two or more cells are connected in series, the result is a *battery*.

Primary and Secondary Cells

Some electrical cells, once their potential (chemical) energy has all been changed to electricity and used up, must be thrown away. They are no good anymore. This type of cell is called a *primary cell*. Other kinds of cells can get their chemical energy back again. Such a cell is a *secondary cell*.

Primary cells include the ones you usually put in a flashlight, in a transistor radio, and in various other consumer devices. They use dry electrolyte pastes along with metal electrodes. They go by names such as *dry cell, zinc-carbon cell, alkaline cell*, and others. Go into a department store and find the panel of "batteries," and you'll see various sizes and types of primary cells, such as "AAA batteries," "D batteries," "camera batteries," and "watch batteries." These are usually cells, not batteries. You will also see real batteries, such as miniature 9-V "transistor batteries" and large 6-V "lantern batteries."

Secondary cells can also be found, increasingly, in consumer stores. *Nickel-cadmium (Ni-Cd or NICAD) cells* are fairly common. They are available in some of the same sizes as nonrechargeable dry cells. The most common sizes are AA, C, and D. These cost several times as much as ordinary dry cells, and a charging unit also costs a few dollars. But if you take care of them, these rechargeable cells can be used hundreds of times, and will pay for themselves several times over if you use a lot of "batteries" in your everyday life.

The battery in your car is made from secondary cells connected in series. These cells recharge from the alternator or from an outside charging unit. It is extremely dangerous to short-circuit the terminals of such a battery, because the acid (sulfuric acid) can boil out, sometimes violently. As a matter of fact, it is never a good idea to directly short-circuit any cell or battery, because it might burst or explode.

Storage Capacity

The most commonly used unit of electrical energy is the *watthour* (Wh) or the *kilowatthour* (kWh). Any electrochemical cell or battery has a certain amount of electrical energy that can be drawn from it, and this can be specified in watthours or kilowatthours. Energy can also be indirectly expressed in *ampere-hours* (Ah); this figure can be multiplied by the battery voltage, in volts, to obtain watthours.

A battery with a rating of 2 Ah can provide 2 amperes (A) for 1 hour, or 1 A for 2 hours. Or it can provide 100 milliamperes (mA) for 20 hours. Within reason, the product of the current in amperes, and the use time in hours, can be as much as, but not more than, 2. The limitations are the *shelf life* at one extreme and the *maximum safe current* at the other. Shelf life is the length of time the battery will last if it is sitting on a shelf without being used; this might be months or even years. The maximum safe current is represented by the lowest load resistance (heaviest load) that the battery can work into before its voltage drops because of its own *internal resistance*. A battery should never be used with loads that are too heavy, because this will cause a voltage drop, resulting in poor equipment performance and shortening the battery life.

Small cells have storage capacity of a few milliampere-hours (mAh) up to 100 or 200 mAh. Medium-sized cells might supply 500 mAh or 1 Ah. Large automotive or truck batteries can provide upward of 50 Ah.

When an *ideal cell* or battery is used, it delivers a fairly constant current for awhile, and then the current starts to diminish (Fig. 5-8). Some types of cells and batteries approach this ideal behavior, called a *flat discharge curve*, and others have current that declines gradually, almost right from the start. When the current that a battery can provide has tailed off to about half of its initial value, the cell or battery is said to be "weak" or "low." At this time, it should be replaced. If it is allowed to run all the way out until the current drops to zero, the cell or battery is "dead." Some rechargeable cells and batteries, especially the nickel-cadmium

Figure 5-8
A flat battery discharge curve. This is considered ideal.

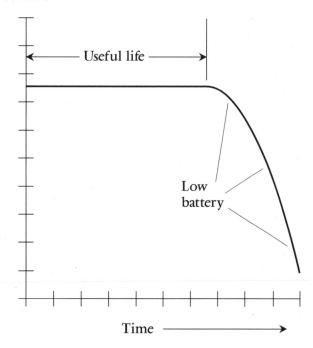

and nickel-metal-hydride type, should never be used until they are completely dead; this can ruin them permanently.

The area under the curve in Fig. 5-8 is the *total energy capacity* of the cell or battery in ampere-hours. This area is always pretty much the same for any particular type and size of cell or battery, regardless of the amount of current drawn while it is in use. Nevertheless, there are infinitely many possible discharge curves for any cell or battery. Higher current demands tend to shorten the time span before the deliverable current begins to decline. Conversely, lower current demands tend to lengthen the useful life.

Dime-Store Cells and Batteries

The cells you see in general stores, and that are popular for use in household convenience items like flashlights and transistor radios, are usually of the zinc-carbon or alkaline variety. These provide 1.5 V, and are available in sizes known as AAA (very small), AA (small), C (medium large), and D (large).

Zinc-carbon cells have a fairly long shelf life. The zinc forms the outer case or shell, and is the negative electrode. A carbon rod serves as the positive electrode. The electrolyte is a paste of manganese dioxide and carbon. Zinc-carbon cells are inexpensive and will work at moderate temperatures, and in applications where the current drain is moderate to high. They are not very good in extreme cold.

Alkaline cells employ granular zinc for the negative electrode, potassium hydroxide as the electrolyte, and a device called a *polarizer* as the positive electrode. The geometry of construction is similar to that of the zinc-carbon cell. An alkaline cell can work at lower temperatures than a zinc-carbon cell. Its shelf life is much longer than that of a zinc-carbon cell. As you might expect, it costs more. Zinc-carbon and alkaline cells and batteries have current output that tails off in a more or less uniform fashion, as shown in Fig. 5-9. This is a *declining discharge curve,* and is less than ideal for electronic applications. Nevertheless, alkaline cells are popular for use in calculators, transistor radios, and cassette tape and compact-disk players.

Transistor batteries consist of six tiny zinc-carbon or alkaline cells in series. Each of the six cells supplies 1.5 V. Even though these batteries

Figure 5-9
A declining discharge curve.

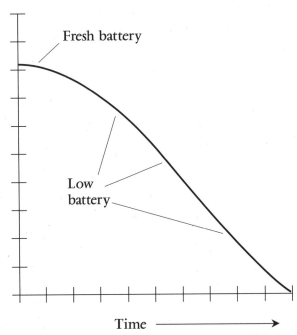

have more voltage than individual cells, the total energy available from them is less than that from a C cell or D cell. This is because the electrical energy that can be drawn from a cell or battery is directly proportional to the amount of chemical energy stored in it, and this, in turn, is a direct function of the physical mass of the cell. A C or D size cell has more mass than a "transistor battery" and therefore contains more stored energy, assuming it is of the same chemical type. The ampere-hour capacity of a "transistor battery" is very small. But transistor radios do not need much current. These batteries are also used in other low-current electronic devices, such as remote-control garage-door openers, TV channel changers, remote video-cassette recorder (VCR) controls, and electronic calculators.

Lantern batteries get their name from the fact that they find much of their use in lanterns. They are massive, so they last a long time. One type has spring contacts on the top. The other type has thumbscrew terminals. Besides keeping a lantern lit for awhile, these big batteries, usually rated at 6 V and consisting of four zinc-carbon or alkaline cells, can provide enough energy to operate a low-power radio transceiver. Two of them in series make a 12-V battery that can power a citizens band (CB) or amateur radio station. These batteries are also good for scanner radio receivers in portable locations, for camping lamps, and for other low-power appliances.

Miniature Cells and Batteries

In recent years, cells and batteries (but especially cells) have become available in many different sizes and shapes besides the old cylindrical cells, transistor batteries, and lantern batteries. These are used in watches, cameras, and other microminiature electronic devices.

Silver-oxide cells are usually made into a buttonlike shape and can fit inside a small wristwatch. They come in various sizes and thicknesses, all with similar appearance. They supply 1.5 V, and offer excellent energy storage for the weight. They also have a flat discharge curve, like the one shown in the graph of Fig. 5-8. Silver-oxide cells can be stacked to make batteries. Several of these miniature cells, one on top of the other, can provide 6 or 9 V for a transistor radio or other light-duty electronic device. The resulting battery is about the size of an AAA cylindrical cell.

Mercury cells, also called *mercuric oxide cells*, have advantages similar to those of silver-oxide cells. They are manufactured in the same general form. The main difference, often not of significance, is a somewhat lower voltage per cell: 1.35 V. If six of these cells are stacked to make a battery, the resulting voltage will be about 8.1 rather than 9 V. One additional cell can be added to the stack, yielding about 9.45 V. There has been a decrease in the use of mercury cells and batteries in recent years, because mercury is toxic. When mercury cells and batteries are dead, they must be discarded. Eventually the mercury, a chemical element, leaks into the soil and groundwater. Mercury pollution has become a significant health concern in some places.

Lithium cells and batteries first became popular during the 1980s. There are several variations in the chemical makeup of these cells. They all contain lithium, a light, highly reactive metal. Lithium cells can be made to supply 1.5 to 3.5 V, depending on the particular chemistry used. These cells, like their silver-oxide cousins, can be stacked to make batteries. The first application of lithium batteries was in memory backup for electronic microcomputers. Lithium cells and batteries have superior shelf life, and they can last for years in very-low-current applications such as memory backup or the powering of liquid-crystal-display (LCD) devices. These cells also provide energy capacity per unit volume that is vastly greater than the above-mentioned types of electrochemical cells.

Lead-Acid Cells and Batteries

A *lead-acid cell* has a solution of sulfuric acid, along with a lead electrode (negative) and a lead-dioxide electrode (positive). These batteries are rechargeable. *Automotive batteries* are made from sets of lead-acid cells having a free-flowing liquid acid. You cannot tip such a battery on its side, or turn it upside-down, without running the risk of having some of the acid electrolyte get out.

Lead-acid batteries are also available in a construction that uses a semisolid electrolyte. These batteries are popular in consumer electronic devices that require a moderate amount of current. They are also used in uninterruptible power supplies (UPSs) for personal computers.

A large lead-acid battery, such as the kind in your car, can store several tens of ampere-hours. The smaller ones, like those in UPSs, have less capacity but more versatility. Their main advantage is reasonable cost, considering that they can be charged and discharged many times.

Nickel-Cadmium Cells and Batteries

You've probably seen, or at least heard of, NICAD cells and batteries. They are fairly common in consumer devices such as the portable headphone radios and cassette players you can wear while doing aerobics. (This type of device is dangerous to wear while walking, jogging, or bicycling on roads where motor vehicles travel.) You can buy two sets of cells and switch them every couple of hours of use, charging one set while using the other. Plug-in charger units cost a few dollars.

NICAD cells are made in several types. *Cylindrical NICADs* look like dry cells. *Button NICADs* are used in cameras, watches, memory backup applications, and other places where miniaturization is important. *Flooded NICADs* are used in heavy-duty applications and can have a charge capacity of as much as 1000 Ah. *Spacecraft NICADs* are made in packages that can withstand the vacuum and temperature changes of a spaceborne environment. Most orbiting satellites are in darkness half the time, and in sunlight half the time. Solar panels can be used while the satellite is in sunlight, but during the times that the earth eclipses the sun, batteries are needed to power the electronic equipment on board the satellite. The solar panels can charge a set of NICADs, in addition to powering the satellite, for half of each orbit. The NICADs can provide the power during the dark half of each orbit.

NICAD batteries are available in packs of cells. These packs can be plugged into the equipment, and might even form part of the case for a device. An example of this is the battery pack for a handheld amateur radio transceiver. Two of these packs can be bought, and they can be used alternately, with one installed in the "handie-talkie" (HT) while the other is being charged.

NICAD cells and batteries should never be discharged all the way until they "die." This can cause the polarity of a cell, or of one or more cells in a battery, to reverse. Once this happens, the cell or battery is ruined. But it is important to discharge them almost all the way as often as possible. A phenomenon peculiar to this type of cell and battery is called *memory drain* or *NICAD battery memory.* If a NICAD is used over and over, and is discharged to about the same extent every time (say, two-thirds of the way), it might start to "go to sleep" at that point in its discharge cycle. Memory drain is probably less common than is rumored, but it can give the illusion that the cell or battery has lost some of its storage capacity. If you think a NICAD has a problem with memory drain, use the cell or battery almost all the way up, and then fully charge it. Repeat the process two or three times, and the memory will be "erased."

NICAD cells and batteries work best using chargers that take several hours to fully replenish the cells or batteries. There are high-rate or quick chargers available, but these can sometimes force too much current through a NICAD. It is best if the charger is made especially for the cell or battery type being charged. An electronics dealer should be able to tell you which chargers are best for which cells and batteries.

Photovoltaics

A *photovoltaic cell* is completely different from electrochemical cells. This type of device is also known as a *solar cell*. It converts visible light, infrared, and/or ultraviolet directly into electric current.

Several, or many, photovoltaic cells can be combined in series-parallel to make a *solar panel*. Such an array might consist of, say, 50 parallel sets of 20 series-connected cells. The series scheme boosts the voltage to the desired level, and the parallel scheme increases the current-delivering ability of the panel. It is not unusual to see hundreds of solar cells combined in this way to make a large panel.

The construction of a photovoltaic cell is shown in Fig. 5-10. The device is a flat semiconductor P-N junction, and the assembly is made

Figure 5-10
Cross section of a silicon photovoltaic cell.

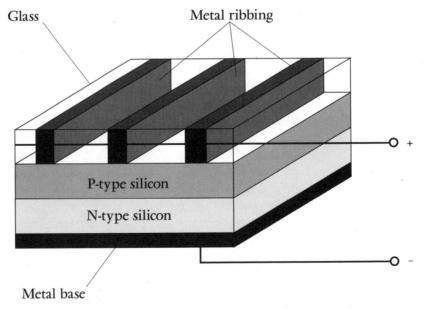

Glass

Metal ribbing

P-type silicon

N-type silicon

Metal base

+

−

transparent so that light can fall directly on the P-type silicon. The metal ribbing, forming the positive electrode, is interconnected by means of tiny wires. The negative electrode is a metal backing, placed in contact with the N-type silicon. Most solar cells provide about 0.5 V. If there is very low current demand, dim light will result in the full output voltage from a solar cell. As the current demand increases, brighter light is needed to produce the full output voltage. There is a maximum limit to the current that can be provided from a solar cell, no matter how bright the light. This maximum current limit can be increased by connecting solar cells in parallel.

Solar cells have become cheaper and more efficient in recent years, as researchers have looked to them as a long-term alternative energy source. Solar panels are used in satellites. They are commonly used in conjunction with rechargeable batteries, such as the lead-acid or NICAD types, to provide power independent of the commercial utilities.

Power Inverters

A *power inverter*, sometimes called a *chopper power supply*, is a circuit that delivers high-voltage AC from a DC source. The input is typically 12 V DC, and the output is usually 117 V rms AC. Inverters were once used in the operation of mobile or portable equipment containing vacuum tubes. Power inverters facilitate the use of small appliances such as computers, television sets, and communications radios in portable and mobile environments.

A simplified block diagram of a power inverter is shown in Fig. 5-11. The *chopper* consists of a low-frequency oscillator that opens and closes a high-current switching transistor. This interrupts or modulates the battery current, producing pulsating DC. The transformer converts the pulsating DC to AC, and also steps up the voltage. If the battery is rechargeable, solar panels can be used to replenish its charge and provide a long-term source of utility power.

The output of a low-cost power inverter is generally not a good sine wave; square-wave choppers are inexpensive and easy to manufacture. For this reason, low-cost power inverters do not always work well with appliances that need a nearly perfect, 60-Hz sine-wave source of power. More sophisticated inverters produce fairly good sine waves, and have a frequency close to 60 Hz. Although such inverters are expensive, they are a good investment if they are to be used with sensitive equipment such as computers.

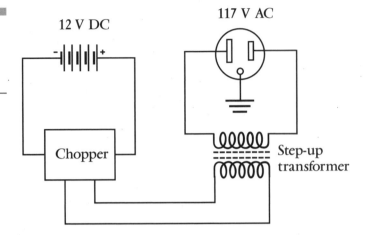

Figure 5-11
A power inverter for
converting low-volt-
age DC into high-
voltage AC.

If the transformer shown in the block diagram is followed by a rectifier and filter, the device becomes a *DC transformer,* also called a *DC-to-
DC converter.* Such a circuit can provide hundreds of volts DC from a
12-V battery source.

Uninterruptible Power Supplies

When a piece of electronic equipment is operated from utility power,
there is always a possibility of a power failure. This can place the user at
risk of inconvenience. If the power to a computer fails, for example, all
data in the random-access memory (RAM) will be lost. This is a nuisance
at best, and a disaster at worst. To prevent problems in case of a power
outage, a UPS can be used.

Figure 5-12 is a block diagram of a simple UPS. Under normal conditions, the equipment gets its power via the transformer and regulator. The
regulator eliminates transients, surges, and dips in the utility current. A
rechargeable battery, such as a lead-acid or NICAD type, is kept charged by
a small current via the rectifier and filter. If the power goes out, a power-
interrupt signal causes the switch to disconnect the equipment from the
regulator and connect it to the power inverter, which converts the battery
DC output to AC. When utility power returns, the switch disconnects the
equipment from the battery and reconnects it to the regulator.

If you leave a computer powered up most of the time, as is often the
case in a business, a UPS can lend you peace of mind. But it's a good
idea to save data in a nonvolatile memory such as the hard disk or
diskette whenever you leave the workstation, even if only for a few

Figure 5-12
An uninterruptible power supply keeps AC flowing to equipment during a momentary power blackout.

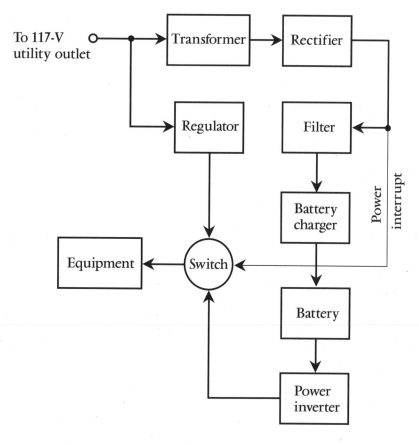

moments. While working, get into the habit of saving data on disk every 5 to 10 minutes. This should become an automatic reflex on your part, whether or not you have a UPS.

If power to a computer fails and you have a UPS, save all your work immediately on the hard disk, and also on a diskette if possible. Then switch the main unit, and all peripherals (including the UPS), off until normal power returns. If the outage has occurred because of an electrical storm, follow the aforementioned procedure and then unplug the main power cord from the wall outlet until the storm has passed.

Solar-Electric Energy Systems

Electric energy can be obtained directly from sunlight by means of photovoltaic cells. Most modern solar cells can produce about 0.1 W of power for each square inch of surface area exposed to bright sunlight.

This is about 15 watts per square foot, or 150 watts per square meter, of cell surface area. Large solar panels, consisting of series-parallel combinations of solar cells, can have hundreds of square feet of surface area and produce thousands of watts of power in direct sunlight. This makes the sun a viable source of energy for residences and businesses. Electric motor vehicles can also be powered by sunlight.

Solar cells produce DC, and batteries deliver DC electricity. But most household appliances require AC at 117 V and 60 Hz. Solar electric energy systems intended for general home and business use must therefore employ power inverters. In some situations, inverters are not necessary. For example, if you plan to operate a notebook computer from a battery and to keep the battery charged via solar panels, you will not need a power inverter. But the charging circuit must be carefully designed so the battery is not damaged during the charging process.

There are two basic types of solar electric energy systems capable of powering homes and small businesses. These are called the *stand-alone system* and *the interactive system*. Each scheme has its assets and limitations.

A stand-alone system uses large banks of rechargeable electrochemical batteries, such as the lead-acid type, to store electric energy as it is supplied by photovoltaics during hours of bright sunshine. The energy is released by the batteries at night or in gloomy daytime weather. This system does not depend on the electric utility companies in any way. It is the sort of system you might construct for a house on a remote island not supplied by conventional utilities. Although this scheme offers independence from the utility companies, you must endure a blackout if the system goes down.

An interactive system operates in conjunction with the utility companies. Energy is sold to the companies during times of daylight and minimum usage, and is bought back from the companies at night or during times of heavy usage. The big advantage of this system is that you can keep using electricity (by buying it all from the utilities) if the solar-energy system breaks down. You do not have to worry about blackouts or brownouts. But such a system is designed to function with the help of the utility companies, and will not provide total independence.

Large solar panels and battery banks are rather expensive; a typical stand-alone or interactive solar electric energy system costs several thousand dollars. But there are some advantages in using this source of energy. Its main asset is the fact that solar energy is dependable (although variable) throughout the civilized world.

Hazards and Precautions

Power supplies can be dangerous. This is especially true of high-voltage circuits, but anything over 12 V should be treated as potentially lethal. In all AC-operated electronic apparatus, high voltage exists at the input to the supply (where the 117-V AC appears). A cathode-ray-tube (CRT) computer display has even higher voltages that operate its deflecting coils. Power inverters might get their current from low-voltage batteries, but their output voltages can be lethal.

Electric shock is the harmful flow of electric current through living tissue. The effects of such current can be injurious, and can sometimes cause death. All living cells contain electrolyte matter, and therefore are capable of conducting electric current. Harmful currents are produced by varying amounts of voltage, depending on the resistance of the path through the tissue.

In the human body, the most susceptible organ, as far as life-threatening effects are concerned, is the heart. A current of 100 to 300 mA for a short time can cause the heart muscle to stop its rhythmic beating and begin to twitch in an uncoordinated way. This is called *heart fibrillation*. An electric shock is much more dangerous when the heart is part of the circuit, as compared to when the heart is not in the circuit.

Under certain conditions, low voltages can produce sufficient current to cause a lethal electric shock. When working on equipment carrying more than 12 V, strict precautions must be taken to minimize the chances of electric shock. The 117-V utility lines present an extreme hazard, because people often treat it casually although it is more than sufficient to cause electrocution.

For details about assisting victims of electric shock, consult your local chapter of the American Red Cross or your local fire department. They offer courses in *cardiopulmonary resuscitation* (CPR), first aid, and other emergency procedures. *If you have any doubts about your ability to work on a piece of equipment safely, then don't.* Call a professional technician or engineer, or take the equipment to a repair shop.

A high-voltage power supply is not necessarily safe after it has been switched off. Filter capacitors hold a charge for a long time. In high-voltage supplies of good design, *bleeder resistors* are connected across each filter capacitor, so the capacitors will discharge in a few minutes after the supply is turned off. But it is unwise to bet your life on components that might not exist in a piece of hardware and that can sometimes fail even when they are provided.

6

Transmitter
Basics

A *transmitter* converts data in the form of digital impulses, voices, and/or images, into electromagnetic (EM) waves in the radio spectrum. Such a device is a specialized, sophisticated transducer. Transmitters are usually designed to work only at specific frequencies, or within relatively narrow ranges (bands) of frequencies. A few transmitters are capable of working over a wide range of frequencies.

Components of a Transmitter

In all radio services, operators must strictly observe the limits of their assigned frequency bands when transmitting. In addition, radio amateurs must use only those frequencies and modes that their license classes allow.

Bands, Modes, and Power

High-frequency (HF) communications transmitters cover various bands between about 1.5 and 30 megahertz (MHz). Some transmitters work one or more very-high- and ultra-high-frequency (VHF and UHF) bands. A few are designed for use at low frequencies (LF) and medium frequencies (MF).

Some radio transmitters can work in several different modes, while others are designed for one specific mode. At HF, transmitters can commonly generate *continuous-wave* (CW) *radiotelegraphy, radioteletype* (RTTY), *single-sideband* (SSB), and *slow-scan television* (SSTV) signals when used with appropriate peripheral input equipment. For RTTY, *frequency-shift keying* (FSK) is usually done by inputting audio tones at the microphone jack in SSB mode. At VHF, transmitters can generate the above-mentioned types of signals and, in addition, are usually capable of *audio-frequency-shift keying* (AFSK), *frequency modulation* (FM), *pulse modulation* (PM), *amplitude modulation* (AM), and *fast-scan television* (FSTV).

Typical fixed-station communications transmitters provide radio-frequency (RF) output power between about 25 and 200 W. A *linear amplifier* can boost the output to several kilowatts if necessary. The most common specification is the *peak envelope power* (PEP). Mobile transmitters, designed for operation from batteries backed up by alternators or generators, generally produce from 5 to 200 W PEP RF output. Portable transmitters, usually incorporated into handheld transceivers ("handie-talkies"), provide from about 1 to 10 W PEP RF output, and

Figure 6-1
Block diagram of an
HF radio transmitter.

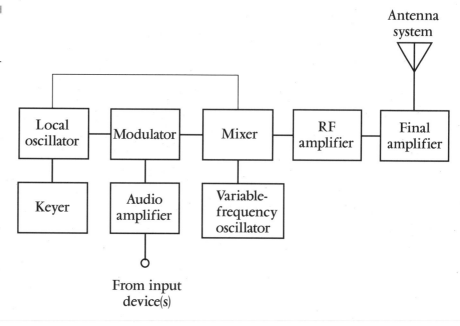

employ rechargeable batteries such as nickel-cadmium (NICAD), nickel-metal-hydride (NiMH), or lithium-ion types.

Stages

A block diagram of an HF SSB radio transmitter is shown in Fig. 6-1. The various stages are defined in this section. Further details are in individual sections later in this chapter.

Local Oscillator The *local oscillator* (LO) is where the signal is generated. This signal is "tailored" by other stages, for ultimate conversion to the transmitting frequency, followed by amplification to the final power output level. The LO is crystal-controlled for optimum stability. Its frequency can be switchable at increments of several hundred kilohertz (500 and 1000 kHz are typical), facilitating band changes when mixed with the output of the *variable-frequency oscillator* (VFO). The LO might operate at a single fixed frequency if the VFO tunes over an exceptionally wide range.

Keyer A radiotelegraphy transmitter might have an internal electronic keyer, with a jack for an external paddle-type key. Most transmitters simply have a key jack to which an external keyer can be connected to switch the LO on and off, or to block its output. Among amateur radio operators, keying devices are a specialty in themselves; there are keyboard keyers, "straight keys," semiautomatic keys ("bugs"), and various paddle-actuated devices.

Microphone Amplifier   The audio from a microphone must be amplified for use by the transmitter modulator. This is done by a simple, low-distortion, low-noise amplifier. This amplifier usually has a tailored frequency response, passing audio between about 300 and 3000 hertz (Hz), and attenuating audio outside this range. The range 300 to 3000 Hz is sufficient to convey intelligible voice signals, and also allows for AFSK and SSTV audio input.

Modulator   The *modulator* in an SSB transmitter is of the balanced variety. This cancels out the carrier from the LO, leaving only the sidebands. Also included (but not shown in the diagram) is a *filtering circuit* that gets rid of either the *lower sideband* (LSB) or the *upper sideband* (USB), producing SSB. This filter also provides further attenuation of sideband energy arising from audio outside the range 300 to 3000 Hz that might have leaked through the microphone amplifier. If FM is used, the modulator works by varying the frequency or phase of the LO. Frequency-shift keying can be accomplished by varying the reactance in the LO; more often, especially in HF transceivers, it is done indirectly via SSB mode. Two audio sine-wave tones are fed into the microphone input, producing two different sideband frequencies that come out as carrier waves.

VFO   The variable-frequency oscillator is either of two types: a tuned *inductance-capacitance* (LC) *oscillator* or a *frequency synthesizer*. The LC type tunes over a span of frequencies, usually 500 or 1000 kHz wide. It is carefully designed and sealed so it maintains its frequency despite mechanical shock, temperature changes, and other variables in the environment. A frequency synthesizer is more stable than an LC-tuned oscillator. A synthesizer can be tuned over a wide range, doing away with the need for band switching at the LO. Or it can cover a specific range equal to the switched-frequency increment of the LO.

Mixer   The *mixer* combines the LO and VFO signals to obtain energy at the actual transmitting frequency. This is a nonlinear circuit that produces sum and difference signals, either of which can be selected for amplification by later stages. The operation of a mixer is basically the same in a transmitter as in a receiver. Details concerning mixer operation are found in Chap. 7.

RF Amplifier   From one to several stages of RF amplification follow the mixer. The last of these stages, prior to the *final amplifier,* is called the *driver,* because it produces the signal that drives the final amplifier. In a typical transmitter with 200 W PEP output, the driver produces approximately 5 W. All the amplifying stages must be linear to ensure that the SSB *modulation envelope* (waveform) is not distorted.

Final Amplifier   The final amplifier, also called the *power amplifier* (PA), boosts the signal power to the level to be transmitted. The final amplifier in an SSB transmitter is always linear. Sometimes "outboard" linear amplifiers are used to obtain higher PEP output than the transmitter alone can provide. Some transmitters have "no-tune" final amplifiers; others require adjustment of the output circuitry for resonance and/or coupling to the antenna system. In transmitters intended for use only in digital modes, or for FM, the final amplifier need not be linear.

Transceivers

A *transceiver* is a combination transmitter/receiver with a frequency control common to both units. The entire system is usually housed in a single cabinet. Transceivers are extensively used in two-way radio communication at all frequencies, and in all modes.

The main advantage of a transceiver over a separate transmitter and receiver is economic. Many of the components and circuits can do double duty, being used in both transmit and receive modes. Another asset is that a transceiver is more easily tuned than a separate transmitter and receiver if the operating frequency must be changed often.

The primary limitation of a transceiver is that it can be difficult to carry out *split-frequency communication* or *split-band communication* on two frequencies that differ greatly. However, some transceivers have a provision for use of two different VFOs that can be independently tuned for this type of operation.

The principal components in a transceiver are a VFO or frequency synthesizer, a transmitter, a receiver, and an antenna-switching circuit. Most transceivers use local oscillators and mixers (superheterodyne operation) in both the transmitting and receiving sections.

190

Chapter 6

Oscillators

For a circuit to oscillate, the gain must be high, the feedback must be positive, and the coupling from output to input must be good. The oscillation frequency is controlled via tuned, or resonant, circuits. These can be LC or resistance-capacitance (RC) combinations, or they can be quartz crystals. Crystals and LC circuits are common at RF; the RC method is more often used at audio frequencies (AF). The tuned circuit makes the feedback path easy for a signal to follow at one frequency, but hard to follow at all other frequencies. As a result, oscillation takes place at a predictable and stable frequency.

Armstrong Oscillator

A common-emitter or common-source amplifier can be made to oscillate by coupling the output to the input through a transformer that reverses the phase of the fed-back signal. The phase at a transformer output can be inverted by reversing the secondary terminals.

The schematic diagram of Fig. 6-2A shows a common-source field-effect-transistor (FET) oscillator whose drain circuit is coupled to the gate circuit via a transformer. In practice, getting oscillation is easy. If the circuit will not oscillate with the transformer secondary hooked up one way, you can switch the wires.

The frequency of this oscillator is controlled by means of a capacitor across either the primary or the secondary winding of the transformer, or both. The inductance of the winding, along with the capacitance, forms a resonant circuit.

The circuit shown in Fig. 6-2A is called an *Armstrong oscillator.* A bipolar transistor can be used in place of the FET. It would need to be biased, using a resistive voltage-divider network, like a class A RF amplifier.

Hartley Oscillator

A method of obtaining controlled feedback at RF is shown in Fig. 6-2B and C. At B, an NPN bipolar transistor is used; at C, an N-channel FET is employed. The analogous PNP bipolar and P-channel FET circuits are identical, but the power supply is negative instead of positive.

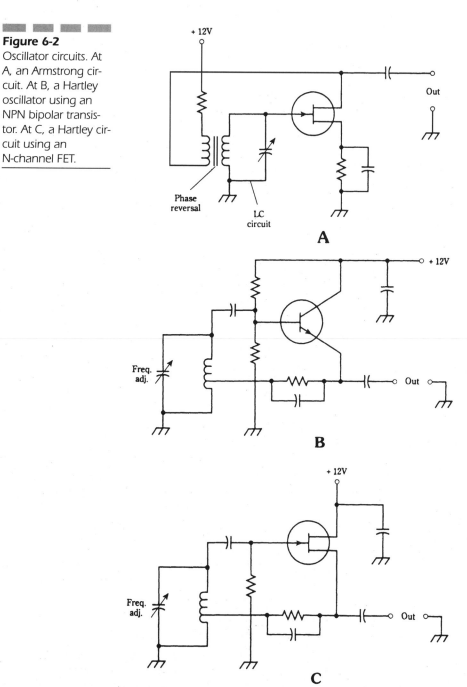

Figure 6-2
Oscillator circuits. At
A, an Armstrong cir-
cuit. At B, a Hartley
oscillator using an
NPN bipolar transis-
tor. At C, a Hartley cir-
cuit using an
N-channel FET.

The circuit uses a single coil with a tap on the windings to provide the feedback. A variable capacitor in parallel with the coil determines the oscillating frequency, and allows for frequency adjustment. This circuit is called a *Hartley oscillator.*

The Hartley oscillator uses about 25 percent of its output power to produce feedback. The other 75 percent of the power can be delivered to external circuits or devices. Oscillators do not, in general, produce more than a fraction of a watt of useful signal output power. If more power is needed, the signal can be boosted by one or more stages of amplification.

In a Hartley oscillator, it is important to use only the minimum amount of feedback necessary to get oscillation. The amount of feedback is controlled by the position of the coil tap.

Colpitts Oscillator

Another way to provide RF feedback is to tap the capacitance instead of the inductance in the tuned circuit. In Fig. 6-2D and E, NPN bipolar and N-channel FET *Colpitts oscillator* circuits are diagramed.

The amount of feedback is controlled by the ratio of capacitances. The coil, rather than the capacitors, is variable in this circuit. This is a matter of convenience. It is difficult to find a dual variable capacitor with the right capacitance ratio between sections to build a Colpitts oscillator circuit. Even if you find one, you cannot change the ratio of capacitances. But it is easy to set the capacitance ratio using a pair of fixed capacitors. Unfortunately, finding a good variable inductor is not easy either. A permeability-tuned coil can be used, but ferromagnetic cores impair the frequency stability of an RF oscillator. A roller inductor might be employed, but these are bulky and expensive. An inductor with several switch-selectable taps can be used, but this does not allow for continuous frequency adjustment. Despite these difficulties, a Colpitts circuit offers exceptional stability and reliability when properly designed.

As with the Hartley circuit, the feedback should be kept to the minimum necessary to sustain oscillation.

In the oscillators shown in Fig. 6-2D and E, the outputs are taken from the emitter or source, rather than the collector or drain as you might expect. The reason for this is that the stability is enhanced when the output of an oscillator is taken from the emitter or source portion of the circuit. To prevent the output signal from being short-circuited to ground, an RF choke (RFC) is connected in series with the emitter or

Figure 6-2 (cont.)
Oscillator circuits. At
D, a Colpitts circuit
using an NPN bipolar
transistor. At E, a Col-
pitts circuit using an
N-channel FET. At F, a
Clapp circuit using an
N-channel FET.

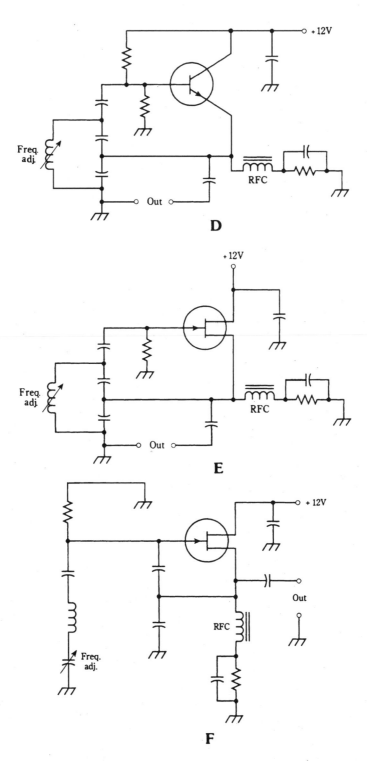

source. The choke lets DC pass while blocking AC at the signal frequency. Typical values for RF chokes range from about 100 microhenrys (μH) at high frequencies (such as 15 MHz) to 10 millihenrys (mH) at low frequencies (such as 150 kHz).

Clapp Oscillator

A variation of the Colpitts oscillator makes use of series resonance, instead of parallel resonance, in the tuned circuit. Otherwise the circuit is basically the same as the parallel-tuned Colpitts oscillator. A schematic diagram of an N-channel FET *Clapp oscillator* circuit is shown in Fig. 6-2F. The P-channel circuit is identical, except the power supply polarity is reversed. The bipolar-transistor Clapp circuit is almost exactly the same as the FET circuit, with the emitter in place of the source, the base in place of the gate, and the collector in place of the drain. The only difference is the addition of a resistor between the base and the positive supply voltage (for NPN) or the negative supply voltage (for PNP).

The Clapp oscillator offers excellent stability at RF. Its frequency will not change much when high-quality components are used. The Clapp oscillator is a reliable circuit; it is easy to get it to oscillate. Another advantage of the Clapp circuit is that it allows the use of a variable capacitor for frequency control, while still accomplishing feedback through a capacitive voltage divider.

Stability

The term *stability* is used often by engineers when they talk about oscillators. In an oscillator, this term has two meanings: constancy of frequency and reliability of performance. Both of these considerations are important in the design of a good oscillator circuit.

The above-described oscillator types allow for frequency adjustment using variable capacitors or variable inductors. The component values are affected by temperature, and sometimes by humidity. When designing a VFO, it is crucial that the components maintain constant values, as much as possible, under all anticipated conditions.

Some types of capacitors maintain their values better than others, when the temperature goes up or down. Polystyrene capacitors are among the best. Silver-mica capacitors also work well when polystyrene units cannot be found.

Inductors are the most temperature-stable when they have air cores. They should be wound, when possible, from stiff wire with strips of plastic to keep the windings in place. Some air-core coils are wound on hollow cylindrical cores, made of ceramic or phenolic material. Ferromagnetic solenoidal or toroidal cores are not as good for VFO coils, because these materials change their permeability as the temperature varies. This changes the inductance, in turn affecting the oscillator frequency.

An oscillator should always start working as soon as power is supplied. It should keep oscillating under all normal conditions, not quitting if the load impedance changes slightly or if the temperature rises or falls. A "finicky" oscillator is a great annoyance. The failure of a single oscillator can cause a transmitter to stop working.

When an oscillator is built and put to use in a practical system, *debugging* is always necessary. This is a trial-and-error process of getting the flaws, or "bugs," out of the circuit. Rarely can an engineer build something from the drawing board and have it work perfectly the first time. In fact, if two oscillators are built from the same diagram, with the same component types and values in the same geometric arrangement, one circuit might work fine, and the other be unstable. This usually happens because of differences in the quality of components that do not show up until they are put to work in actual circuits.

Oscillators are generally designed to work into high load impedances. It is important that the load impedance not be too low. (You need never be concerned that it might be too high. In general, the higher the load impedance, the better.) If the load impedance is too low, the load will try to draw power from an oscillator. Then, even a well-designed oscillator might become unstable. Oscillators are not meant to produce powerful signals. High power should, if needed, be obtained via amplifier stages following an oscillator circuit.

Crystal-Controlled Oscillators

Quartz crystals can be used in place of tuned LC circuits in RF oscillators, if it is not necessary to change the frequency often. Crystal oscillators offer frequency stability far superior to that of LC-tuned VFOs.

There are several ways that crystals can be connected in bipolar or FET circuits to get oscillation. One common circuit is the *Pierce oscillator.* An N-channel FET and quartz crystal are connected in a Pierce configuration as shown in the schematic diagram of Fig. 6-2G. The crystal frequency can be varied slightly (about plus or minus 0.1 percent) by

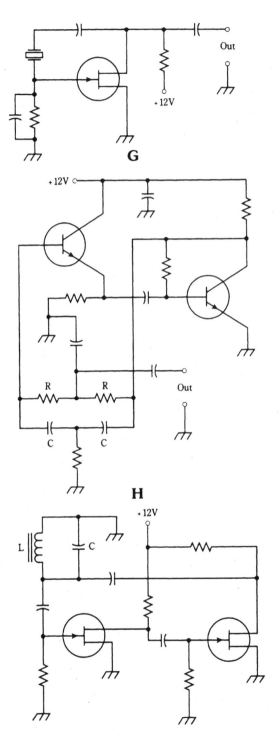

Figure 6-2 (cont.)
Oscillator circuits. At
G, a Pierce circuit
using an N-channel
FET. At H, a twin-T
audio oscillator circuit
using two bipolar
transistors. At I, a
multivibrator using
two N-channel FETs.

means of an inductor or capacitor in parallel with the crystal. But the frequency is determined mainly by the thickness of the crystal and by the angle at which it is cut from the quartz rock.

Crystals change slightly in frequency as the temperature changes. Some crystal oscillators are housed in temperature-controlled chambers called *crystal ovens*. They maintain their frequency so well that they are often used as frequency standards against which other oscillators are calibrated. The accuracy can be within a few hertz at working frequencies of several megahertz.

Voltage-Controlled Oscillator

The frequency of a VFO can be adjusted via a *varactor* in the tuned LC circuit. A varactor, also called a *varicap,* is a semiconductor diode that works as a variable capacitor when it is reverse-biased. The capacitance depends on the reverse-bias voltage. The greater this voltage, the lower the value of the capacitance.

The Hartley and Clapp oscillator circuits lend themselves well to varactor frequency control. The varactor is placed in series or parallel with the tuning capacitor, and is isolated for DC by blocking capacitors. The resulting oscillator is called a *voltage-controlled oscillator* (VCO). There are several reasons why varactor frequency control is often preferable to the use of mechanically variable capacitors or inductors. But it all comes down to the bottom line: Varactors are cheap. They are also less bulky than mechanically variable capacitors and inductors.

Phase-Locked Loop

One type of oscillator that combines the flexibility of a VFO with the stability of a crystal oscillator is known as a *phase-locked loop* (PLL). This scheme is extensively used in modern radio transmitters and receivers.

In a PLL, the output of a VCO is passed through a *programmable divider,* a digital circuit that divides the VCO frequency by any of hundreds or even thousands of numerical values chosen by the operator. The output frequency of the programmable divider is locked, by means of a *phase comparator,* to the signal from a crystal-controlled reference oscillator.

As long as the output from the programmable divider is exactly on the reference-oscillator frequency, the two signals are in phase, and the output of the phase comparator is zero volts DC. If the VCO frequency begins to drift, the output frequency of the programmable divider will drift, too (although at a different rate). But even the tiniest frequency change—a fraction of 1 Hz—causes the phase comparator to produce a DC error voltage. This error voltage is either positive or negative, depending on whether the VCO has drifted higher or lower in frequency. The error voltage is applied to a varactor in the VCO, causing the VCO frequency to change in a direction opposite to that of the drift. This forms a DC feedback circuit that maintains the VCO frequency at a precise multiple of the reference-oscillator frequency, that multiple having been chosen by the programmable divider.

The key to the stability of the PLL frequency synthesizer lies in the fact that the reference oscillator is crystal-controlled. When you hear that a radio receiver, transmitter, or transceiver is "synthesized," it usually means that the frequency is determined by a PLL frequency synthesizer.

The stability of a synthesizer can be enhanced by using an amplified signal from the National Bureau of Standards, transmitted on the shortwave radio bands by time-and-frequency stations WWV and WWVH at 5, 10, and 15 MHz, directly as the reference oscillator. These signals are frequency-exact to a tiny fraction of 1 Hz, because they are controlled by atomic clocks. Most people do not need precision of this caliber, so you will not often see advertisements for consumer devices like ham radios and shortwave receivers with primary-standard PLL frequency synthesis. But it is employed by some corporations and government agencies, such as the military.

AF Oscillators

All AF oscillators work in the same way, consisting of amplifiers with positive feedback. One form of AF oscillator that is popular for general-purpose use is the *twin-T oscillator* (Fig. 6-2H). The frequency is determined by the values of the resistors R and capacitors C. The output is a near-perfect sine wave. The small amount of distortion helps to alleviate the aural irritation produced by an absolutely pure sinusoid. This particular circuit uses two NPN bipolar transistors. However, FETs will work also.

Another audio-oscillator circuit uses two identical common-emitter or common-source amplifier circuits, hooked up so that the signal goes around and around between them. This is sometimes called a *multivibrator* circuit, although that is technically a misnomer, the term being more appropriate to various digital signal-generating circuits. Two N-channel FETs can be connected to form a multivibrator as shown in Fig. 6-2I. Each FET amplifies the signal and reverses the phase by 180 degrees. Thus the signal goes through a 360-degree shift each time it gets back to any particular point. A 360-degree shift results in positive feedback, being effectively equivalent to no phase shift. The frequency is set by means of an LC circuit. The coil uses a ferromagnetic core because such a core is necessary to obtain the large inductance needed for resonance at these low frequencies. The value of L is typically from 10 mH to as much as 1 henry (H). The capacitance is chosen to obtain an audio tone at the frequency desired.

Modulators

Some characteristic of a signal must be varied in a systematic way if the signal is to convey any information. There are several different *modulation* methods via which a signal can be made to change in a controlled way, so that data is "impressed" on it. Modulation can be accomplished by varying the amplitude, the frequency, or the phase of a signal. Another method is to transmit a series of pulses, whose duration, amplitude, or spacing is made to change in accordance with the data to be conveyed.

The Carrier

The "heart" of most communications signals is a sine wave, usually of a frequency far above the range of human hearing. This is called a *carrier wave*. The lowest carrier frequency normally used for EM signal transmission is 9 kHz. The highest frequency is less well defined; some wireless devices make use of visible light waves.

For modulation to work effectively, the carrier must have a frequency many times the highest frequency of the modulating signal. For example, if you want to modulate a radio wave with hi-fi music, which has a frequency range from a few hertz up to 20 kHz or so, the carrier wave must have a frequency well above 20 kHz. A good rule is that the carrier must have a frequency of at least 10 times the highest modulating frequency. So for good hi-fi music transmission, a radio carrier should be at 200 kHz or higher.

This rule holds for all kinds of modulation, whether it be of the amplitude, the phase, or the frequency of the carrier. If the rule is violated, the efficiency of transmission will be degraded, resulting in less than optimum data transfer.

On-Off Keying

The simplest, and oldest, form of modulation is *on-off keying*. Early telegraph systems used direct currents that were keyed on and off, and were sent along wires. The first radio transmitters employed spark-generated "signals" (more closely resembling *noise* by today's standards) that were keyed using the *Morse code*. The impulses from the sparks, like ignition noise from a car, could be heard in crystal-set receivers several miles away. In the early years of the twentieth century, the "wireless telegraph" was a miracle of technology with significance comparable to that of today's most advanced wireless systems. It was the first glimmer of what was to become mass electromagnetic media.

Keying is usually accomplished at the oscillator of a continuous-wave (CW) radio transmitter. This is the basis for a mode of communications that is, and always has been, popular among amateur radio operators and radio experimenters. While the Morse code is archaic, a CW transmitter is simple to build. A human operator, listening to Morse code and writing down the characters as they are sent, is an efficient (if slow) digital data receiver. Morse code can be broken down into bits, each having a length of one *dot*. A *dash* is three bits long. The space between dots and dashes, within a single character, is one bit. The space between characters in a word is three bits. The space between words is seven bits. Punctuation marks are sent as characters attached to the ends of words.

Morse code is a tedious way to send and receive data. Human operators typically use speeds ranging from about five *words per minute* (5 wpm) to 40 or 50 wpm. A few expert radiotelegraphers can follow along by ear (without writing the data down) at speeds of around 90 or 100 wpm. This is a tiny fraction of modern digital communications speeds.

Frequency-Shift Keying

Morse keying is the most primitive form of amplitude modulation (AM). The amplitude is varied between two extreme conditions: fully on and fully off. The trouble with using this scheme for automatic data transmission, such as RTTY, is that noise pulses can be interpreted as signal pulses by the receiving apparatus. There is no problem if a noise burst takes place during the full-on, or *mark*, part of the signal; but if it happens during a pause or *space* interval, the receiver might interpret it as a mark instead. This problem can be practically eliminated by sending a signal during spaces, but at a different frequency from the marks. The easiest way to do this is to send the mark (logic 1) part of the signal at one carrier frequency and the space (logic 0) part at another frequency a few hundred hertz higher or lower. This is frequency-shift keying (FSK), the most primitive form of frequency modulation (FM). The difference between the mark and space frequencies, called the *shift*, is usually between 100 and 1000 Hz.

There are several text data teleprinter codes, two of which are known as the *Baudot* (pronounced "baw-DOE") and *ASCII* (pronounced "ASK-ee") codes. A carrier wave can be keyed on and off using such codes at speeds ranging from 60 to many thousands of words per minute. In recent years, ASCII has replaced Baudot as the most common code. A special circuit, called a *terminal unit* (TU), converts RTTY signals into electrical impulses to work a teleprinter or to display the characters on a monitor screen. The terminal unit also generates the signals necessary to send RTTY as the operator types on the keyboard. A computer can be made to work as an RTTY terminal by means of *terminal emulation software*. This software is available in several different forms, and is popular among radio amateurs and electronics hobbyists.

Audio-Frequency-Shift Keying

The scheme used by a computer *modem*, which is a special kind of TU, is audio frequency-shift keying (AFSK). It is a simple type of FM, and is in fact the same thing as FSK, except that it is done at audio frequencies rather than at radio frequencies. Two tones are transmitted, one for the mark signals (logic 1) and the other for the space signals (logic 0). This is illustrated in Fig. 6-3A.

You might think that the types of data that can be sent via AFSK must be limited to text and software. How, you might ask, can anything complicated, such as a visual scene, a human voice, a musical tune, or a robot command, be conveyed by using two audio tones? The answer is that all variable quantities can be represented as sequences of digital bits. In fact, digital methods work better than older analog schemes for transmitting and receiving analog information, because the digits are sent at

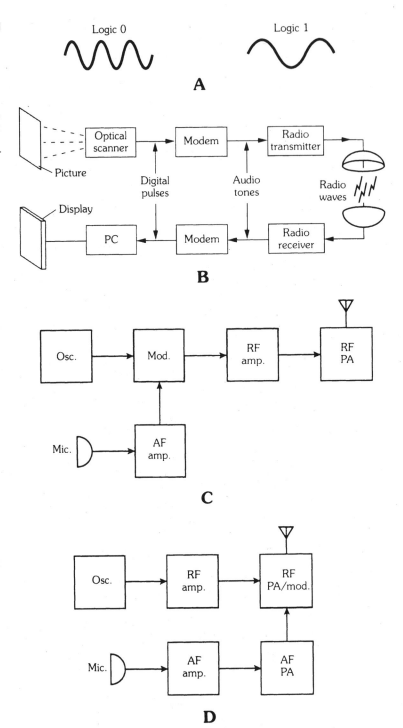

high speeds. Data can be changed from one form to the other by means of *analog-to-digital (A/D) conversion* and *digital-to-analog (D/A) conversion.*

Even an intricate color photograph can be converted into digital highs and lows (logic states 1 and 0) at a *source* computer, sent over wires, radio waves, or light beams, and then converted back into the original picture at the *destination.* If you've done much downloading of data online, you have probably witnessed this for yourself. In the future, expect to see control signals, such as the kind used for personal robots, transferred via telephone lines, television cable systems, and wireless satellites.

Figure 6-3B shows a system for digitally sending a picture by radio. At the source, an *optical scanner* converts the picture into digital pulses. These pulses are changed into AFSK by the D/A converter in a modem. The resulting AFSK is sent to the radio transmitter, which contains a special modulator that impresses the audio onto radio waves. These waves travel through space, perhaps being retransmitted by *repeaters* along the way. At the destination, the receiver extracts the AFSK from the radio waves. The audio is then fed to a modem, in which an A/D converter changes the signal into digital pulses. These pulses are then sent to the computer, which has software that changes the pulses back into the original picture.

Even at fairly high data speeds, it takes awhile to send a picture with excellent image resolution (detail). It can also take a long time to send complex software programs. However, the intended data can be conveyed faster by using fewer logic bits to represent the same information. This is done via a scheme called *data compression.*

Text and software can be compressed somewhat without losing any of the detail, but the *compression ratio* can rarely be greater than 2:1. Images are just the opposite. There's always at least a slight sacrifice in *image resolution* when a picture's digital representation is squeezed down. But images can be compressed by factors of 25:1 or more, and the degradation in image quality is difficult or impossible to notice. To speed up the transmission of information, compressed data can be used to modulate the signals sent over a communications circuit. With a compression ratio of, say, 20:1, an image that takes an hour to send in uncompressed form will take only 3 minutes to send in compressed form, assuming the same raw data speed in bits per second along the transmission medium.

Amplitude Modulation

A voice signal is a complex waveform with frequencies mostly in the range 300 Hz to 3 kHz. DC can be modulated by these waveforms, thereby transmitting voice information over wires. This is how early telephones worked.

Around the year 1920, when CW oscillators were developed to replace spark-gap transmitters, engineers wondered if a radio wave might be modulated with a voice, like DC in a telephone. If this were possible (as it proved to be), voices could be sent by "wireless." Radio communications via Morse code was being done over long distances, but the mode was agonizingly slow.

An RF amplifier was built with the intent of making its gain variable at audio frequencies. Actually, this is not what happens in an AM transmitter (see the next paragraph), but the concept led to a system that did what the engineers hoped. Vacuum tubes were used as amplifiers back then, because transistors had not been invented yet. But the principle of AM is the same, whether the active devices are electron tubes, bipolar transistors, field-effect transistors, or integrated circuits.

In AM, the carrier wave itself does not rise and fall in amplitude, as early engineers imagined. Instead, the audio energy appears at *sideband frequencies* above and below, and very close to, the carrier frequency. It is these sideband signals that carry the voice information. The carrier does not contribute anything to the intelligence being conveyed, although it does make the amplifying device, whatever it might be, work hard.

Two complete AM transmitters are shown in block-diagram form in Fig. 6-3C and D. At C, modulation is done at a low power level. This is *low-level AM*. All the amplification stages after the modulator must be linear. If a class C power amplifier (PA) is used, the signal will be distorted. In some broadcast transmitters, AM is done in the final PA, as shown at D. This is *high-level AM*. The PA is the modulator as well as the final amplifier. As long as the PA is modulated correctly, the output will be a "clean" AM signal; RF linearity is of no concern.

If bipolar transistors had existed in 1920, the first amplitude modulator would have resembled the circuit shown in Fig. 6-3E. This circuit is in fact an RF/audio amplifier/mixer that impresses the voice data on the carrier signal. The voice signal affects the instantaneous voltage between the emitter and base, varying the instantaneous bias. Thus the RF output signal is modulated in a way that duplicates the waveform of the voice signal. The circuit of Fig. 6-3E will work well as an AM voice modulator, provided that the audio-frequency (AF) input is not too great. If the AF is excessive, *overmodulation* will occur. This will result in a distorted signal.

The extent of amplitude modulation is expressed as a percentage, from 0 percent, representing an unmodulated carrier, to 100 percent, representing full modulation. Increasing the modulation past 100 percent

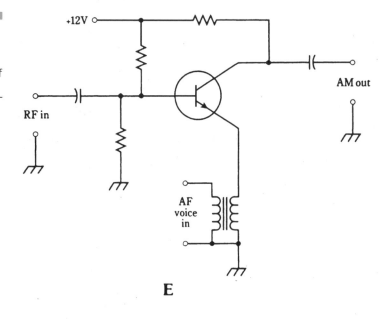

E

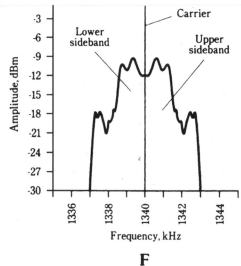

F

will degrade the quality of the transmission. In an AM signal modulated
100 percent, only one-third of the power is actually used to convey the
data; the other two-thirds is consumed by the carrier wave. For this reason, AM is rather inefficient. There are voice modulation techniques that
make better use of available transmitter power. Perhaps the most widely
used is single sideband (SSB), to be discussed shortly.

Suppose you could get a graphic display of an AM signal, with frequency on the horizontal axis and amplitude on the vertical axis. This is in fact done using an instrument called a *spectrum analyzer.* In Fig. 6-3F, the spectral display for an AM voice radio signal at 1340 kHz is illustrated. On a spectrum analyzer, unmodulated radio carriers look like vertical lines, or *pips,* of various heights depending on how strong they are. The carrier wave at 1340 kHz shows up as a strong pip. The horizontal scale of the display is calibrated in increments of 1 kHz per division. This is an ideal scale for looking at an AM signal. The vertical scale is calibrated in decibels (dB) below the signal level that produces 1 milliwatt (mW) at the input terminals. Each vertical division represents 3 dB. Decibels relative to 1 mW are abbreviated dBm by engineers. Thus, in Fig. 6-3F, the top horizontal line is 0 dBm; the first line below it is −3 dBm and the second line, −6 dBm.

The audio components of the voice signal show up as sidebands on either side of the carrier. All of the voice energy in this example is at or below 3 kHz. This results in sidebands within the range 1340 kHz plus or minus 3 kHz, that is, the band 1337 to 1343 kHz. The frequencies between 1337 and 1340 kHz are the *lower sideband* (LSB); those from 1340 to 1343 kHz are the *upper sideband* (USB). The *bandwidth* of the RF signal is the difference between the maximum and minimum sideband frequencies, and is twice the highest audio modulating frequency. In the example of Fig. 6-3F, the voice energy is at or below 3 kHz, and the bandwidth of the RF signal is 6 kHz.

At 3 kHz above and below the carrier, the frequency cutoffs are abrupt; the transmitter uses an *audio lowpass filter* that cuts out the audio above 3 kHz. Audio above 3 kHz contributes nothing to the intelligibility of a human voice. It is important to keep the bandwidth of a signal as narrow as possible, to maximize the number of signals that can be transmitted within a given frequency band.

Single Sideband

As mentioned previously, AM is not efficient. Most of the power is used up by the carrier; only one-third of it carries data. Besides that, the two sidebands are mirror-image duplicates. An AM signal is not only inefficient but needlessly redundant as well.

During the 1950s, engineers began to work on an alternative to conventional AM. They wanted to put all the transmitter output energy into the voice, that is, the meaningful part of the signal. This could in theory represent a threefold effective increase in transmitter power. And what if the bandwidth could be cut to 3 rather than 6 kHz, by getting rid of the sideband redundancy? That would put all the energy into half the spectrum space, so that twice as many voice signals could fit into a given band.

These improvements were realized by means of circuits that cancel out, or suppress, the carrier in the modulator circuit, and that filter out, or phase out, one of the two sidebands. The remaining voice signal has a spectrum display that looks like the graph at Fig. 6-3G. Either LSB or USB can be used, and either mode works as well as the other.

The heart of an SSB transmitter is a *balanced modulator.* This circuit works like an ordinary AM modulator, except that the carrier wave is phased out. This leaves only the LSB and USB. One of the sidebands is removed by a filter that passes only the RF within a 3-kHz-wide band. A block diagram of an SSB transmitter is shown in Fig. 6-3H.

High-level modulation does not work for SSB. The balanced modulator is in a low-power part of the transmitter. Therefore, the RF amplifiers that follow the modulator must all be linear amplifiers. If nonlinear amplification is used with an SSB signal, the modulation envelope will be distorted. This will degrade the quality of the signal. It can also cause the bandwidth to exceed the nominal 3 kHz, resulting in interference to other stations using the band. Engineers and technicians refer to this as *splatter.*

Frequency and Phase Modulation

Both AM and SSB work by varying the overall signal amplitude, or strength. This can be a disadvantage when there are *sferics* (also called "static") caused by thundershowers in the vicinity. Sferics are recognizable as crashing or crackling sounds in an AM or SSB receiver. *Ignition noise* can also be a problem; this sounds like a buzz or whine. Sferics and ignition noise are both predominantly amplitude-modulated. Frequency-modulated signals can be received by circuits designed for immunity to these kinds of noise. In FM, the overall amplitude of the signal remains constant, and the instantaneous frequency or phase is made to change. Because the carrier is always full-on, class C power amplifiers can be used. Linearity is of no concern when the signal level does not change.

The most direct way to get FM is to apply the audio signal to a varactor in a VFO. An example of this scheme, known as *reactance modulation,* is shown by Fig. 6-3I. The fluctuating voltage across the varactor causes its capacitance to change in accordance with the audio waveform. The

Figure 6-3 (cont.)
At G, the spectral display of an SSB signal. At H, a simplified block diagram of an SSB transmitter. At I, reactance modulation to obtain FM.

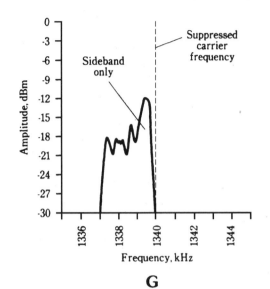

G

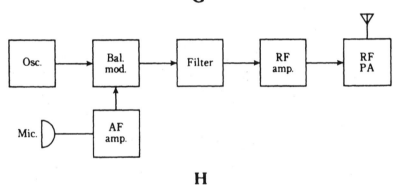

H

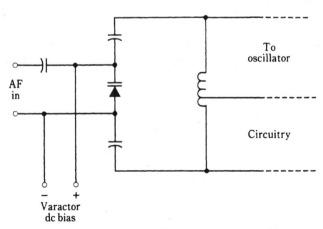

I

changing capacitance results in an up-and-down swing in the frequency generated by the VFO. In the illustration, only the tuned circuit of the oscillator is shown.

Another way to get FM is to modulate the phase of the oscillator signal. This causes small fluctuations in the frequency as well, because any instantaneous phase change shows up as an instantaneous frequency change (and vice versa). This is called *phase modulation*. The circuit is more complicated than the reactance modulator. When phase modulation is used, the audio signal must be processed, adjusting the amplitude-versus-frequency response of the audio amplifiers. Otherwise the signal will sound muffled in an FM receiver.

The amount by which the FM carrier frequency varies will depend on the relative audio signal level, and also on the degree to which the audio is amplified before it is applied to the modulator. The *deviation* is the maximum extent to which the instantaneous carrier frequency differs from the unmodulated-carrier frequency. For most FM voice transmitters, the deviation is standardized at plus or minus 5 kHz (Fig. 6-3J). The deviation obtainable by means of direct FM is greater, for a given oscillator frequency, than the deviation that can be realized with phase modulation. But the deviation of a signal can be increased by a *frequency multiplier*, which is an RF amplifier circuit whose output is tuned to some harmonic (multiple frequency) of the input. A multiply-by-two circuit is called a *frequency doubler*; a multiply-by-three circuit is a *frequency tripler*.

When an FM signal is passed through a frequency multiplier, the deviation is multiplied along with the carrier frequency. If a modulator provides plus or minus 1.6-kHz deviation, the frequency can be doubled and the result will be a deviation of plus or minus 3.2 kHz. If the frequency is tripled, the deviation increases to plus or minus 4.8 kHz, which is just about standard for narrowband FM voice communications. In FM hi-fi broadcasting, and in some other applications, the deviation is much greater than plus or minus 5 kHz. These transmitters employ *wideband FM*, allowing for faithful reproduction of stereo music and complex or high-speed data.

The deviation of an FM signal should be equal to the highest modulating audio frequency, if optimum fidelity is to be obtained. Thus plus or minus 5 kHz is more than enough for voice (3 kHz would probably suffice). For music, a deviation of about plus or minus 15 or 20 kHz is needed for excellent hi-fi reception. The ratio of the frequency deviation to the highest modulating audio frequency is called the *modulation index*. For good fidelity, it should be at least 1:1. But it should not be much greater, because that would waste spectrum space.

Figure 6-3 (cont.)
At J, frequency-ver-
sus-time graph of an
FM signal. At K, pulse
amplitude modula-
tion. At L, pulse width
modulation. At M,
pulse interval modu-
lation. At N, pulse
code modulation.

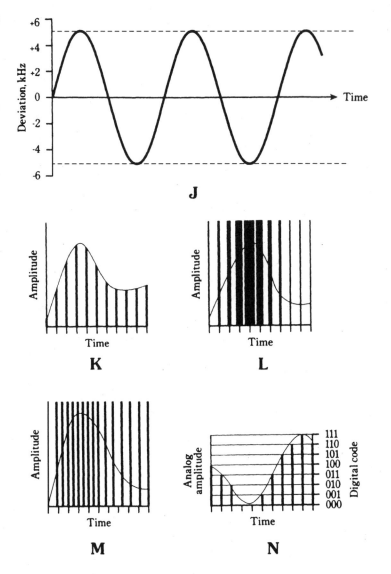

Pulse Modulation

Still another method of modulation works by varying some aspect of a
stream of signal bursts or pulses. Several types of pulse modulation (PM)
are briefly described here. They are diagramed in Fig. 6-3K, L, M, and N
as amplitude-versus-time graphs. The modulating waveforms are shown
as curves, and the pulses as vertical lines.

In *pulse amplitude modulation* (PAM), the strength of each individual
pulse varies according to the modulating waveform. In this respect, PAM

is very much like ordinary amplitude modulation. An amplitude-versus-time graph of a hypothetical PAM signal is shown in Fig. 6-3K. Normally, the pulse amplitude increases as the instantaneous modulating-signal level increases. But this can be reversed, so that higher audio levels cause the pulse amplitude to go down. Then the signal pulses are at their strongest when there is no modulation. Either positive or negative PAM will provide good results. The transmitter works a little harder if the negative modulation method is used. In PAM, the pulses all last for the same length of time; they are equally "wide."

Another way to change the transmitter output is to vary the duration of the individual pulses. This is called *pulse duration modulation* (PDM) or *pulse width modulation* (PWM), shown in Fig. 6-3L. Normally, the pulse duration increases as the instantaneous modulating-signal level increases. This is *positive PDM.* But, as with PAM, this can be reversed. The transmitter must work harder to accomplish *negative PDM.* Regardless of whether positive or negative PDM is employed, the peak pulse amplitude remains constant.

Even if all the pulses have the same amplitude and the same duration, modulation can still be accomplished by varying how often they occur. In PAM and PDM, the pulses are always sent at the same time interval, such as 0.0001 second (s). But in *pulse interval modulation* (PIM), also called *pulse frequency modulation* (PFM), pulses can occur more or less frequently than the zero-modulation interval. This is shown in Fig. 6-3M. Every pulse has the same amplitude and the same duration, but their rate fluctuates with the modulating waveform. When there is no modulation, the pulses are evenly spaced with respect to time. An increase in the instantaneous data amplitude might cause pulses to be sent more often, as is depicted in Fig. 6-3M. Or, an increase in instantaneous data level might slow down the rate at which the pulses are sent. Either method will work equally well.

In recent years, communication has been done more and more by digital means. In digital signals, the modulating data attains only certain defined states, rather than continuously varying in an analog way. Digital transmission offers better efficiency than analog transmission. With digital modes, the signal-to-noise ratio is better, the bandwidth is narrower, and there are fewer errors. Teleprinter and computer data is always sent digitally, as is Morse code. Voices and images can be sent digitally, too. The only drawback that early digital experimenters faced was somewhat degraded fidelity. But today that has been overcome to the extent that digitized music recordings and transmissions sound better than the best analog reproductions. *Pulse-code modulation* (PCM) represents a digital method of signal modulation. In PCM, any of the previously described aspects—amplitude, duration, or frequency—of a pulse sequence can be varied. But rather than having infinitely many possible states, there are only a few. The greater the number of states, the better the fidelity. The transmitting and receiving equipment must be made more sophisticated as the number of digital states increases. An example of eight-level PAM/PCM is shown in Fig. 6-3N.

Amplification in Transmitters

There are several ways in which transmitter amplifiers can be categorized. One way amplifier circuits are distinguished involves the manner in which the active device, usually a bipolar transistor or FET, is biased. The *class of amplification* is an expression of the biasing scheme, and therefore of the general behavior, of an analog amplifier circuit.

Class A

In a *class A amplifier*, the bias is such that, with no signal input, the device is near the middle of the straight-line portion of the *characteristic curve* of the active amplifying device. Figure 6-4A shows the general nature of the characteristic curve for an NPN bipolar transistor; Fig. 6-4B shows a characteristic curve typical of an N-channel FET. For PNP bipolar transistors, reverse the polarity signs in Fig. 6-4A; for P-channel FETs, reverse the polarity signs in Fig. 6-4B.

In bipolar transistors, class A operation results from bias at the point marked "Class A" in the curve of Fig. 6-4A. In FETs, class A operation is obtained with bias at the point marked "Class A" in the curve of Fig. 6-4B.

With class A amplifiers, it is important that the input signal not be too strong. Otherwise, during part of the cycle, the device will be driven outside of the straight-line part of the characteristic curve. When this occurs, the output waveshape will not be a faithful reproduction of the input waveshape, and the amplifier will no longer be linear. This will cause distortion of the signal waveform. In an AF amplifier, the resulting sound will have a raspy or muddled quality. In an RF circuit, the output signal will contain considerable energy at harmonic frequencies (whole-number multiples of the fundamental frequency).

When a class A amplifier is working properly, it produces almost no waveform distortion. But class A operation is inefficient; considerable power is dissipated as heat in the active device and its peripheral components. This is mainly because the bipolar transistor or FET draws a fair amount of current whether there is signal input or not. Even with zero

Figure 6-4
At A, a characteristic
curve for an NPN
bipolar transistor. At
B, a characteristic
curve for an N-chan-
nel FET. At C, a sim-
ple class B push-pull
amplifier circuit.

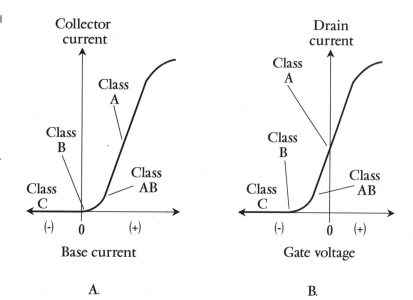

A.

B.

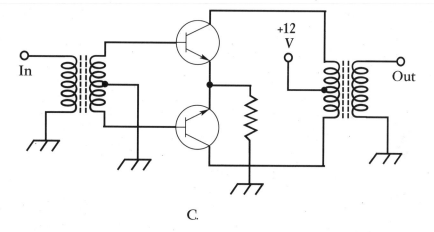

C.

signal, the device is working hard. Class A amplifiers typically convert
only about 25 to 35 percent of the DC input power into actual signal
output power.

Class AB

In amplifiers designed for transmitters, efficiency is a significant consideration. Any power not used toward generating a strong output signal
will end up as heat in the circuit components. If an amplifier is

designed to produce high power output, inefficiency translates to a lot of heat.

When a bipolar transistor is biased close to *cutoff* under no-signal conditions (shown by the point marked "Class AB" in Fig. 6-4A), or when an FET is near *pinchoff* (shown by the point marked "Class AB" in Fig. 6-4B), the input signal will drive the device into the nonlinear part of the characteristic curve during part of the cycle. A small collector or drain current will flow when there is no signal input, but it will be less than the no-signal current that flows in a class A amplifier. This is a *class AB amplifier.*

With class AB operation, the input signal might or might not cause the device to go into cutoff or pinchoff for a small part of the cycle. Whether or not this happens depends on the actual bias point, and also on the strength of the input signal. You can visualize this by imagining the dynamic operating point oscillating back and forth along the curve, in either direction from the static (no-signal) operating point. If the bipolar transistor or FET is never driven into cutoff/pinchoff during any part of the signal cycle, the circuit is a *class AB_1 amplifier.* If the device goes into cutoff/pinchoff for any part of the cycle (up to almost half), the circuit is a *class AB_2 amplifier.*

In a class AB amplifier, the output waveshape is not identical with the input waveshape. But if the wave is amplitude-modulated in any way, as is the case in some voice radio transmitters, the modulation envelope will come out undistorted. That is, the circuit is not linear with respect to the actual signal waveform, but it is linear relative to the modulation envelope. Class AB operation is useful in linear RF power amplifiers, such as those used by amateur radio operators in single-sideband (SSB) communications. The efficiency ranges from about 40 to 60 percent.

Class B

When a bipolar transistor is biased exactly at cutoff, or an FET is biased exactly at pinchoff under zero-input-signal conditions, an amplifier is working in class B. These operating points are labeled "Class B" on the characteristic curves in Fig. 6-4A and B.

In a *class B amplifier,* there is no collector or drain current when there is no signal. This saves energy, because the circuit is not consuming any power unless there is a signal going into it. (Class A and class AB amplifiers draw current even when the input is zero.) When there is an input signal, current flows in the device during exactly half of the cycle. The output waveshape is greatly different from the input waveshape; in fact, it is half-wave rectified. But in RF work, this is not normally a problem.

Sometimes two bipolar transistors or FETs are used in a class B circuit, one for the positive half of the cycle and the other for the negative half. In this way, distortion is eliminated. This is called a *class B push-pull amplifier.* A class B push-pull circuit using two NPN bipolar transistors is illustrated in Fig. 6-4C. This configuration combines the efficiency of class B with the low distortion of class A. Its main disadvantage is that it needs two center-tapped transformers, one at the input and the other at the output. This translates into large physical bulk and mass; it also drives up the cost.

The class B scheme lends itself well to RF power amplification. Although the output waveshape is distorted, resulting in harmonic energy, this problem can be overcome by a resonant LC circuit in the output. If the signal is amplitude-modulated in any way, the modulation envelope will not be distorted. If a push-pull circuit is used, the even harmonics of the fundamental frequency are canceled in the output circuit.

You will sometimes hear of class AB or class B linear amplifiers, especially with respect to amateur-radio RF power amplification. The term *linear* refers to the fact that the modulation envelope is not distorted by the amplifier. The carrier waveform itself is affected in a nonlinear fashion, because the amplifiers are not biased in the straight-line part of the operating curve.

Class AB_2 and class B amplifiers draw some power from the input signal source. Engineers say that such amplifiers require a certain amount of *drive* or *driving power* to function. Class A and class AB_1 amplifiers theoretically need no driving power, although there must be an input voltage. Class B amplifiers are typically 55 to 65 percent efficient.

Class C

A bipolar transistor or FET can be biased past cutoff or pinchoff, and it will still work as a power amplifier if the driving power is sufficient to overcome the bias during at least some portion of the input signal cycle. You might think, at first, that this bias scheme could never result in amplification. Intuitively, it seems as if this would produce a marginal signal loss, at best. But in fact, if there is significant driving power, a *class C amplifier* can be more efficient than class A, class AB, or class B circuits. In some cases, class C amplifiers are 75 percent efficient. The operating bias zones for this method of operation are labeled "Class C" in Fig. 6-4A and B.

Class C power amplifiers are never linear for AM or SSB signals. Because of this, a class C circuit is used mainly in applications where the signals are either full-on or full-off. Continuous-wave (CW) and radioteletype (RTTY) are examples of such signals. Class C RF power amplifiers work well with FM, because the instantaneous signal amplitude never varies. Some AM broadcast stations amplify the carrier wave alone with a class C circuit and accomplish the modulation by applying a high-power AF signal directly to the final amplifier output circuit. This circumvents the envelope distortion that would occur if the modulated signal were actually passed through the amplifier.

A class C power amplifier requires considerable drive. The gain, or amplification factor, is fairly low. It might take 300 W of RF drive to get 1000 W of RF power output, a gain of only a little more than 5 dB. Nevertheless, because the efficiency is excellent, class C power amplifiers are commonly employed in CW, RTTY, and FM radio transmitters.

Broadband versus Tuned

Some RF power amplifiers offer the convenience of requiring little or no tuning on the part of the radio operator. Such amplifiers are especially popular in amateur radio communications, where operating frequencies change often. Figure 6-5A is a simple schematic diagram of a broadband RF PA.

Broadband PAs have certain advantages over tuned types. In a tuned amplifier, unless the collector or plate current is "dipped" (minimized) by adjusting the output capacitor, the efficiency will be low, and most of the power will be dissipated by the collector or plate of the output device. This can destroy the transistor(s) or tube(s). A broadband PA is easier to use. Changing frequencies or bands requires that the operator adjust only the tuning dial and/or band switch on the radio. Broadband amplifiers are less likely to oscillate than tuned amplifiers.

There are some advantages to tuned RF amplifiers. Broadband amplifiers offer no attenuation of spurious signals or harmonics within the entire design range of a radio transmitter. Tuned amplifiers are also more efficient than broadband types. Figure 6-5B illustrates a generic tuned RF PA circuit.

Figure 6-5
At A, a broadband RF
power amplifier. At B,
a tuned RF power
amplifier.

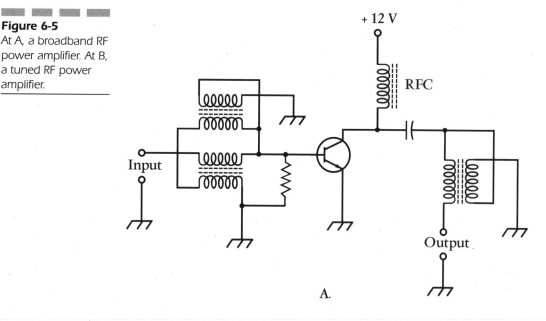

A.

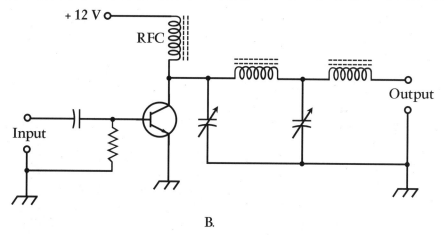

B.

Frequency Considerations

Obtaining high RF transmitter power becomes more difficult as the frequency increases, particularly at UHF and above (300 MHz and higher frequencies). Vacuum tubes are far easier to work with than transistors for kilowatt-plus amplifiers at these frequencies. *Parasitic oscillation* can also be a problem; a UHF or microwave power amplifier can generate and emit signals at HF unless precautions are taken to prevent it.

As digital modes have become increasingly common, class C PAs are being used more widely at VHF and above. These circuits are easier to design, especially at VHF and above, than linear amplifiers.

Tuned circuits in PAs at frequencies well into UHF, and at microwavelengths, take the form of strips of transmission line rather than coils and capacitors. These strips can be etched onto a printed circuit. Alternatively, *resonant cavities* can be used.

Linear Amplifiers

A *linear amplifier,* sometimes called simply a *linear,* is an analog amplifier in which the output varies in direct proportion to the input. Such an amplifier exhibits linearity of the modulation envelope. All properly operating class A amplifiers are linear; some class AB and class B amplifiers are linear. High-fidelity audio amplifiers are designed for exceptional linearity. Class C amplifiers are nonlinear.

Applications

A linear amplifier can be used with any type of emission. The circuit will not, when operated properly, introduce any significant distortion into the signal. A linear is generally used when the power output from a transmitter or transceiver is not enough to provide reliable communications. A linear should not be used when an amplifier is not necessary to maintain contact.

A linear is usually a self-contained unit, sometimes with a built-in power supply and sometimes with an outboard supply. The most rugged amateur-radio linears can provide 1500 W continuous (100 percent duty cycle) RF output, the maximum legal limit for amateur service in the United States. Some linears are designed to supply this output for only a 50 percent duty cycle, typical of CW and SSB. Other linears cannot provide the maximum legal limit but something less, such as 1000 W output. A typical amateur linear needs about 100 W of drive power to produce the maximum output. Some linears can work with as little as 10 W of drive.

Types

The term *linear* means that the output signal waveform has, in a specific respect, the same shape as the input signal waveform. In engineering terms, the *amplification curve* is a straight line. The instantaneous output is a linear function of the instantaneous input. In amateur-radio service, the modulating signal waveform is duplicated by a linear amplifier, but the actual carrier waveform is distorted until it passes through tuned circuits in the amplifier output.

Class A amplifiers are always linear. Such circuits are used in high-fidelity sound systems, and also as weak-signal amplifiers. This type of circuit is not often used for RF power amplification because it is inefficient; RF linear power amplifiers are usually operated in class AB or B.

Push-pull, class B circuits work very well as RF linear power amplifiers, offering attenuation of the even-numbered harmonics as well as reasonable efficiency. But these circuits have the disadvantage of requiring center-tapped transformers, and the balance between the two amplifying transistors or vacuum tubes is critical.

Class C amplifiers are never linear. They have high efficiency and are useful for FM and for digital signals such as CW and FSK, but they introduce distortion into emissions such as SSB, where the instantaneous amplitude varies.

Commercially manufactured amateur linears have features that make them versatile. These features include built-in wattmeters, built-in standing-wave-ratio (SWR) meters, full break-in option, automatic shutdown in case of overheating, automatic level control (ALC), and automatic tune-up or broadband operation.

Television Transmission

Television (TV) is the transfer of moving visual images from one place to another. Television is used not only for broadcasting but for two-way communications, monitoring, and security applications. The video signals of TV are often transmitted along with audio signals. Systems can be categorized as *fast-scan television* (FSTV) or *slow-scan television* (SSTV), and as *National Television Systems Committee* (NTSC) *television* or *high-definition television* (HDTV).

Analog FSTV Transmission

In order to get a realistic impression of motion, it is necessary to transmit at least 20 complete *frames* (stationary images) per second, and the detail must be adequate. An FSTV system usually provides 30 frames per second. There are 525 or 625 lines in each frame in a conventional FSTV picture. In HDTV, there are more lines; in SSTV there are fewer lines. The lines run horizontally across the picture, which usually has a horizontal-to-vertical dimensional ratio, or aspect ratio, of 4:3. Each line contains shades of brightness in a *gray-scale TV system*, and shades of brightness and hue in a *color TV system*. In FSTV broadcasting, the image is generally sent as an AM signal, and the sound is sent as an FM signal. In SSTV communications, SSB is the most often used mode.

Because of the large amount of information sent, an FSTV channel is wide. A standard FSTV channel in the North American system takes up 6 MHz of spectrum space. All FSTV broadcasting is done at VHF and above for this reason.

An FSTV transmitter consists of a *camera tube*, an oscillator, an amplitude modulator, and a series of amplifiers for the video signal. The audio system consists of an input device (such as a microphone), an oscillator, a frequency modulator, and a feed system that couples the RF output into the video amplifier chain. There is also an antenna or cable output. A simplified block diagram of an analog FSTV transmitter is shown in Fig. 6-6.

Analog SSTV Transmission

As previously mentioned, a standard FSTV signal, of the kind used in broadcasting, requires 6 MHz of spectrum space to be sent. This is because of the rapid signal fluctuations required for realistic full-motion video. Thus FSTV is practical only at VHF and above, where 6 MHz is a reasonable percentage of the carrier frequency.

If some aspects of the signal are compromised, it is possible to send a video image in a band far narrower than 6 MHz. Slow-scan television (SSTV) accomplishes this by greatly reducing the rate at which the *frames* (individual still images) are sent. The *image resolution* (detail) is also reduced somewhat, compared with conventional FSTV. An SSTV signal is sent within 3 kHz of spectrum, the same as needed by an SSB voice signal. This makes SSTV practical for transmission over telephone lines. Amateur radio operators use this mode on portions of their HF bands. An SSTV signal typically contains one frame every 8 s. There are 120 lines per frame.

The modulation for SSTV in radio communications is obtained by inputting audio signals into an SSB transmitter. An audio frequency of 1500 Hz corresponds to black; an audio frequency of 2300 Hz corresponds to white. Intermediate audio frequencies produce shades of gray. Synchronization signals are sent at 1200 Hz. These are short bursts, last-

■■■ ■■■ ■■■ ■■■
Figure 6-6
Block diagram of an
analog fast-scan
television transmitter.

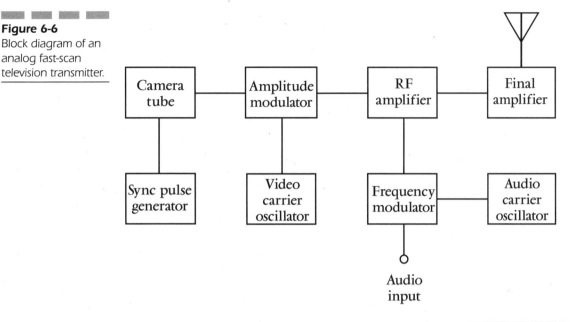

ing 0.030 s [30 milliseconds (ms)] for vertical synchronization and 0.005 s (5 ms) for horizontal synchronization.

Sometimes, SSTV is sent along with a voice. The video is sent on one sideband (say the USB), while the audio is sent on the other sideband (say the LSB). The resulting signal is a continuous voice narration, with still frames that change every 8 s, something like a slide show that proceeds at a brisk and constant pace.

Color SSTV is used by many radio amateurs. The differences between color SSTV and *gray-scale SSTV* are similar to the differences between color FSTV and gray-scale ("black-and-white") FSTV.

HDTV Transmission

The standard TV broadcast scheme was originally developed by the NTSC in 1953. When it was first devised, it was regarded as a breakthrough. But some engineers and consumers are no longer satisfied with it.

HDTV refers to any of several similar methods for getting more detail into a TV picture, and for obtaining better audio quality, compared with NTSC TV. Perhaps you've used a personal computer with a high-caliber display such as an *SVGA (super video graphics array) monitor.* If so, you have noticed the crispness of the image—considerably sharper than that of an ordinary TV set—and wondered why your color TV does not provide as much detail. With HDTV, you see a picture whose *definition,* also called the *image resolution,* is comparable to that of SVGA computer monitors. These terms refer to the amount of detail that the screen can portray.

While there are several different schemes for HDTV, they all share some common features. The most important, and most immediately noticeable, is the crispness of the picture. This is vividly apparent in big-screen TV installations, which have traditionally suffered from image blurring. While an NTSC TV picture has 525 lines per frame, HDTV systems have between 787 and 1125 lines per frame. The image is scanned about 60 times per second.

Some HDTV systems use a technique called *interlacing* to double the number of lines in each frame, without the need for extremely sophisticated scanning apparatus. An interlaced system scans the screen twice for each complete video image, so there are actually two overlapping images instead of one image. The two images, called *rasters,* are "meshed" together. This effectively doubles the image resolution without doubling the cost of manufacture. But it can cause problems with fast-moving pictures. For this reason, many engineers prefer that interlacing not be used. This increases the cost for a given image resolution, but it optimizes the picture quality for fast-changing images, a common characteristic of popular action-oriented TV programming.

There are other, more technical aspects of HDTV that make it superior to NTSC TV. One important difference is that HDTV is digital, while NTSC TV is analog. Digital signals propagate better, are easier to deal with when they are weak, and can be processed in various ways that analog signals cannot.

Some HDTV developers and marketers believe that consumers want *interactive technology* with TV. Translated into everyday terms, that means shopping, responding to polls, taking university courses, banking, and voting. The big-screen HDTV set can double as a computer monitor for education, presentations, and enjoyment. It will be optimal if the telephone, the computer, and the TV set can be made compatible, so they can be used together or separately.

Digital TV Transmission

The signals in conventional satellite TV, traditional cable TV, and broadcast TV are analog in nature. Analog data can be converted into binary digital signals, resulting in improved signal-to-noise ratio and superior fidelity. The term *digital television* refers to the transmission of moving video images in digitized form, especially via *active communications satellites.*

Until the early 1990s, a satellite television installation required a *dish antenna* several feet in diameter. Many such systems are still in use. The antennas are expensive, they attract attention (sometimes unwanted), and they are subject to damage from ice storms, heavy snows, and high winds. Digitization has changed this situation. In any communications system, digitization allows the use of smaller receiving antennas, smaller transmitting antennas, and/or lower transmitter power levels. Manufacturers and broadcasters realize that large dish antennas are a problem for many consumers. Therefore, in the development of digital television, primary emphasis has been placed on reducing the size of the subscriber antenna as much as possible. Engineers have managed to get the diameter of the receiving dish down to about 2 ft.

A pioneer in digital TV was RCA (Radio Corporation of America), which developed the *digital satellite system* (DSS). Figure 6-7 is a simplified block diagram of a DSS link. The analog signal is changed into digital pulses at the transmitting station via *analog-to-digital* (A/D) *conversion.* This signal resembles the binary data used by computers. The digital

Figure 6-7
Block diagram of a digital satellite television link.

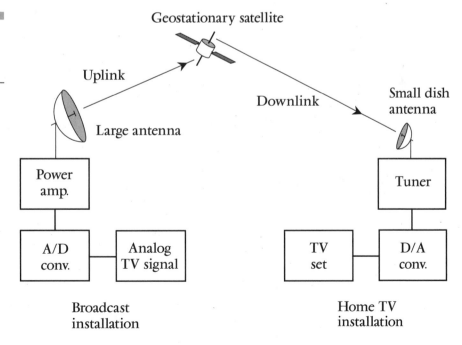

signal is amplified and sent up to a *geostationary satellite*. This signal is called the *uplink*. The satellite has a *transponder* that receives the signal, converts it to a different frequency, and retransmits it back toward the earth. The return signal is called the *downlink*. The downlink is picked up by a portable dish that can be placed on a balcony or patio, on a rooftop, or in a window. The *tuner* selects the channel that the subscriber wants to watch. The digital signal is amplified. If necessary, *digital signal processing* (DSP) can be used to improve the quality of reception under marginal conditions. The digital signal is changed back into analog form, suitable for viewing on a conventional TV set, via *digital-to-analog* (D/A) *conversion*.

Receiver Basics

A *receiver* is a device or system that converts electromagnetic (EM) waves, having come from some transmitter, into the original messages that were sent by that transmitter. In the broadest sense, a receiver is a form of transducer. This chapter is an overview of common receivers. Some specialized communications techniques are also discussed.

Simple Designs

A receiver need not be a complicated system. For some applications, especially in hobby radio and for instructional purposes, simple circuits can function effectively as EM signal receivers.

Crystal Set

One of the earliest diode devices was a semiconductor component, developed before the first vacuum tubes. Known as a *cat whisker*, it consisted of a fine piece of wire in contact with a small piece of the mineral *galena*. This strange-looking thing had the ability to act as a rectifier of weak radio-frequency (RF) currents. When the cat whisker was connected in a circuit like that shown in Fig. 7-1, the result was a receiver capable of picking up amplitude-modulated (AM) radio signals.

Figure 7-1
A crystal set radio receiver. All the power is supplied by incoming signals.

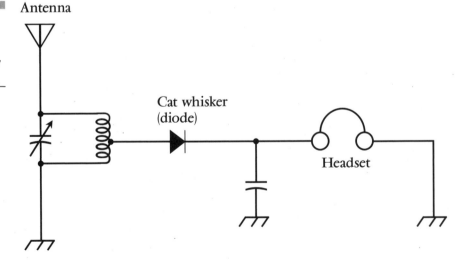

The cat whisker was finicky. Engineers had to adjust the position of the fine wire to find the best point of contact with the galena. A tweezers and magnifying glass were invaluable in this process. A steady hand was essential. The galena, sometimes called a "crystal," gave rise to the nickname *crystal set* for this low-sensitivity radio. You can still build a crystal set today, using a simple RF diode, a coil, a tuning capacitor, a headset, and a long-wire antenna. There is no battery or other source of power for this circuit. All of its energy comes from the antenna.

The diode acts to recover the audio from the radio signal. This is called *detection;* the circuit is a *detector.* If the detector is to be effective, the diode must be of the right type. It should have low capacitance, so that it works as a rectifier at RF, passing current in one direction but not in the other. Some modern RF diodes are actually microscopic versions of the old cat whisker, enclosed in a glass case with axial leads. You have probably seen these in electronics hobby stores. Most of these devices, called *point-contact diodes,* have low capacitance and are excellent for use in RF devices.

An interesting feature of the crystal set is that it requires no power source. It is a passive device. All the power is supplied by the EM signal energy in the antenna. Nevertheless, the audio output is sufficient to drive a common earphone or headset on the AM broadcast band, if the station is within about 25 miles of the receiver, and if the antenna is sufficiently large.

Direct-Conversion Receiver

A *direct-conversion receiver* is a system that derives audio output by beating, or mixing, incoming signals with the output of a variable-frequency *local oscillator* (LO). The received signal is fed into a *mixer,* along with the output of the LO. As the oscillator is tuned across the frequency of an unmodulated carrier, a high-pitched audio beat note is heard; this pitch becomes lower until it vanishes at *zero-beat,* when the LO frequency is exactly equal to the signal frequency. Then the beat note rises in pitch again as the oscillator frequency moves away from the signal frequency. A diagram of a direct-conversion receiver is shown in Fig. 7-2.

For reception of continuous-wave (CW) radiotelegraphy signals, the LO is set slightly above or below the signal frequency. The audio tone will then have a frequency equal to the difference between the oscillator and signal frequencies. For reception of AM or single-sideband (SSB) signals, the oscillator should be set to zero beat with the carrier frequency of the incoming signal.

Figure 7-2
Block diagram of a
direct-conversion
receiver.

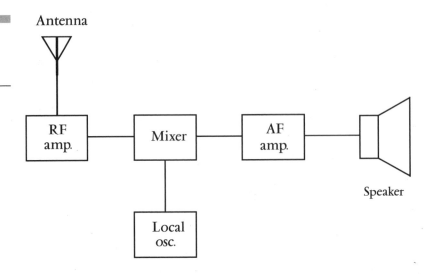

A direct-conversion receiver normally cannot provide optimum selectivity. This is because signals on either side of the LO frequency tend to be heard "on top of" one another. For example, if the LO is tuned to 7.050 megahertz (MHz) and there are signals at 7.449 and 7.051 MHz, both signals will cause a 1-kilohertz (kHz) beat note in the receiver. The result will be severe interference between the two signals. A selective filter in the RF amplifier stage can, in theory, eliminate this problem. But this type of filter must be designed for a fixed frequency if it is to have sufficiently steep cutoff curves (or "skirts"), and the RF amplifier in a direct-conversion receiver must accommodate signals having many different frequencies.

Superheterodyne Receiver

A *superheterodyne receiver* uses one or more local oscillators and mixers to obtain a constant-frequency signal. A fixed-frequency signal is more easily processed than a signal that changes in frequency. The incoming signal is first passed through a tunable, sensitive amplifier called the *front end*. The output of the front end is mixed with the signal from a tunable, unmodulated LO. Either the sum or the difference between the frequencies of the two signals is then amplified. The sum or difference signal is called the first *intermediate-frequency* (IF) signal. The IF signal is filtered to obtain a high degree of selectivity.

If the first IF signal is detected, the radio is called a *single-conversion receiver*. Some receivers use a second mixer and second LO, converting the *first IF* signal to a much lower frequency called the *second IF.* This type of radio is known as a *double-conversion receiver.*

Superheterodyne receivers offer advantages over direct-conversion units. The design makes it possible to optimize the selective characteristics. The RF bandpass filter can be constructed for use on a fixed frequency (the IF), allowing superior skirt selectivity and facilitating adjustable bandwidth. The sensitivity is also enhanced, because the fixed IF amplifiers are easy to keep in optimum tune. One problem with the superheterodyne circuit is that it can receive or generate unwanted signals. The signals might be external to the receiver, in which case they are called *images,* or they might come from the local oscillator or oscillators, in which case they are called *birdies.* If the LO frequencies are judiciously chosen, false signals are generally not a problem.

The Modern Receiver

Communications receivers operate over specific ranges of the radio spectrum. The range covered depends on the application for which the receiver is designed. For example, a "general coverage" communications receiver works from about 1.5 through 30 MHz; a 2-meter (m) very-high-frequency (VHF) amateur-band receiver operates from approximately 144 to 148 MHz. Some receivers work from the very-low-frequency range well into the ultra-high-frequency (UHF) spectrum. These are *all-band communications receivers.*

Specifications

Any communications receiver, whether analog or digital, audio or video, must do certain basic things well. The *specifications* of a radio indicate how well it can perform the functions it is supposed to perform.

Sensitivity   The sensitivity of a receiver is its ability to recover weak signals and process them into readable data. The most common way to express receiver sensitivity is to state the number of signal microvolts that must exist at the antenna terminals to produce a certain *signal-to-noise ratio* (S/N). Sometimes, the *signal-plus-noise-to-noise ratio,* abbreviated S+N/N, is given. When you look at the specifications table for a radio receiver, you might see, for example, "better than 0.2 microvolt for 10 dB

S/N." This means that a signal of 0.2 microvolt or less, at the antenna terminals, will result in a S/N ratio of 10 decibels (dB). A poor sensitivity figure for one application, or on one frequency band, might be acceptable for some other intended use, or on some other frequency band. The *front end,* or first RF amplifier stage, of a receiver is the most important stage with regard to sensitivity. Sensitivity is directly related to the gain of this stage, but the amount of noise the stage generates is even more significant. A good front end should produce the best possible S/N or S + N/N ratio at its output. All subsequent stages will amplify the front-end noise output as well as the front-end signal output.

Selectivity   Selectivity is the ability of a receiver to respond to a desired signal, but not to undesired ones. A receiver must have a frequency "window" within which it is sensitive, but outside of which it is not sensitive. This "window" is established by a *preselector* in the early RF amplification stages of the receiver, and is honed to precision by bandpass filters in later amplifier stages. Suppose you want to receive a lower-sideband signal whose suppressed-carrier frequency is 3.885 MHz. The signal information is contained in a band from about 3.882 to just under 3.885 MHz, a span of a little less than 3 kHz. The preselector makes the receiver sensitive in a range of about plus or minus 10 percent of the signal frequency; other frequencies are attenuated. This reduces the chance for a strong, out-of-band signal to impair the performance of the receiver. The narrow-band filter responds only to signals having frequencies in the range of 3.882 to 3.885 MHz; signals in nearby channels are rejected. The better this filter works, the better the *adjacent-channel rejection.*

Dynamic Range   The signals at a receiver input vary over several orders of magnitude (powers of 10) in terms of their absolute voltage. If you are listening to a weak signal and a strong one appears on the same frequency, you don't want to be startled by a loud noise. But you want the receiver to work well if a strong signal appears at a frequency near, but not exactly at the frequency of the weak one. Dynamic range is the ability of a receiver to maintain a fairly constant output, and yet to maintain its rated sensitivity in the presence of signals ranging from very weak to extremely strong. Dynamic range is specified in decibels, and is typically 100 dB or more.

Noise Figure   All RF amplifier circuits generate some wideband noise. The less *internal noise* a receiver produces, the better the S/N ratio. This becomes more and more important as the frequency increases. It is paramount at VHF, UHF, and microwave frequencies (30 MHz and above). The noise figure of a receiver is specified in various different ways. The lower the noise figure, the better the sensitivity. You might see a specification such as "NF: 6 dB or better." Noise figure depends on the type of active amplifier device used in the front end of a receiver. *Gallium-arsenide field-effect transistors* (FETs) are well known for the low levels of noise they generate, even at quite high frequencies. Other types of FETs can be used at lower frequencies. *Bipolar transistors* tend to be rather noisy.

Stages

A block diagram of a superheterodyne communications receiver is illustrated in Fig. 7-3. Each stage shown is discussed briefly below. Individual receiver designs can vary somewhat from this basic form.

Front End   The front end of a receiver consists of the first RF amplifier, and often includes bandpass filters between the amplifier and the antenna. The dynamic range and sensitivity of a receiver are determined by the performance of the front end. These two characteristics are among the most important for any receiver. Low-noise, high-gain amplifiers are the rule. FETs are commonly used.

Figure 7-3
Block diagram of a typical super-heterodyne receiver.

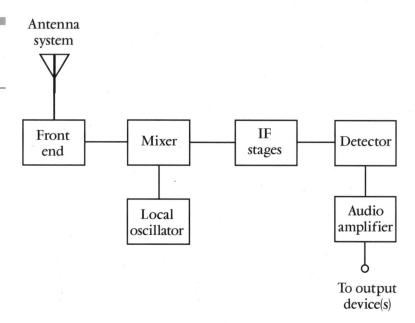

Mixer   The superheterodyne receiver uses one or two mixer stages that convert the variable signal frequency to a constant IF. The mixer is a nonlinear circuit that combines the signal with a carrier from an LO. The output is either the sum or the difference of the signal frequency and the LO frequency. Mixing always produces images and birdies. In a well-designed receiver, these signals are weak and are not on frequencies that cause interference to desired signals.

IF Stages   The IF stages of a receiver are where much, if not most, of the gain takes place. These stages are also where the best possible RF selectivity is obtained. The IF is a constant frequency. This simplifies the design of the amplifiers to produce optimum gain and selectivity. *Crystal-lattice filters* or *mechanical filters* are commonly used in these stages to obtain the desired bandwidth and response.

Detector   The detector extracts the information from the signal. The circuit details depend on the type of emission to be received. Common detectors are the *product detector* for CW radiotelegraphy, frequency-shift keying (FSK), and SSB; and the *discriminator,* the *phase-locked loop* (PLL), or the *ratio detector* for frequency modulation (FM).

Postdetector Stages   Following the detector, one or two stages of audio amplification are generally employed to boost the signal to a level suitable for listening with a speaker or headset. Alternatively, the signal can be fed to a printer, facsimile machine, slow-scan television picture tube, or computer via a *terminal unit* (TU). The TU is a decoding circuit similar to a *modem* that converts the detector output to impulses suitable for various display devices. In amateur fast-scan television, for example, the detector output is a video signal. The fast-scan television receiver is usually a consumer-type television (TV) set or a personal computer with a special converter between the antenna and the front end.

The following several sections discuss the components of a superheterodyne receiver, also known as a *superhet,* in greater detail.

Predetector Stages

The stages preceding the first mixer must be designed so they provide reasonable gain but produce as little noise as possible. In some receivers, these stages must be capable of operating effectively over a wide range of frequencies.

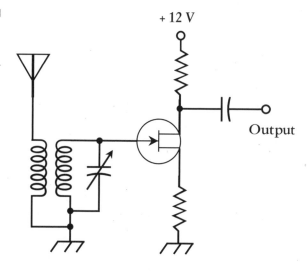

Figure 7-4
A tunable RF
preamplifier.

+ 12 V

Output

Preamplifiers

An RF *preamplifier* is a low-noise amplifier that boosts the strength of
weak signals and/or enhances the S/N ratio in a communications system. All preamplifiers operate in class A. Preamplifiers are used when it
is necessary to improve the sensitivity of a communications or television
receiver, especially at frequencies above 30 MHz. Most RF preamplifiers
employ FETs. An FET has a high input impedance that is ideally suited
to weak-signal work.

Figure 7-4 shows a simple RF preamplifier. Input tuning is used to
reduce noise and provide some selectivity against signals on unwanted
channels. This circuit will produce about 5 to 10 dB gain, depending on
the frequency, and also depending on the choice of FET.

In the design of any preamplifier, regardless of the intended frequencies of operation, it is important that *linearity* be as good as possible.
Nonlinearity in a preamplifier results in *intermodulation distortion* at radio
frequencies. This can wreak havoc in a receiver to which the preamplifier
is connected.

The Front End

When people talk about the front end of a receiver, they usually mean
the first RF amplifier stage. The front end is one of the most important
parts of any receiver, because the sensitivity of the front end dictates the

sensitivity of the entire receiver (unless a preamplifier is used). A low-noise, high-gain front end provides a good signal for the rest of the receiver circuitry. A poor front end results in a high amount of noise or distortion.

The front end of a receiver becomes more important as the frequency gets higher. At low and medium frequencies (below about 30 MHz), there is considerable atmospheric noise, and the design of a front-end circuit is rather simple. Above 30 MHz, the amount of atmospheric noise becomes small, and the main factor that limits the sensitivity of a receiver is the noise generated within the receiver itself. Noise generated in the front end of a receiver is amplified by all of the succeeding stages, so it is most important that the front end generate very little noise.

Another front-end requirement is low distortion. The amplifier circuit must be as linear as possible; the greater the degree of nonlinearity, the more susceptible the front end is to the generation of *mixing products*. Mixing products result from the heterodyning, or beating, of two or more signals in the circuits of a receiver; this causes false signals to appear in the output of the receiver. The front end should also have the greatest possible dynamic range; it must be capable of handling very strong as well as very weak signals without being driven into nonlinearity. If the front end cannot handle strong signals, *desensitization* will occur when a strong signal is present at the antenna terminals. Mixing products might also appear. At best this is an annoyance; at worst it renders a receiver useless.

Preselectors

A tuned circuit in the front end of a radio receiver is called a *preselector*. Most receivers incorporate preselectors. The preselector provides a bandpass response that improves the S/N ratio and also reduces the likelihood of receiver overloading by a strong signal far removed from the operating frequency. The preselector also provides a high degree of image rejection in a superheterodyne circuit. Most preselectors have a 3-dB bandwidth that is a few percent of the received frequency.

A preselector can be tuned by means of *tracking* with the tuning dial. This eliminates the need for continual readjustment of the preselector control, thereby making the receiver more convenient to operate. The tracking type of preselector, however, requires careful design and alignment. Some receivers incorporate preselectors that must be adjusted independently. Although this is less convenient from an operating standpoint, the control need be reset only when a large frequency change is made. The independent preselector simplifies receiver design.

Mixers

A mixer is a nonlinear circuit that combines two signals of different frequencies to produce a third signal, the frequency of which is either the sum or difference of the input frequencies. Mixers are widely used in radio and TV receivers and transmitters.

Suppose there are two signals with frequencies f_1 and f_2, where f_1 is the lower frequency and f_2 is the higher frequency. If these signals are combined in a nonlinear circuit, new signals will result. One of these new signals will have the frequency $f_2 - f_1$, and the other will occur at the frequency $f_2 + f_1$. These new frequencies are known as *beat frequencies*. The signals are called *mixing products*.

A mixer requires some kind of nonlinear circuit element in order to function properly. The nonlinear element can be a diode or combination of diodes. This forms a *passive mixer*. The diode mixer does not require an external source of power, so there is some insertion loss. An *active mixer* employs one or more transistors or integrated circuits to produce gain, in addition to acting in a nonlinear manner to generate mixing products.

The mixer operates on a principle similar to that of an amplitude modulator. But there is one major difference: The output of the mixer circuit can be tuned to either the sum frequency or the difference frequency, as desired.

The IF Chain

A high IF, such as several megahertz, is preferable to a low IF for purposes of image rejection. However, a low IF is better for obtaining sharp selectivity. This is why double-conversion receivers are common: They provide the advantages of both a high *first IF* and a low *second IF*.

IF amplifiers generally are cascaded two or more in a row, with tuned-transformer coupling. The amplifiers follow the mixer stage and precede the detector stage. Double-conversion receivers have two sets of IF amplifiers; the first set follows the first mixer and precedes the second mixer, and the second set follows the second mixer and precedes the detector.

The IF amplifier chain serves two main purposes: to provide high gain and to provide excellent selectivity. Gain and selectivity are much easier to obtain with amplifiers that operate at a single frequency, as compared with tuned RF amplifiers.

The selectivity of the IF chain in a superheterodyne receiver is often expressed mathematically. The bandwidths are compared for two voltage-attenuation values, usually 6 and 60 dB. For example, after you purchase a new receiver, you might read in the specifications:

Selectivity: 8 kHz or more at -6 dB

16 kHz or less at -60 dB

This gives you an idea of the shape of the bandpass response.

A receiver might have several different bandpass filters, each with a different degree of selectivity. The ideal amount of selectivity depends on the mode of communication. The above figures are typical of a VHF radio for FM communications. For AM signals, the plus-or-minus bandwidth at the 6-dB points is usually 3 to 5 kHz. For SSB, it is roughly 1 to 1.5 kHz, representing an overall 6-dB bandwidth of 2 to 3 kHz. For CW reception, the overall bandwidth figure can be as small as 30 to 50 Hz at lower speeds, and perhaps 100 Hz at higher speeds.

The ratio of the 60-dB selectivity to the 6-dB selectivity is called the *shape factor.* It is of interest, because it denotes the degree to which the response is rectangular. A *rectangular response* is the most desirable response in most receiving applications. The smaller the shape factor, the more rectangular the response. Filters such as the ceramic, crystal-lattice, and mechanical types can provide a good rectangular response in the IF chain of a receiver.

Detectors

Detection, also called *demodulation,* is the recovery of information such as audio, a visible image (either still or moving), or printed data from a signal. The way this is done depends on the *modulation mode.* Detection schemes for several common modulation modes are discussed in this section.

Detection of AM

Amplitude modulation is accomplished by combining data with a carrier wave in a manner similar to the mixing of signals. This produces sum and difference frequency components within two bands of frequencies, one just above the carrier and the other just below it. These frequency bands, usually a small fraction of the carrier frequency in width, are called *sidebands*.

One way to eliminate the modulating waveform from an AM signal is to "cut off" half of the carrier wave cycle. This is a somewhat oversimplified view of the situation but can be useful to illustrate the practical effects of "rectifying" an AM signal. The result (Fig. 7-5A) is fluctuating, pulsating DC. The rapid pulsations occur at the carrier frequency; the slower fluctuation is a duplication of the modulating audio or video. This form of AM detection can be done by passing the signal through a diode with small *junction capacitance*. A bipolar transistor biased at *cutoff*, or an FET biased at *pinchoff*, can also be used.

The rapid pulsations are smoothed out by passing the output of the diode or transistor through a capacitor. The capacitance should be large enough so that it holds the charge for one carrier current cycle, but not so large that it smooths out the cycles of the modulating signal. This scheme is known as *envelope detection*.

Detection of CW

If you tune in an unmodulated carrier with an envelope detector, you won't hear anything. A keyed radiotelegraph carrier might produce barely audible thumps or clicks, but it will be impossible to read the signals.

For detection of CW radiotelegraph signals, it is necessary to inject a signal into the receiver a few hundred hertz from the carrier. This injected signal beats against the carrier, producing a tone whose frequency is the difference between the carrier and injected-signal frequencies. The injected signal is produced by a specialized type of LO called a *beat-frequency oscillator* (BFO). The beating occurs in a signal combiner or mixer. The BFO is tunable.

Suppose there is a CW signal at 3.550 MHz. As the BFO approaches 3.550 MHz from below, a high-pitched tone will appear at the output. When the BFO reaches 3.549 MHz, the tone will be 3.550 MHz − 3.549 MHz, or 1 kHz. This is a comfortable listening pitch for most people.

Figure 7-5
At A, envelope
detection. At B, slope
detection.

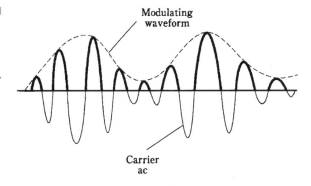

Modulating
waveform

Carrier
ac

A

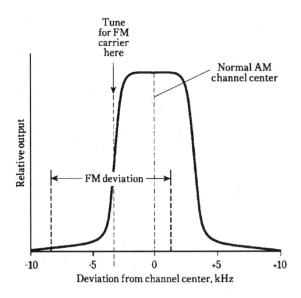

Tune
for FM
carrier
here

Normal AM
channel center

Relative output

FM deviation

-10   -5   0   +5   +10

Deviation from channel center, kHz

B

The BFO setting is not too critical; in fact, it can be changed to get a different tone pitch if you get tired of listening to one pitch. As the BFO frequency passes 3.550 MHz, the pitch will descend to a rumble, then to silence. As the BFO frequency continues to rise, the tone will reappear and its pitch will increase, eventually rising beyond the range of human hearing.

To receive a keyed Morse-code CW signal, the BFO is tuned to a frequency that results in a comfortable listening pitch for the audio-frequency (AF) beat note. For most people this is approximately 700 Hz. The audio tone follows the keying of the signal being received. This demodulation scheme is called *heterodyne detection.*

Detection of FSK

Frequency-shift keying can be detected using the same method as CW detection. The carrier beats against the BFO in the mixer, producing an audio tone that alternates between two different pitches.

With FSK, the BFO frequency is always set a few hundred hertz above or below both the mark and the space carrier frequencies. The *frequency offset,* or difference between the BFO and signal frequencies, determines the audio tone frequencies and is set so that certain standard pitches result. There are several sets of standard tone frequencies, depending on whether the communications is amateur, commercial, or military. To get the proper pitches, the BFO must be set to precisely the right frequency. There is little tolerance for error.

Detection of FM

Frequency-modulated or phase-modulated signals can be detected in several ways. The best receivers for these modes respond to frequency and phase fluctuations, but not to amplitude variations.

Slope Detection   An AM receiver can be used to detect FM. This is done by setting the receiver frequency near, but not exactly at, the FM unmodulated-carrier frequency. An AM receiver has a narrowband filter with a passband of about 6 kHz. This gives a selectivity curve such as that shown in Fig. 7-5B. If the FM unmodulated-carrier frequency is near the *skirt,* or slope, of the filter response, modulation will cause the signal to move in and out of the passband. This will make the receiver output vary with the modulating data. This scheme has two shortcomings. First, the receiver will respond to amplitude variations (as it is designed to do). Second, there will be nonlinearity in the received signal, producing distortion, because the slope of the bandpass response curve is not a straight line.

Phase-Locked Loop   If an FM signal is injected into a PLL circuit, the loop will produce an error voltage that is a duplicate of the modulating waveform. The frequency fluctuations might be too fast for the PLL to lock onto, but the error voltage will still appear. Many modern receivers take advantage of this effect to achieve FM detection. A circuit called a *limiter* can be placed ahead of the PLL so that the receiver does not respond to AM. Thus, one of the major advantages of FM over AM is realized. Atmospheric noise and ignition noise causes much less interference in a good FM receiver than in AM, CW, or SSB receivers, provided that the signal is strong enough. Weak signals tend to appear and disappear, rather than fading, in an FM receiver that employs limiting.

Discriminator A *discriminator* produces an output voltage that depends on the instantaneous signal frequency. When the signal is at the center of the discriminator passband, the output voltage is zero. If the instantaneous signal frequency decreases, a momentary phase shift results, and the output voltage becomes positive. If the frequency rises above center, the output becomes negative. The instantaneous voltage level (positive or negative) is directly proportional to the instantaneous frequency. Therefore, the output voltage is a duplicate of the modulating waveform. A discriminator is sensitive to amplitude variations in the signal, but this problem can be overcome by the use of a limiter.

Ratio Detector A discriminator with a built-in limiter is known as a *ratio detector*. This type of FM detector was developed by RCA (Radio Corporation of America) and is used in high-fidelity receivers and in the audio portions of TV receivers. A simple ratio detector circuit is shown in Fig. 7-5C. A transformer splits the signal into two components. A change in signal amplitude causes equal level changes in both halves of the circuit. These effects cancel because they are always 180 degrees out of phase. This nullifies amplitude variations on the signal. A change in signal frequency causes a phase shift in the circuit. This unbalances it, so the outputs in the two halves of the circuit become different. This produces an output in direct proportion to the instantaneous phase shift. The output signal is a duplication of the modulating waveform on the FM signal.

Detection of SSB

An SSB signal is basically an AM signal without the carrier, and with only one of the sidebands. If the BFO frequency of a CW receiver is set exactly at the suppressed-carrier frequency of an SSB signal, the sideband

Figure 7-5 (cont.)
At C, a ratio detector.
At D and E, product
detectors.

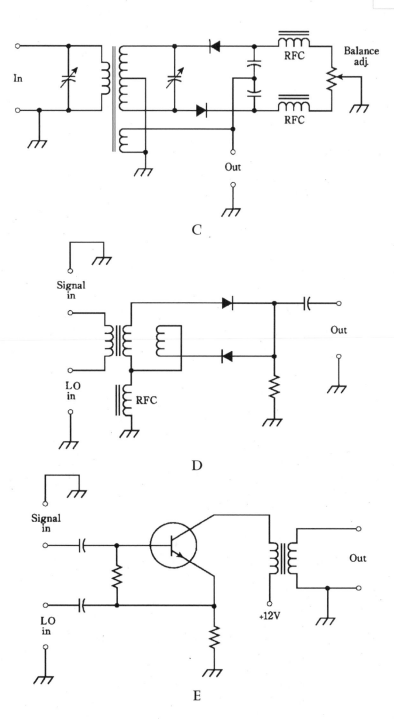

(either upper or lower) will beat against the BFO signal and produce audio output. In this way, a CW/FSK receiver can be used to detect SSB. The BFO frequency is critical for good SSB reception. If it is not set precisely at the frequency of the suppressed carrier, the voice will sound muffled or distorted.

For the reception of CW, FSK, and SSB signals, a *product detector* is generally used. It works according to the same basic principle as the BFO and mixer. The incoming signal combines with the signal from an unmodulated LO, producing audio or video output. But product detection is done at a single frequency, rather than at a variable frequency as in direct-conversion reception. The single, constant frequency is obtained by mixing the incoming signal with the output of the LO.

Two product-detector circuits are shown in Fig. 7-5D and E. At D, diodes are used; there is no amplification. At E, a bipolar transistor is employed; this circuit provides some gain. The essential characteristic of either circuit is the nonlinearity of the semiconductor devices. This is responsible for producing the sum and difference frequencies that result in audio or video output.

Detection of PM

The common forms of pulse modulation (PM) usually operate at a low *duty cycle*. This means that the transmitter is "on" only a small proportion of the time. The pulses are far shorter in duration than the intervals between them. The ratio of average signal power to peak signal power is sometimes less than 1 percent. When the amplitude or duration of the pulses changes, the average transmitter power also changes. Stronger or longer pulses increase the effective signal amplitude; weaker or shorter pulses result in decreased amplitude. Therefore, PM can be detected in the same way as AM.

A major advantage of PM is that it is "mostly empty space." With *pulse amplitude modulation, pulse duration modulation,* or *pulse code modulation,* the time interval is constant between pulse centers. Even at maximum modulation, the ratio of "on" time to "off" time is low. Two or more signals can be intertwined on a single carrier. A PM receiver can pick out one of these signals and detect it, ignoring the others. This is a form of *multiplexing,* which will be discussed in more detail later in this chapter.

Audio Stages

In a receiver, enhanced selectivity can be obtained by tailoring the audio response of the receiver in stages following the detector. *Audio filters* can be of the bandpass, band-rejection (notch), or low-pass type. High-pass audio filters are not often used in communications receivers.

Filtering

A voice signal occupies a band ranging from about 300 to 3000 Hz. Although the human voice contains components outside this range, the 2.7-kHz-wide band is enough for good intelligibility under most conditions. Therefore, an *audio bandpass filter,* with a passband of about 300 to 3000 Hz, can improve the quality of reception with some voice receivers. Sometimes an *audio low-pass filter,* with a cutoff frequency of about 3000 Hz, is used instead. The results are similar for either type of audio filter in voice communications use.

A Morse code, or CW, signal requires only a few hundred hertz of bandwidth in order to be clearly read. Most high-frequency communications-receiver IF stages provide a passband of 2.7 kHz, suitable for SSB. Those with CW filters can narrow this down to 500 Hz or so, and some can provide IF selectivity as narrow as approximately 200 Hz. A narrower audio bandpass filter can sometimes improve the quality of reception when a band is crowded.

Passbands narrower than 100 Hz are not used often, because *ringing* can take place. Ringing is the tendency for an audio bandpass filter to produce *damped oscillations* at its center frequency. This effect reduces the speed at which a Morse signal can be clearly received.

An ideal audio passband filter has a relatively rectangular response. This means that there is very little, or no, attenuation within the passband range, and very great attenuation outside the range. The skirts are steep. These are the same features that are desirable in IF filtering stages. There are various reasons why a flat passband response and steep skirts are advantageous in filtering systems. These features are more important in voice communications than they are in CW. Some CW operators prefer a less flat bandpass response for audio filtering. Some CW bandpass audio filters have center frequencies that are adjustable from approximately 300 to 3000 Hz. Comfortable CW listening frequencies, for most operators, are between 400 and 1000 Hz.

An *audio notch filter* is a band-rejection filter with a sharp, narrow response. The idea is that a *heterodyne*, or an interfering carrier that produces a tone of constant frequency in the receiver output, can be nulled out by means of such a filter. Some receivers have IF notch filters. These work better than any audio notch filter, because IF filtering gets rid of *automatic-gain-control* (AGC) effects caused by interfering heterodynes, and audio filters do not. Nonetheless, for receivers that lack IF notch filters, an audio notch can be useful.

Audio notch filters are usually tunable from roughly 300 to 3000 Hz. Some tune wider ranges. A few tune automatically; when an interfering heterodyne appears and remains for more than a few tenths of a second, the notch is activated and centers itself on the frequency of the heterodyne.

Squelching

When it is necessary to listen to a radio receiver for long periods, and signals are present infrequently, the constant hiss or roar becomes fatiguing to the ears. A *squelch circuit* is used to silence a receiver when no signal is present, while allowing reception of signals when they appear.

Squelch circuits are most often used in channelized units, such as citizens band and VHF communications transceivers. Squelch circuits are less common in continuous-tuning radio receivers. Most FM communications receivers use squelching systems. A squelch circuit operates in the AF stages. The squelch is actuated by the incoming audio signal. When no signal is present, the rectified hiss produces a negative DC voltage, cutting off a transistor and keeping the noise from reaching the output. When a signal appears, the hiss level is greatly reduced, and the transistor conducts, allowing the audio to reach the output. The cutoff squelch circuit is said to be closed; the conducting squelch is said to be open.

In most receivers, the squelch is normally closed when no signal is present. A potentiometer facilitates adjustment of the squelch, so that it can be opened if desired, allowing the receiver hiss to reach the speaker. This control also provides for adjustment of the *squelch sensitivity*. Most squelch controls are open when the knob is turned fully counterclockwise. As the knob is rotated clockwise, the receiver hiss abruptly disappears; this point is called the *squelch threshold*. At this point, relatively weak signals will open the squelch. As the squelch knob is turned farther clockwise, stronger and stronger signals are required to open the squelch.

In some receivers, the squelch will not open unless the signal has certain characteristics. This is called *selective squelching*. It is used in some repeaters and receivers to prevent undesired signals from being heard. The most common methods of selective squelching use *subaudible-tone generators* or *tone-burst generators*. Selective squelching is sometimes called *tone squelching*. This scheme is common in congested areas. It is used by some amateur radio clubs to control access to repeaters.

Television Reception

A television receiver contains an antenna, or an input having an impedance of either 75 or 300 ohms, a tunable front end, an oscillator and mixer, a set of IF amplifiers, a *video demodulator*, an audio demodulator and amplifier chain, a picture tube with associated peripheral circuitry, and a speaker.

Fast-Scan

A receiver for analog fast-scan television (FSTV) is shown in simplified block form in Fig. 7-6. In order for the picture to appear normal, the

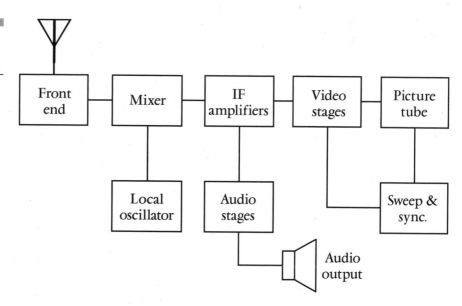

Figure 7-6
Block diagram of an analog FSTV receiver.

transmitter and the receiver must be exactly synchronized. The studio equipment generates pulses at the end of each line and at the end of each complete frame. These pulses are sent along with the video signal. In the receiver, the demodulator recovers these *sync pulses* and sends them to the picture tube. The electron beam in the picture tube thus moves in exact synchronization with the scanning beam in the camera tube. If the frame sync is upset, the picture appears to *roll*. If the line sync is not right, the picture seems to *tear*. Most FSTV receivers have controls that allow adjustment of the frame and line sync, usually called *vertical hold* and *horizontal hold*.

In the United States, FSTV broadcasts are made on 68 different channels in the VHF and UHF bands of the EM spectrum. Each channel is 6 MHz wide, including video and audio information. There is no channel 1. Channels 2 through 13 comprise the *VHF TV broadcast channels* (see Table 7-1). Channels 14 through 69 are the *UHF TV broadcast channels* (Table 7-2).

When cable is not used, and receiving installations employ autonomous, rotatable outdoor antennas, *ionospheric propagation* can dramatically affect reception on channels 2 through 6. Long-distance propagation can take place because of dense ionization in the E layer on

TABLE 7-1

VHF Television Broadcast Channels

Channel	Frequency, MHz
2	54—60
3	60—66
4	66—72
5	76—82
6	82—88
7	174—180
8	180—186
9	186—192
10	192—198
11	198—204
12	204—210
13	210—216

TABLE 7-2

UHF Television
Broadcast
Channels

Channel	Frequency, MHz	Channel	Frequency, MHz
14	470—476	42	638—644
15	476—482	43	644—650
16	482—488	44	650—656
17	488—494	45	656—662
18	494—500	46	662—668
19	500—506	47	668—674
20	506—512	48	674—680
21	512—518	49	680—686
22	518—524	50	686—692
23	524—530	51	692—698
24	530—536	52	698—704
25	536—542	53	704—710
26	542—548	54	710—716
27	548—554	55	716—722
28	554—560	56	722—728
29	560—566	57	728—734
30	566—572	58	734—740
31	572—578	59	740—746
32	578—584	60	746—752
33	584—590	61	752—758
34	590—596	62	758—764
35	596—602	63	764—770
36	602—608	64	770—776
37	608—614	65	776—782
38	614—620	66	782—788
39	620—626	67	788—794
40	626—632	68	794—800
41	632—638	69	800—806

these channels. *Tropospheric propagation* is occasionally observed on all VHF channels. The UHF channels are unaffected by the ionosphere, although tropospheric effects can occur to a small extent.

Slow-Scan

A slow-scan television (SSTV) communications station needs a transceiver with SSB capability, a device called a *scan converter,* a TV set, and a *video camera.* The scan converter consists of two data converters (one for receiving and the other for transmitting), a memory, a tone generator, and a detector. When you listen to an SSTV signal with an SSB receiver, you hear a rapid sequence of audio tones. These tones contain the video information. The scan converter changes the tones into signals suitable for viewing on a TV set.

Scan converters are commercially available and are advertised in amateur radio and electronics magazines. If you are interested in buying one, it is best to look through the most recent periodicals obtainable, because the technology is rapidly advancing and the quality is constantly improving. Converters are available for both gray-scale SSTV and color SSTV. If you are an amateur radio operator, you might want to consider building the scan converter yourself. Further information can be found in publications of the American Radio Relay League, 225 Main Street, Newington, CT 06111.

Problems with Reception

Reception of EM signals can be an uncertain business. Trouble can result from poor conditions, congestion on the airwaves, marginal equipment design, hardware failure, or a combination of these things.

Desensitization

Receiver *desensitization,* also called *desensing,* is a reduction in the gain of RF stages that occurs when a strong signal comes in. In some cases this effect is slight; in other instances it can be so severe that *blocking* (complete loss of reception) takes place. Desensitization generally occurs in the front end of a receiver. This is the first RF amplifier stage that signals encounter when they arrive from the antenna system. The front end is designed to produce as little noise as possible; it also usually contains some selective circuitry to allow passage of signals in a particular band, attenuating signals at frequencies far removed from the band of interest.

When an extremely strong signal appears at the input of the receiver front end, the amplifying device, usually a bipolar transistor or FET, might be driven into nonlinearity. The effective bias at the base or gate of the transistor is altered by the signal, reducing the gain of the stage. If the signal is intermittent in nature, or if it fluctuates in amplitude, the gain of the front end will constantly vary. The effect might not be noticed if the strong signal happens to be the one to which the receiver operator is listening. But if the signal is at some frequency other than that to which the receiver is tuned, all the received signals will get fainter every time the offending signal comes in. This is annoying, and in bad cases it renders the desired signals unreadable.

The risk of desensitization is greatest when the offending signal is near in frequency to the desired one(s). This is because the tuned circuits in the front ends of most receivers are not especially narrow-banded. While a receiver tuned to 7.000 MHz might not be desensed by a strong signal at 5.000 MHz, the same signal might cause slight desensing if it is at 6.700 MHz, severe desensing if it is at 6.900 MHz, and total blocking if it is at 6.980 MHz.

The best way to eliminate desensitization depends on the cause. Overall, the best protection against this problem is good engineering of the receiver front end, with special attention given to maximizing the dynamic range.

A potentiometer can be installed between the antenna and the receiver input terminals to reduce the total signal energy in the front end when necessary. But this reduces the overall sensitivity of a receiver and might make it impossible to receive weak signals when an offending strong signal is present on a nearby frequency. This is, however, an improvement for reception of moderately strong signals; while it will weaken them, it will also eliminate the intermittent nature of desensitization that can be so distracting.

Sometimes, desensing problems can be greatly reduced, or even overcome, by changing the configuration of the receiving antenna. Many offending signals are local, originating within a few miles of the receiver. As a result, their polarization is constant, and the wavefronts arrive from a defined and consistent direction. If the offending signal is vertically polarized, a horizontally polarized receiving antenna can be used, or vice versa. Another technique is to use a receiving antenna with a sharp null, such as a small loop, and orient the antenna so the null exists in the direction from which the offending signal is coming. Still another approach, which will often work if a unidirectional or bidirectional parasitic or phased array is used for receiving, is to turn the antenna so it points in a compass direction (azimuth) 90 degrees away from the azimuth of the offending signal source. Then the interfering signal will arrive from the side of the antenna, where its gain is the smallest.

If desensitization is taking place within a repeater (the transmitter interferes with the receiver), various steps might be necessary. The transmitter power can be decreased; the frequency offset can be increased; the receiver front-end filtering can be improved; the entire receiver front end can be reengineered.

Intermodulation

In a receiver, an undesired signal sometimes interacts with the desired signal. This usually, but not always, occurs in the front end. The desired signal appears to be modulated by the undesired signal, as well as having its own modulation. This is known as *intermodulation,* often called "intermod." It is sometimes mistaken for spurious emissions from a transmitting station. Intermodulation might be comparatively mild, or it might be so severe that reception becomes impossible.

Intermod can take place for any or all of several reasons. Inadequate selectivity in the receiver input can result in unwanted signals passing into the front-end transistor. Nonlinear operation of the front-end transistor increases the chances of intermod. An excessively strong signal can drive the front end into nonlinear operation, even when the circuit is well designed. A poor electrical connection in the antenna system can cause intermod in a receiver used with that antenna system.

Intermod is never desirable, and modern receivers are designed to keep this form of distortion to a minimum. Manufacturers and designers speak of a specification called *intermodulation spurious-response attenuation.* This is usually expressed in decibels. Two signals are applied simultaneously to the antenna terminals of the receiver under test. The amplitude ratio is prescribed according to preset specifications. The strength of the unwanted signal required to produce a given amount of intermodulation is then compared with the strength of the desired signal, and the ratio is calculated in decibels. Alternatively, the strength of the unwanted signal can be held constant, and the extent of the distortion measured as a modulation percentage in the desired signal. This percentage is called the *intermodulation-distortion percentage*. Standards for intermodulation spurious-response attenuation are set in the United States by the *Electronics Industries Association* (EIA). When testing a receiver for intermodulation response, the procedures and standards of this association should be consulted.

Cross modulation is a form of intermodulation caused by the presence of a strong signal, and also by the existence of a nonlinear component or electrical junction near the receiver but external to it. Cross modulation causes all desired signal carriers to appear modulated by the undesired signal. This modulation can usually be heard only if the undesired signal is amplitude-modulated, although a change in receiver gain might occur in the presence of extremely strong unmodulated signals. When a problem-causing nonlinearity exists external to the receiver and antenna system, cross modulation can be eliminated only by locating the nonlinear junction and getting rid of it. Marginal electrical bonds between two wires, or between water pipes, or even between parts of a metal fence, can be responsible. Sometimes, eliminating cross modulation can be extremely difficult.

Ghosting

A *ghost* is a false image in a television receiver, caused by reflection of a signal from some object just prior to its arrival at the antenna. The ghost appears in a position different from the actual picture because of the delay in propagation of the ghost signal with respect to the main signal. In especially severe cases, the ghost is almost as prominent as the actual picture, which makes it difficult or impossible to view the TV signal.

Ghosting can often be minimized simply by reorienting an indoor television antenna or rotating an outdoor antenna. In some cases, ghosts can be difficult to eliminate. The object responsible for the signal reflection must then be moved far away from the antenna. This is not always practical.

With cable and satellite TV, ghosting is seldom a problem. Ghosting can sometimes take place in these types of systems if *impedance discontinuities* are present in the feed line near the TV receiver. Fortunately, eliminating such discontinuities is not difficult. Replacing the feed line and/or connectors will usually solve such problems. In satellite systems, relocating the dish antenna might be necessary.

External Noise and Interference

Atmospheric static, noise from electrical appliances and internal-combustion engines, and interference from other people sharing a frequency band represent long-standing (and probably eternal) nuisances to EM reception. The nature of these problems, and some solutions, are covered in Chap. 15.

Specialized Communications Techniques

In all communications systems, the objective is to get data from one place to another with the fewest possible errors. Many schemes, some simple and others complex, some obvious and others obscure, have been devised to this end. Here are some examples of specialized wireless communications techniques that have proved effective.

Diversity Reception

Diversity reception is a scheme for minimizing the effects of fading in radio reception or communication via ionospheric propagation. Usually, two receivers are used. Both receivers are tuned to the same signal. They employ separate antennas, spaced several wavelengths apart. The outputs of the receivers are fed into a common audio amplifier, as shown in Fig. 7-7.

At high frequency (above 3 MHz), ionospheric signal fading occurs over small areas at irregular rates, because numerous wave components arrive in constantly varying relative phase. When two antennas are separated by at least several wavelengths (nominally a few hundred feet) at the receiving location, the probability is low that a deep fade will occur at both antennas simultaneously. At least one antenna usually receives a fair signal. When the antennas are connected to independent receivers and the receiver outputs are combined, the result is the best of both signals.

While diversity reception provides some immunity to fading, the tuning procedure is critical and the equipment is expensive. In some diversity receiving installations, three or more antennas and receivers

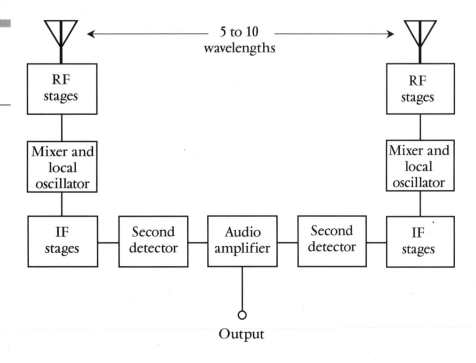

Figure 7-7
Diversity reception combines signals from two different antennas.

are employed. This provides superior immunity to fading, but it compounds the tuning difficulty and further increases the expense.

Synchronized Communications

One of the most important problems in communications is the maximization of the number of signals that can be accommodated within a given band of frequencies. This has traditionally been done by attempting to minimize the bandwidth of a signal. But there is a limit to how small the bandwidth in a specific communications mode can be if information is to be effectively received.

Digital signals generally require less bandwidth than analog signals to convey a given amount of information per unit time. Digital signals consist of finitely many well-defined states with precise timing, while analog signals can have infinitely many states and generally have no timing scheme. *Synchronized communications* refers to a specialized digital mode, in which the transmitter and receiver operate from a common time standard to optimize the amount of data that can be sent in a communications channel or band. This mode is impossible with analog communications methods.

Consider the example of Morse code (radiotelegraphy). Perfectly timed radiotelegraphy consists of defined bits, each bit having the duration of one dot. The length of a dash is three bits; the space between dots and/or dashes in a single character is one bit; the space between characters in a word is three bits; the space between words and sentences is seven bits. This makes it possible to identify every bit by number, even in a long message, provided the starting time and data speed are both precisely agreed upon by the receiver and transmitter. The receiver and transmitter can be synchronized so the receiver, in a sense, knows which bit of the message is being sent at any instant in time.

In synchronized digital communications, also called *coherent communications,* the receiver and transmitter operate in lock step. The receiver evaluates each transmitted bit as a unit, for a block of time lasting from the bit's exact start to its exact finish. This makes it possible to use a receiving filter having extremely narrow bandwidth. The synchronization requires the use of an external frequency or time standard. The broadcasts of the National Bureau of Standards time and frequency stations, such as WWV or WWVH, can be used for this purpose. Frequency dividers are employed to obtain the necessary synchronizing frequencies. A tone or pulse is generated in the receiver output for a particular bit if, but only if, the average signal voltage exceeds a certain value over the duration of that bit. False signals, such as might be caused by filter ringing, sferics, or other noise, are generally ignored because they rarely produce sufficient average bit voltage.

Synchronization can be used with all digital codes, including Morse, Baudot, ASCII, digitized voice, and digitized video. Experiments with synchronized communications have shown that the improvement in S/N ratio, compared with nonsynchronized systems, is several decibels at low to moderate data speeds. Further improvement can often be obtained by the use of *digital signal processing* at the receiving end of the communications circuit. The reduced bandwidth of coherent communications allows proportionately more signals to be placed in any given band of frequencies, compared with conventional modulation and detection methods.

Digital Signal Processing

Digital signal processing (DSP) was first used with radio and TV receivers to produce a clear voice and/or picture from a marginal signal. Some of the earliest experimenters with DSP were amateur radio operators.

In analog modes such as voice or video, the signals are first changed into digital form by *analog-to-digital* (A/D) *conversion*. Then the digital data is "cleaned up," so that the pulse timing and amplitude adhere strictly to the protocol for the type of digital data being used. Noise is eliminated. Finally, the digital signal is changed back to the original voice or video via *digital-to-analog* (D/A) *conversion*.

Digital signal processing can vastly extend the workable range of almost any communications circuit, because it allows reception under much worse conditions than would be possible without it. Digital signal processing also improves the quality of fair signals, so that the receiving equipment or operator makes fewer errors. In circuits that use only digital modes, A/D and D/A conversion are irrelevant, but DSP can still be used to "clean up" the signal. This improves the accuracy of the system, and also makes it possible to copy data over and over many times (that is, to produce multigeneration duplicates).

Digital signals have discrete, well-defined states. It is easier for a machine to process a digital signal than to process an analog signal, which has a theoretically infinite number of possible states. In particular, digital signals have well-defined patterns that are easy for microprocessors to recognize and clarify. Binary (two-state) signals are the easiest for machines to work with.

In an analog signal, the patterns are infinitely varied and complicated. Even the most sophisticated digital computers cannot deal directly with the complex curves of an analog function. Although the modulation envelope of an SSB signal is easy for an analog communications receiver to process, a digital machine sees such a waveform the way a typical American sees a newspaper in Chinese. When analog data is changed to simple, high/low binary digital format, a sophisticated digital electronic circuit can understand it and modify it. The most powerful mainstream microprocessors are binary digital devices.

The DSP circuit works primarily by getting rid of the noise and interference components in a received digital signal, as shown in Fig. 7-8. A hypothetical signal before processing is shown at the top; the signal after processing is shown at the bottom. Relative amplitude is shown on the vertical scale; time is shown on the horizontal scale. The electronic circuit makes its decision for defined time intervals. If the incoming signal is above a certain level for an interval of time, the DSP output will be high (logic 1). If the level is below the critical point for a time interval, then the output will be low (logic 0).

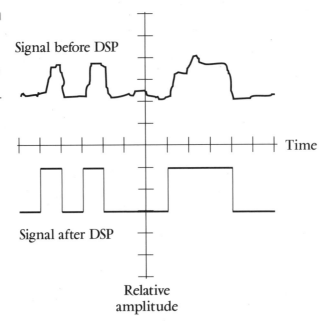

Figure 7-8

A digital signal with noise (top), and the same signal after DSP (bottom).

Signal before DSP

Time

Signal after DSP

Relative
amplitude

Errors can occur with this scheme. A sudden burst of noise, such as atmospheric "static" (sferics) from a nearby thunderstorm, might fool the DSP circuit into thinking the signal is high when it is really low. But this does not happen often unless the signal is exceedingly weak. All other factors being equal, errors are far less frequent with DSP than without it.

Multiplexing

Two or more signals can occupy the same medium or channel at the same time. This is called *multiplexing,* and it can be done in various ways. The most common methods are *frequency-division multiplexing* and *time-division multiplexing.* Multiplexing requires a special encoder at the transmitting end and a special decoder at the receiving end. The medium must be such that interference does not occur among the various channels.

Any EM communications or broadcast channel can be broken down into subchannels. Suppose a channel is 24 kHz wide, so it can contain eight signals, each 2.7 kHz wide. The frequencies of the signals must be evenly spaced so they do not overlap. There is a little extra space on either side of each subchannel to ensure that overlapping does not occur. This is an example of frequency-division multiplexing. Each of these eight signals can be entirely independent of all the others. In digital communications, eight-bit *bytes* can be transmitted over the multiplexed circuit in the same amount of time it takes one bit to go over a single channel. This is an example of *parallel data transfer*. If multiplexing were not used, the circuit would have to employ *serial data transfer*, in which case the bits would have to be transmitted one by one.

Suppose there is a communications channel carrying signals from three different digital sources. Call these signals XXXX, YYYY, and ZZZZ. They are sent in little bits, in a rotating sequence: XYZXYZXYZXYZ... and so on. This is an example of time-division multiplexing. A time-division-multiplex communications circuit requires that the receiver be synchronized with the transmitter. The receiver and transmitter can be clocked from a single independent primary time standard such as the broadcasts of radio station WWV. The receiver detector is blocked off during intervals between transmitter pulses and opens up only during "windows" lasting as long as the longest transmitter pulses. The received data is selected by adjusting the "windows" to correspond with the desired pulse train. Because the duty cycle of any single signal is so low, it is possible to multiplex dozens or even hundreds of signals on one transmitted carrier.

Spread Spectrum

In most communications systems, efforts are made to minimize the bandwidth occupied by each signal. This is because the S/N ratio improves, all other things being equal, as the bandwidth of a signal is reduced. But the signal frequency need not be constant; it can be rapidly varied at the transmitter, and the receiver can follow along. This dilutes the energy over a wide band of frequencies, but the receiver still "thinks" it is picking up a narrowband signal. This is called *spread-spectrum communications* because the signal energy is spread out over a portion of the radio spectrum. If a signal frequency is varied according to a mathematical function, it can be received if, but only if, the receiver "knows" that function and synchronizes its frequency with that of the transmitter. Thus the probability of *catastrophic interference*, in which one strong interfering signal obliterates the desired one accidentally, is near zero. This is the first major asset of spread-spectrum communications.

In EM communications, the random seeking and establishment of communications, in which someone calls "CQ" (meaning "Calling anyone") and waits for an answer, requires that another operator be listening on the same frequency at the same time. With spread spectrum, this is unlikely. It is also unlikely that anyone could eavesdrop on a contact in progress. Privacy is the second major asset of spread-spectrum communications. The frequency-spreading function can be extremely complex. If the transmitting and receiving operator do not divulge the code to anyone, and if they do not tell anyone about the existence of their contact, then no one else on the band will know the contact is taking place, much less be able to listen in.

During a spread-spectrum contact between a given transmitter and receiver, the operating frequency is changing all the time. It might fluctuate all over a band, a range of several megahertz or tens of megahertz. As a band becomes occupied with more and more of these types of signals, the overall noise level in the band appears to increase. This is the result of larger numbers of signals "sweeping across" each other. Ultimately, there is a limit to the number of spread-spectrum contacts that a band can handle. This limit is about the same as it would be if all the signals were constant in frequency.

The *modulation* of the signal whose frequency is made to fluctuate should concentrate energy into the narrowest possible band. That is, SSB is preferable to AM or FM for this signal, and digital modes are better still. Although the signal is in effect "wideband frequency-modulated" by the spreading function, the receiver, which follows the spreading function precisely, "hears" the signal as if it uses ordinary narrowband modulation. Thus, from the receiver's point of view, once the signal gets into the amplification and selective circuits, the applicable engineering principles are the same as if the communications were fixed in frequency.

A common way of getting the frequency to change, thereby spreading the spectrum, is known as *frequency hopping*. In this scheme, the transmitter has a list of channels that it follows in a certain order. The receiver must be programmed with this same list, in the same order, and must be synchronized with the transmitter. The list is a *pseudorandom sequence* that repeats for as long as the communication takes place. An example might be 420.4, 420.2, 420.7, 420.1, 420.8, 420.5, 420.9, 420.6, 421.0, and 420.3 MHz, with each frequency being sent for 0.005 second (s), and the whole sequence repeated again and again by the transmitter, synchronized with a clock that has been initialized against some primary standard such as WWV/WWVH.

The *dwell time* is the interval at which the frequency changes occur. It should be short enough so that a signal will not be noticed, and not cause interference, on any single frequency. In the foregoing example the dwell time is 0.005 s, or 5 milliseconds, and there are 10 different frequencies. Therefore, the energy is diluted on any one frequency by a factor of 10, as compared with the way it would be if its frequency were constant. This is a difference of only 10 dB below the level of the signal if it were on a constant, fixed frequency. In practice, there should be hundreds or thousands of *dwell frequencies,* so the signal energy is diluted to the extent that, if someone tunes to any frequency in the sequence, the signal will not be noticeable.

Another method of getting spread spectrum is to modulate the fundamental signal at a slow rate, say 20 Hz, with a sine wave that guides it up and down over the range of the band. The receiver will be able to pick the signal up only if it follows at the same frequency, and in the same phase. The frequency of this sine wave might itself be gradually increased and decreased. The wave might have some distorted, complicated shape. This can lend various dimensions to the sweeping function. Almost anything will work, as long as the receiver is programmed to follow the transmitter.

Radio Telescopes

In the 1930s, Karl Jansky, an employee of Bell Laboratories, discovered that the center of our galaxy, the Milky Way, produces RF energy at a wavelength of 15 m (a frequency of 20 MHz). Jansky was the first scientist to discover that radio waves come from distant objects in space. A few years after Jansky's discovery, a radio amateur, Grote Reber, built a 31-ft dish antenna for the purpose of receiving RF energy from space. This was a primitive *radio telescope.* Reber found radio waves coming from the Milky Way at wavelengths down to about 2 m (a frequency of 150 MHz). With his modest apparatus and a lot of hard work, Reber was able to make a rough *radio map* of our galaxy. By 1940, astronomers in general had become interested, and the science of *radio astronomy* was born.

Design

Modern radio telescopes use advanced antennas, receiver preamplifiers, and sophisticated signal-processing techniques. Radio astronomy is generally carried out at wavelengths shorter than 10 m (frequencies above 30 MHz), and especially at ultra-high and microwave frequencies (wavelengths less than 1 m, or frequencies greater than 300 MHz). A typical radio telescope installation consists of an antenna, a feed line, a pream-plifier, the main receiver, a signal processor, and a signal recorder.

A radio-telescope antenna can consist of a large dish or an extensive phased array. The dish antenna offers the advantage of being easily steered without affecting the *resolution,* or the ability to separate different radio sources in the sky. Some radio-telescope dish antennas can be point-ed to any part of the sky. A phased array, which can consist of a set of Yagi, dipole, or dish antennas, provides better resolution because it can be made very large. However, the steerability is not as flexible; phased arrays can usually be steered only along the meridian (north and south). The rotation of the earth must then be used to obtain full coverage of the sky.

Two or more large antennas, spaced a great distance apart, can be used to obtain extreme resolution with a radio telescope. This apparatus is called an *interferometer.* It allows the radio astronomer to probe details of celestial radio sources. In some cases, what was originally thought to be a single source has been found to contain multiple components.

The antenna feed line is usually a coaxial cable or waveguide. The lowest possible loss and the highest possible *shielding continuity* must be maintained. This minimizes interference from earth-based sources and ensures that the antenna directivity and sensitivity are optimum.

A radio-telescope receiver differs from a communications receiver. Sen-sitivity is of the utmost importance. A special preamplifier is needed. The components can be cooled down to a few degrees above absolute zero to reduce the noise in the circuit. The receiver in a radio telescope must have a relatively large response bandwidth, in contrast to a commu-nications receiver, which usually exhibits a narrow bandwidth. The wave-length must be adjustable over a wide range. A speaker can be used for effect, but is of little practical value in most cases, because cosmic radio emissions sound like little more than sferics. A *pen recorder* and/or *oscillo-scope,* and perhaps a *spectrum analyzer,* can be used to evaluate the charac-teristics of the cosmic noise.

Most radio telescopes are located far from sources of humanmade noise. A city is a poor choice as a location to build a radio telescope. Some radio-telescope observatories are legally protected against human-made radio noise.

Applications

Radio telescopes have provided astronomers with knowledge that could not have been discovered by any other means. Radio telescopes can "see" things that cannot be detected with optical telescopes. The hydrogen line at the 21-centimeter (cm) wavelength (approximately 1400 MHz) can be easily observed with a radio telescope, although it is invisible with optical apparatus. Some astronomers believe that the 21-cm line might serve as a marker for interstellar communication. In 1974, radio astronomer Frank Drake used the massive radio telescope at Arecibo, Puerto Rico, to transmit the first message to the cosmos. This transmission has already passed some of the nearer stars. A binary code was used, telling of our world and our society.

With specialized radio telescopes aboard space probes, astronomers have precisely ascertained the distances to, the motions of, and the surface characteristics of most of the known planets and their moons in our solar system. When a radio telescope is used in this way, it is called a *radar telescope*. It contains a transmitter as well as a receiver, and employs a computer to generate graphic displays of the echoes received.

In 1965, two Bell Laboratories employees, Arno Penzias and Robert Wilson, found weak microwave radio noise coming from all directions in the sky. These observations were confirmed by other radio astronomers. The amplitude-versus-frequency curve, or *spectral distribution,* suggested that these EM waves originated in a primordial explosion. It is believed that this energy is leftover EM radiation from the *Big Bang* several billion years ago, from which the entire known universe has since evolved.

CHAPTER 8

Beyond Microwaves

Wireless technology is not confined to the radio spectrum. Infrared (IR) and visible-light systems are common. Acoustic waves are employed in some wireless systems. Fiberoptics are sometimes classified as wireless technology. More esoteric wireless modes, still largely unexploited, include ground currents, ultraviolet, X-rays, gamma rays, and certain sub-atomic particles.

Infrared and Visible-Light Devices

The IR spectrum consists of electromagnetic (EM) fields whose wavelengths lie between those of microwave radio and those of visible light. Infrared is sometimes mistakenly called "heat radiation." The longest IR waves measure a fraction of a millimeter. The shortest IR waves measure 750 nanometers [nm; 7.5×10^{-7} meters (m)]. Visible wavelengths range from 750 nm (7.5×10^{-7} m) for red light to 390 nm (3.9×10^{-7} m) for violet light.

Photoelectric Cells

The term *photoelectric cell* is used to describe a variety of devices in which visible-light or IR rays affect the behavior of materials. This can happen in either of two ways: (1) variation of electrical conductance depending on exposure to illumination, and (2) generation of electric current as a result of illumination. This section is concerned primarily with devices that exhibit the first of these effects.

There are many materials that exhibit electrical conductance that varies with the level of illumination. This effect is called *photoconductivity*; substances that show this effect are known as *photoconductive materials*. A piece of photoconductive material has a certain conductance when there is no illumination. As the illuminance increases, the conductance improves. There is a limit to the extent that the conductance will continue to increase as the illumination becomes more intense. As the brightness of an IR or visible source increases beyond this limit, the conductivity of the material will not improve further. This condition is called *saturation*, and the conductance in this state is called the *saturation conductance*. Photoconductivity occurs in many substances, but it is most pronounced in semiconductor materials such as silicon, gallium arsenide (GaAs), germanium, and various sulfide compounds. Photoconductive semiconductors are used in the manufacture of photoelectric cells.

A *photodiode* is a semiconductor diode that exhibits variable conductance depending on the intensity of illumination that strikes its P-N junction. Most ordinary diodes having P-N junctions are subject to photoconductivity. An ordinary diode will not work as a photodiode because its housing is opaque, and/or because the P-N junction is between two opaque pieces of semiconductor material. A photodiode, therefore, is designed so its P-N junction is readily exposed to illumination. This necessitates the use of a transparent enclosure. Most photodiodes also have relatively large P-N junction surface area.

The conductivity of a photodiode generally increases in the reverse direction (that is, when the device is reverse-biased) as the illuminance increases, but there is always a defined saturation conductance. When the device is operating in the *saturation region,* the conductance will not increase no matter how bright the illumination gets. This effect is illustrated graphically in Fig. 8-1. Assume a constant reverse-bias voltage applied to the diode. The relative level of illumination is shown on the horizontal axis; the relative reverse current is shown on the vertical axis. The saturation region is the flat part of the curve.

A photodiode is always connected in series with the circuit to be controlled, and is always reverse-biased. Care must be exercised to ensure that the photodiode is not forced to carry too much current, because excessive current will destroy the P-N junction. The current can be limited by using resistors in series with the device, and/or by limiting the applied voltage.

Figure 8-1

Current in a photodiode as a function of illumination.

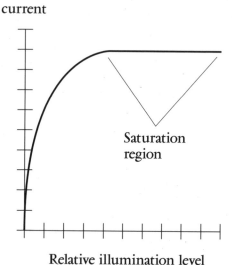

Relative current

Saturation region

Relative illumination level

Phototransistors

A *phototransistor* is a bipolar or field-effect transistor (FET) that exhibits conductance that depends on the illuminance from a source of IR or visible light. Transistors are normally enclosed in opaque packages to eliminate noise that would otherwise be caused by illumination of the semiconductor junctions. Phototransistors have transparent or translucent packages, so that light can reach the internal junctions.

A *bipolar phototransistor* is constructed in such a way that the base-collector P-N junction can receive the maximum possible amount of illumination. The most common bipolar phototransistors are of the NPN type. They are manufactured from silicon. Most do not have terminals at the base electrode, but only at the emitter and the collector. Biasing can be obtained via a *light-emitting diode* (LED) near the device, but base bias is not necessary and can sometimes be detrimental to performance. A bipolar phototransistor is used in much the same way as a photodiode. The effective conductance increases as the illuminance increases, but only up to the saturation point.

A *photoFET* is a field-effect transistor that exhibits photoconductive properties. Most ordinary FETs would be suitable for use as photoFETs, except that their housings and construction are opaque, and it is therefore impossible for illumination to reach the P-N junction inside the device. A photoFET has a transparent window in its housing, and is constructed in such a way that its P-N junction, forming the boundary between the gate and the channel, has the largest possible surface area. The junction is situated so that illumination can fall on it. The main difference between photoFETs and other photoelectric devices is the fact that photoFETs have higher impedance. This can be an advantage in weak-signal modulated-light communications receivers and in fiberoptic systems.

Optoisolator

Optical coupling is a method of transferring a signal from one circuit to another, by converting it to modulated light and then back to its original electrical form. A component specifically designed for this purpose is called an *optoisolator*.

In an optoisolator, the input signal is applied to an LED or *infrared-emitting diode* (IRED). The resulting modulated-light signal propagates across a small gap, where it is intercepted by a photocell, phototransistor, or other light-sensitive device and converted back into electrical impulses. Most optoisolators contain the transmitter and the receptor in a single package with a glass or plastic transmission medium.

Optical coupling is used when it is necessary to prevent impedance interaction between two successive stages in a multistage circuit. Normally, changes in the input signal to an amplifier cause the input impedance to vary, and this in turn affects the preceding stage. Optical coupling avoids this problem by providing essentially complete isolation. Optical coupling also facilitates coupling between circuits that operate at substantially different potentials, for example, a transistorized video amplifier and a cathode-ray tube.

Optical Shaft Encoder

In digital radio receivers, transmitters, and transceivers, frequency adjustment is done in discrete steps rather than continuously. In high-frequency equipment, the usual increment is 1 or 10 hertz (Hz). At very high frequencies and ultra-high frequencies, 10- or 100-Hz increments are used for continuous-wave telegraphy, radioteletype, and single-sideband (SSB); 5 kilohertz (kHz) increments are common for frequency modulation.

For tuning, an alternative to mechanical switches, which corrode or wear out with time, is a device called an *optical shaft encoder.* This consists of an LED, a *photodetector,* and a rotatable disk with alternating transparent and opaque bands called a *chopping wheel.*

A functional diagram of an optical shaft encoder is shown in Fig. 8-2. The LEDs shine on the photodetectors, but their beams must pass through the chopping wheel. The wheel is attached to the tuning shaft (not shown). As the tuning shaft is turned, the light beams are interrupted. Each interruption causes the frequency to change by a specified amount, such as 10 Hz. The difference between the "frequency up" and "frequency down" operations is determined according to the relative timing of the interruptions in the light beams. The two sets of output pulses are out of phase by a certain amount when the shaft is rotated clockwise; they are out of phase in the opposite sense when the shaft is rotated counterclockwise. A microcomputer interprets these phase differences and instructs a voltage-controlled oscillator to increase or decrease its frequency by the proper amount. The same microcomputer

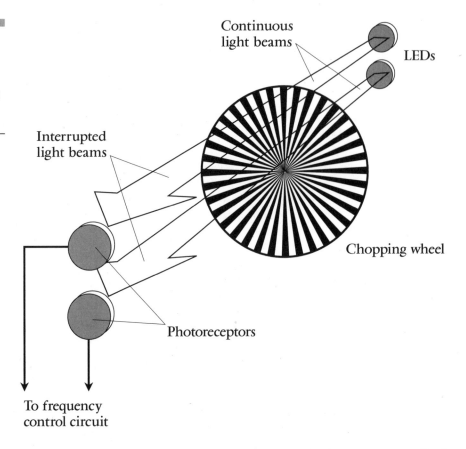

Figure 8-2
An optical shaft encoder. For clarity, the shaft is not shown; it is attached to the center of the chopping wheel.

also sends data to a digital display that shows the frequency, usually in kilohertz or megahertz.

A well-designed optical shaft encoder provides a "feel" similar to that of a mechanical dial, but without most of the moving parts and sliding contacts that cause maintenance trouble in the long term.

Electric Eye

An *electric eye* is a sensing device that uses radiant energy beams to detect objects. Usually, the output of an electric eye is used to control some external machine or system. For example, an electric eye can be set up to detect people passing through a doorway. This can count the number of people entering or leaving a building. Another example is the counting of items on a fast-moving assembly line; each item breaks the light beam once, and a circuit counts the number of interruptions.

Figure 8-3
An electric eye
detects objects that
interrupt the path of
a visible light or
infrared beam.

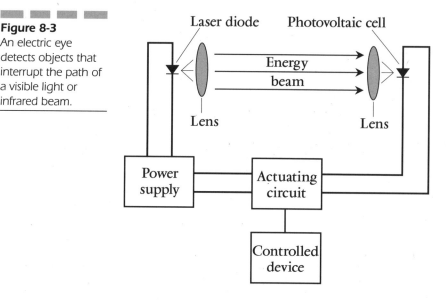

Elevators commonly employ electric eyes in their doors, so the doors will not close on people who hesitate on their way in or out.

The typical electric eye has a light source, usually a laser diode, and a photosensor such as a photoelectric or photovoltaic cell. These are connected to an actuating circuit as shown in Fig. 8-3. When something interrupts the light beam, the voltage or current passing through, or generated by, the sensor changes dramatically. It is easy for electronic circuits to detect this voltage or current change. Using amplifiers, even the smallest change can be used to control large and powerful machines.

Infrared, rather than visible light, is commonly used in electric eyes. This is ideal for use in burglar alarms, because an intruder cannot see the beam and therefore cannot evade it. An IR electric eye operates according to the same principles as its visible-light counterpart; the only difference is that IREDs or IR lasers are used.

Modulated-Light Communications

Electromagnetic energy of any frequency can be modulated for the purpose of transmitting information. The only constraint is that the frequency of the carrier be at least several times the highest modulating frequency. Modulated light has recently become a significant means of conveying data, although the concept is more than a century old.

A Simple Transmitter

A basic modulated-light transmitter, which can be built for only a few dollars using parts available at most electronics stores, is diagramed schematically in Fig. 8-4. It uses a microphone, an audio amplifier, and an incandescent bulb. Complete amplifier modules, capable of producing about 1 watt (W) of audio output, are sold in some stores for very reasonable prices. Alternatively, a simple class A amplifier circuit can be built using a rugged bipolar transistor or operational amplifier. The module or amplifier should have a built-in gain (volume) control.

The direct current (DC) from the battery or power supply provides constant illumination for the light bulb. The bulb should be chosen to match the battery voltage. For example, if a 6-volt (V) battery is used as the power source, then the bulb should be a 6-V lantern or flashlight lamp. The bulb should be illuminated by the DC even when there is no output from the audio amplifier.

When someone speaks into the microphone, the amplifier produces alternating-current (AC) output at audio frequencies (AF). In the light bulb, the AC appears superimposed on the DC from the power supply. The gain of the audio amplifier should be set at minimum to begin with. Then, as the operator speaks into the microphone in a normal voice, the gain should be increased until the light bulb just begins to flicker on voice peaks. Increasing the gain further will cause distortion in the modulated-light signal.

You might think that an incandescent bulb cannot follow the AF of a human voice. This is true in a strict theoretical sense; the instantaneous brilliance of the light bulb changes very little as a result of voice modulation. The visible output from the bulb is not modulated to any-

Figure 8-4
A simple voice-modulated-light transmitter.

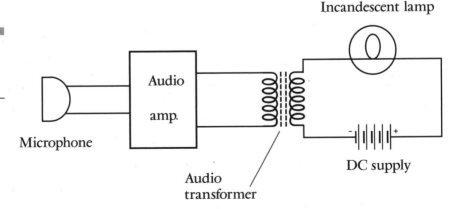

thing near 100 percent. Nevertheless, sufficient modulation can be obtained in this way to provide an excellent demonstration of how light beams can be used to carry information.

An LED might be used in place of the light bulb in the circuit of Fig. 8-4. A resistor would have to be placed in series with the LED to limit the current through it. Most LEDs can follow current variations at much higher frequencies than incandescent lamps. However, the LED illumination output is far lower than that of a flashlight bulb, largely offsetting any advantage gained by the faster response characteristic.

Some types of bulbs follow audio modulation better than flashlight lamps. For example, a mercury-vapor or sodium-vapor lamp, such as the kind used for yard lighting and available in home supply superstores, can be employed if you are willing to build the necessary high-voltage power supply and large audio amplifier, and if you are willing to find and pay for a large enough audio transformer. These lamps follow current fluctuations at much higher frequencies than incandescent lamps.

A Simple Receiver

Modulated light beams can be detected using photodiodes or photo-voltaic cells, as long as the modulating frequency is not too high (less than about 100 kHz). At AF, a circuit such as the one in Fig. 8-5 will provide sufficient output to drive a headset. The same type of audio amplifier module can be used for the receiver as is used for the transmitter. For use with speakers, the output of the module can be fed to the phono or tape input of a high-fidelity (hi-fi) amplifier system. To increase the sensitivity of the receiver (that is, to allow reception of

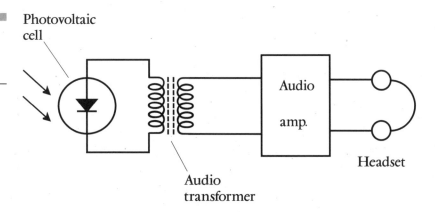

Figure 8-5
A simple voice-modulated-light receiver.

Photovoltaic cell

Audio transformer

Audio amp.

Headset

much fainter modulated-light beams), an FET stage can be added between the photovoltaic cell and the amplifier module.

Using the simple transmitter and receiver depicted in this section, voice-modulated-light communications can be had for distances up to about 50 feet (ft). If a small reflector, such as the type found in lanterns and flashlights, is added at the transmitter, the range can be increased to several hundred feet. This will be true in daylight as well as in darkness. It is important that the photodiode or photovoltaic cell not be exposed to direct sunlight, because this will saturate the device and render it insensitive to the tiny fluctuations in illumination from the transmitter. It is also important that the receiver not be exposed to the light output from lamps using 60-Hz utility AC. The reason for this will be vividly apparent when it is tried. Even ordinary incandescent lamps are modulated by utility AC to the extent that the voice modulation from the transmitter is drowned out. Mercury-vapor or sodium-vapor street lamps present an impediment to communications at night, unless some means is devised to narrow the receiver's "field of vision."

Using this receiver, some interesting things can be discovered. For example, the visible emissions of a cathode-ray tube, such as is used in television sets and computer monitors, produce bizarre sounds when demodulated by this circuit. It's also fun to "listen" to sunlight shining through the leaves of a tree on a windy day, or reflecting off a lake whose water is rippled by a light breeze.

Extending the Range

The communications range of any line-of-sight modulated-light system can be increased in several ways, including *collimation* of the transmitted beam, maximizing the receiver *aperture,* and increasing the transmitter power.

Beam Collimation   A large paraboloidal or spherical reflector can be used to focus the light beam from the transmitter. The larger the reflector is, the narrower the beam will be, and the farther it can propagate on a line of sight. The problem with this scheme is that large reflectors are expensive. An alternative is to employ a *Fresnel lens,* of the type used in overhead projectors. These lenses are made of flat plastic, etched with circular grooves that cause the material to behave like a convex lens. Fresnel lenses are typically about a foot square. They can be obtained from some hobby stores and catalogs. The light bulb or LED should be placed at the focal point of the lens; a fairly narrow beam will be emitted from the other face of the lens.

Receiver Aperture   The same scheme used for collimating the transmitted beam can be used to increase the receiver aperture. However, it is also important that the receiver's "field of vision" be as narrow as possible. This will allow reception of the transmitted beam, while minimizing interference from sources in other directions. One scheme is to insert the photodetector in a telescope in place of the eyepiece. The telescope "finder" can be used to visually zero in on the transmitter. The larger the diameter of the telescope's objective lens or mirror, the larger will be the receiver aperture, and the greater will be the communications range, all other factors being equal. A Fresnel lens can be used to obtain a large receiver aperture, but an opaque box or other enclosure must be used to keep out stray light. A communications range of several miles in clear air can be obtained if Fresnel lenses are used at the transmitter and the receiver, assuming a 6-V lantern bulb is used as the transmitter light source. The best receiving "antenna" is a paraboloidal or spherical reflector 2 ft or larger in diameter.

Transmitter Power   The audio output power of the transmitter depicted in Fig. 8-4 is less than 1 W. Only a small fraction of this actually modulates the light beam. A mercury-vapor or sodium-vapor yard lamp, a high-voltage DC power supply, a massive AF transformer, and a high-power audio amplifier system might produce a light beam modulated with several watts of audio power. Using a reflector such as the type employed in advertising arc lamps (the ones that can create visible spots in the sky on cloudy nights), it should be possible to communicate between mountains over 100 miles (mi) apart if the air is clear.

"Cloudbounce" and Atmospheric Scatter

Designing a high-power transmitter of the sort described above, and getting it to work, could be an interesting experiment for a hobbyist with a lot of extra money. If two suitable paraboloidal or spherical reflectors can be found—one for the transmitter and another for the receiver—it might be possible to communicate by means of light scattering from clouds or within the atmosphere.

Please note that the following "cloudbounce" and atmospheric-scatter experiments have not been carried out by the author, and the author therefore cannot guarantee that they will work. Also, before these experiments are done at night, you should check with the Federal Aviation Administration (FAA) to ensure that the bright light will not interfere with the operation of aircraft in the vicinity.

"Cloudbounce" modulated-light communication is shown in Fig. 8-6A. Of course, the scheme will work only when the sky is overcast. Low, thick clouds will provide better results in general than high, thin clouds. The obtainable range depends on the thickness of the cloud deck, its reflectivity, and its altitude. The transmitter beam is aimed at the sky, in such a way that a spot of light falls on the clouds approximately midway between the receiver and the transmitter. The receiver reflector is aimed so the maximum amount of signal light falls on the photodetector at its focus.

This technique should work in daylight as well as in darkness, as long as the amount of daylight falling on the photodetector is not so great as to cause it to go near, or into, a state of saturation. At night, the background glow from mercury-vapor and sodium-vapor street lamps is likely to cause a 60-Hz "hum" problem. This might be overcome by providing the transmitter lamp with an amplitude-modulated (AM) signal at a frequency of, say, 30 kHz. A special very-low-frequency (VLF) transmitter would have to be built for this purpose. The receiving station could employ a VLF converter connected to the antenna terminals of a general-coverage communications receiver. It would be necessary to exercise care to ensure that the 30-kHz AM signal did not get transmitted over the airwaves (VLF radio waves). Coaxial cable would have to be used between the transmitter and the light bulb, and an excellent electrical ground system would be needed. Another way to get rid of the 60-Hz "hum" would be to use a separate photodetector, without a reflector, to pick up the general lamp glow and introduce its "hum" into the receiver input 180 degrees out of phase with, but at the same amplitude as, the "hum" from the signal photodetector. This would cancel out the "hum" but would not affect the desired voice signals.

Figure 8-6B illustrates how a sensitive receiver might be used to pick up the scattered light from a powerful transmitter when the sky is partly cloudy or cloudless. The receiving and transmitting reflectors must be aimed at a common parcel of air. This scheme should work best in air containing just a little dust or haze. If the air is too clear, not enough light will be scattered; if there is too much dust or haze, beam propagation will be poor.

Figure 8-6
"Cloudbounce"
(A) and atmospheric-
scatter (B) modulated-
light communications.

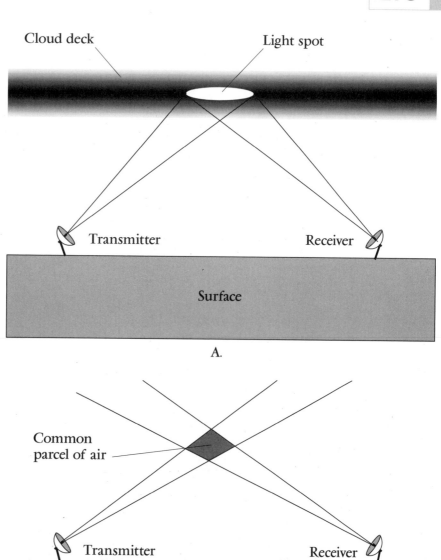

Cloud deck

Light spot

Transmitter

Receiver

Surface

A.

Common
parcel of air

Transmitter

Receiver

Surface

B.

Lasers

The term *laser* is an acronym, derived from the first letters of the words "light amplification by stimulated emission of radiation." Many people think lasers are built for knocking satellites out of orbit or burning holes through walls. Some lasers can do these things, but in wireless communications and electronics, lasers are employed for less spectacular, although perhaps more useful, purposes.

Modulated Lasers

Lasers offer superior performance over long distances for line-of-sight modulated-light communications, because the beam from a typical laser, even an inexpensive one, diverges much more gradually than that from any paraboloidal reflector. The *beam divergence*, or spreading of the rays, can be essentially ignored within a certain distance of the laser source. The emitted rays are almost parallel.

A hobby-type *helium-neon laser* has a beam the diameter of a pencil at close range; this enlarges to the diameter of a half dollar at a distance of several hundred feet. High-gain, low-noise amplifiers can be used for reception. It is within the means of the serious electronics hobbyist to make a system that will work for several miles, or even tens of miles, given clear air and an unobstructed line of sight, and without the need for large reflectors or Fresnel lenses. The addition of a reflector or Fresnel lens to the receiver will multiply the range severalfold.

Amateur radio operators have experimented with laser communication, but the technique has never been put to common use, probably because a clear line of sight usually does not exist between any two randomly chosen locations. But possibilities exist for communication over paths not connected by a single direct line of sight. For example, the beam from a laser might be aimed at a well-placed reflecting mirror, allowing line-of-sight communication between points not "visible" directly to one another. Several such mirrors could be used if necessary. Their alignment would be critical, and earth movements would necessitate frequent realignment. However, once installed, such a point-to-point system would be reliable (except in rain, snow, fog, or thick haze).

Line-of-sight laser communications circuits offer a level of security not possible with radio communications or even with conventional telephone systems. This is because, in order to intercept the information, a detecting device must partially obstruct the light beam. Light beams are not visible to anyone who is not directly in their path. Even if a "laser cracker" were to determine the whereabouts of a beam and intercept its data contents, the data itself could be encrypted. Besides this, the eaves-dropping equipment would produce a sudden decrease in the beam intensity at the receiver. The beam could be left shining at all times, whether communications were taking place or not. A microcomputer at the receiver could alert the operator in the event of a sudden drop in the beam intensity.

Basic Laser Principles

All lasers work because of a phenomenon called *resonance*. A cavity is designed so waves, usually visible light or IR, bounce back and forth inside it, reinforcing each other in phase. This occurs in much the same way as the EM field reinforces itself in a radio-frequency cavity resonator.

The heart of a laser is the region in which resonance takes place. This cavity might contain gaseous matter, a liquid, or a solid. The elements and/or compounds chosen determine the wavelength of the output. Mirrors are placed at each end of the cavity, so energy bounces back and forth many times between them. One mirror reflects 100 percent of the energy that strikes it, and the other mirror is 95 percent reflective. The mirrors might be flat or concave; the cavity can have its ends either perpendicular or angled.

Two cavity laser configurations are shown in Fig. 8-7. These illustrations are vertically exaggerated for clarity. The waves in the drawings represent the EM energy, usually IR or visible light, resonating between the reflectors. Only a few wave cycles are shown; in reality, many thousands of wave cycles occur between each reflection.

Energy is supplied to the cavity by means of a process called *pumping*, so the intensity of the beam builds to higher and higher levels as the energy is repeatedly reflected from the mirrors. The rays finally emerge from the partially silvered mirror in the form of *coherent radiation*. This means that all the waves in the output are in phase with each other and are at exactly the same frequency. This concentrates the radiant energy into the most intense possible beam and also produces the smallest possible beam divergence. Some lasers emit a continuous output beam, while others produce a series of pulses.

Figure 8-7
Two simple lasers. A flat-ended cavity and flat reflectors (A); an angle-ended cavity and concave reflectors (B).

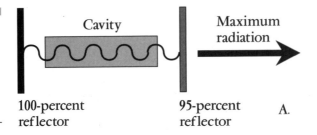

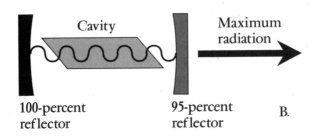

Common Lasers

The helium-neon (He-Ne) laser is the most often used in hobby work, because of its low cost and versatility. It gets its name from the two gases in its resonant cavity. The light from a He-Ne laser is bright red. Other gases can be used. Argon produces a laser with blue light. A mixture of nitrogen, carbon dioxide, and helium provides an IR beam. Hydrogen, xenon, oxygen, and chlorine have also been used to make gas lasers.

The *ruby laser* makes use of solid aluminum oxide in the form of a rod, with a trace of chromium mixed in. The reflectors are silvered onto the ends of the rod. The output is in pulses that last for a fraction of a millisecond. Pumping is done by means of a helical *flash tube* wrapped around the rod. The output is in the red visible range. Low-speed digital modulation is possible with this type of laser.

Semiconductor lasers work differently than the types of lasers mentioned above. Instead of a resonant cavity, a semiconductor laser has a P-N junction that emits coherent radiation when forward-biased. Most *laser diodes* are made with the semiconductor compound *gallium arsenide* (GaAs) and have a primary emission wavelength of about 905 nm in the IR region. Laser diodes are also available for various visible wavelengths. In general, the shorter the wavelength becomes, the more challenging is the task of developing a laser diode to produce emissions at that wavelength. Laser diodes need a certain minimum forward current in order to function. If the current is below the critical level, the laser diode behaves much like an ordinary LED or IRED but when the threshold is reached, charge carriers (electrons and holes) recombine in such a manner that laser action (*lasing*) occurs.

Semiconductor lasers do not have the narrow-beam output typical of larger, *pumped lasers*, although the beam can be collimated using lenses. The output power of a semiconductor laser is very low, so it is not well suited to long-distance communications. Some laser diodes emit continuous energy; others work in short pulses of brief duration. In a pulsed laser, the peak power can be many times the average power. The maximum peak power, in short pulses, can be as great as 100 W, but the pulse duration is so brief that the average power is low.

Laser diodes always produce their emission over an extremely narrow range of wavelengths. This is in contrast to incandescent lamps and fluorescent tubes, which have outputs covering wide bands of EM wavelengths. Laser diodes are used in optical communications and computer systems, especially in fiberoptic data transmission. The coherent output from a laser diode propagates more efficiently, with lower loss per unit distance, than ordinary visible light or IR.

Basic Laser Communications System

A laser-communications transmitter has several components. Any form of data can be sent, including text, voices, music, and images. A system for transmitting computer data via laser is illustrated by the block diagram of Fig. 8-8A. The output of the computer goes to a *modem*, just like the kind used with personal computers to gain access to online services. The audio output of the modem is amplified, and then the data is impressed onto the laser beam via a *modulator*.

A laser-communications receiver uses a photocell, an amplifier, and a signal processor. A simple system for recovering data from a laser beam and feeding it to a computer is shown in Fig. 8-8B. The light falls on the photocell, which converts it to weak audio tones, just like the ones that came from the modem at the transmitting end of the circuit. These tones are amplified if necessary, and then they go to the signal processor, which acts like a modem for received data. The output of this device can be understood directly by the computer.

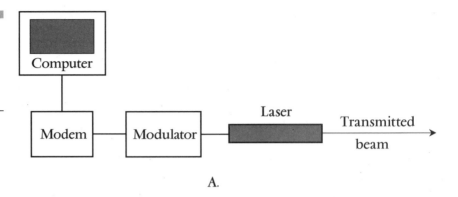

A.

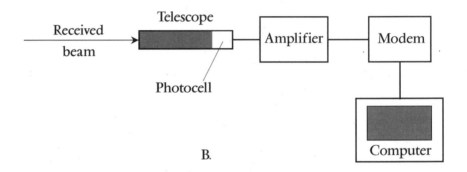

B.

Acoustics

Acoustics is the science of compression waves transmitted through the atmosphere. Acoustics is important to architects and engineers who design and build concert halls, where sound must propagate from a stage or speaker system to a large number of people. Engineers consider acoustics in the construction of speaker enclosures. Some knowledge of acoustics is helpful if you want to set up a hi-fi sound system. *Acoustic waves* can also be used in various wireless applications.

The Sound Spectrum

Sound consists of molecular vibrations at AF, ranging from approximately 20 Hz to 20 kHz. Young people can hear the full range of AF.

Older people lose hearing sensitivity at the upper and lower extremes. An elderly person might only hear sounds from 50 to 5000 Hz.

As the frequency of sound increases, the wavelength becomes shorter. In air, sound travels at about 1100 feet, or 335 meters, per second. The relationship between the frequency f of a sound wave in hertz, and the wavelength w in feet, is as follows:

$$w = 1100/f$$

$$f = 1100/w$$

In the metric system, the relationship between f in hertz and w in meters is given by

$$w = 335/f$$

$$f = 335/w$$

This formula is also valid for frequencies in kilohertz and wavelengths in millimeters. When using the formulas, be careful that you do not get the units confused.

A sound of 20 Hz has a wavelength of 55 ft or 16.8 m in the air. A sound of 1000 Hz produces a wave measuring 13 inches (in) or 33.5 centimeters (cm). At the extreme high-frequency end of the audio spectrum, a wave is just $^{2}/_{3}$ in or 17 mm long. The speed of sound in the atmosphere changes slightly with variations in barometric pressure, temperature, and relative humidity. In substances other than air, such as water or metal, the above formulas do not apply.

Sound waves 55 ft long behave much differently than waves measuring less than an inch, and both of these act differently than a wave 13 inches in length. Acoustics engineers must take this matter into consideration when designing sound systems, especially for music.

Waveforms

The frequency, or *pitch*, of a sound is only one of several variables that acoustic waves can possess. Another important factor is the shape of the wave itself.

The simplest acoustic waveform is a *sine wave*, in which all of the sound energy is concentrated at a single frequency. Sine-wave sounds are rare in nature. A good example of a sine-wave sound is the beat note, or *heterodyne*, produced by a steady carrier in a communications receiver. Certain musical instruments, such as the flute, produce near sine-wave tones.

In music, most of the notes are *complex waveforms*, consisting of energy at some frequency and various multiples, or *harmonics*. Some examples are the *sawtooth wave*, the *square wave*, and the *triangular wave*. An instrument might produce a note at 1 kHz and also at numerous harmonics: 2 kHz, 3 kHz, 4 kHz, and so on. The shape of the waveform depends on the distribution of energy among the *fundamental frequency* (lowest frequency) and the harmonics. There are infinitely many different shapes that a wave can have at a single frequency such as 1 kHz.

Music is usually comprised of many different tones that intermingle to make the overall waveform almost unfathomably complicated. A flute, a clarinet, a guitar, and a piano can each produce a sound at 1 kHz, but the tone quality is different for each instrument. The waveform affects the way a sound is reflected from objects. Acoustics engineers must consider this when designing sound systems and concert halls. The goal is to make sure that all the instruments sound "real" everywhere in the room.

Suppose you have a sound system set up in your living room. You sit on the couch and listen to music. Imagine that, for the particular placement of speakers with respect to your ears, sounds propagate well at 1, 3, and 5 kHz, but poorly at 2, 4, and 6 kHz. This will affect the way musical instruments sound. It will distort the sounds from some instruments more than the sounds from others. Unless all sounds, at all frequencies, reach your ears in the same proportions that they come from the speakers, you will not hear the music the way it originally came from the instruments. For certain instruments and notes, the sound can be altered so much that it doesn't resemble the original at all.

Figure 8-9 shows a listener, a speaker, and three sound reflectors. It is almost certain that the waves traveling along paths X, Y, and Z will add up to something different, at the listener's ears, than the waves that come directly from the speaker. Path X might be most effective for the bass (lowest frequencies), path Y most effective for the midrange frequencies, and path Z most effective for the treble (highest frequencies). This will distort the waveforms produced by various musical instruments. It is almost impossible to design an acoustical room, such as a concert auditorium, to propagate sound perfectly with respect to frequency for every listener. The best the acoustics engineer can do is optimize the situation.

Figure 8-9
Sound propagation
depends on the
paths the waves
follow.

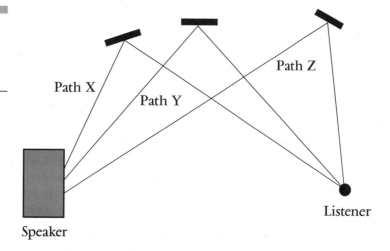

Loudness

People do not perceive the *loudness* or *volume* of sound in direct proportion to the actual energy contained in the waves. Instead, the ears sense sound levels according to the logarithm of the actual intensity. If you change the volume control on a hi-fi set until you can just barely tell the difference, the increment is 1 *decibel* (dB). If you use the volume control to double the actual sound power coming from a set of speakers, you will hear an increase of 3 dB, or three times the minimum detectable difference.

For decibels to have meaning, there must be a reference level against which everything is measured. Sometimes you will hear decibels spoken of as if they are absolute units; for example, a vacuum cleaner might produce 80 dB of sound, and a jet taking off might generate 100 dB as it passes overhead at a low altitude. But these are determined with respect to the *threshold of hearing*, the faintest sound that "good ears" can hear when there is no background noise.

In hi-fi amplifiers, you'll sometimes see meters that show the relative sound level. Such a meter indicates *volume units* (VU), and accordingly, it is called a *VU meter.* These meters are often calibrated in decibels with respect to some reference sound level.

When thinking of decibels as they apply to sound, remember that a doubling of actual acoustic power equals a 3-dB increase in amplitude. A rise in the sound level is indicated by positive decibel values, and a decrease is indicated by negative values. A change of −6 dB means that the sound is cut to one-fourth of its previous level; a change of +9 dB means an eightfold increase. If you hear someone say "zero decibels" (0 dB), he or she means there is no change in the sound level, or that two different sounds have equal amplitudes.

Phase

Another variable in the transmission of sound is the *phase* with which waves arrive. Even if there is only one sound source, acoustic waves reflect from the walls, ceiling, and floor of a room. Some concert halls have flat surfaces called *baffles* strategically placed to reflect the sound from the stage toward the audience.

In the example of Fig. 8-9, the three sound paths X, Y, and Z are almost certain to have different lengths. Therefore, the three reflected waves will usually not arrive in the same phase at the listener's ears, although at certain frequencies and certain listener locations, it is possible that the waves will all arrive in phase. The direct path (not shown), a straight line from the speaker to the listener, is always the shortest path. In this situation, there are at least four different paths via which sound gets from the speaker to the listener.

Suppose that, at a frequency of 700 Hz, the acoustic waves for all four paths happen to arrive in exactly the same phase in the listener's ears. Then sounds at that frequency will be exaggerated in volume. The same phase coincidence might also occur at harmonics of 700 Hz: that is, 1.4 kHz, 2.1 kHz, 2.8 kHz, and so on. But it is not likely to happen at any other frequencies. This is an undesirable effect in hi-fi acoustics applications, because it causes "peaks" at specific frequencies, distorting the original sound. (It can be useful in other applications, however.) At some frequencies, the waves might mix in phase opposition, so that the energy is nearly canceled out. This produces acoustic *dead zones,* a phenomenon that acoustics engineers dread.

If the listener moves a few feet, the phase coincidence will no longer take place at 700 Hz and its harmonics. The frequency might change, or the effect might disappear altogether. The *live zone* for 700 Hz is confined to a single spot in the room.

One of the most important considerations in acoustical design is the avoidance, to the greatest extent possible, of live zones and dead zones. The perfect acoustic chamber delivers sound effectively to all members of the listening audience, without favoring any particular regions of the room or frequencies of the sound.

Acoustic Transducers

An *acoustic transducer* is an electronic component that converts sound waves into some other form of energy, or vice versa. The other form of energy is almost always an electrical signal. Usually, the waveforms of the acoustic energy (sound) and AC electrical energy are identical, or nearly so. Two common acoustic transducers are the *microphone* and the *speaker*. Other such devices include *earphones, headsets, contact microphones, underwater speakers, underwater microphones, ultrasonic emitters*, and *ultrasonic pickups*.

Piezoelectric Transducer

Figure 8-10 is a simplified pictorial diagram of a *piezoelectric transducer*. This device consists of a crystal, such as quartz or ceramic material, sandwiched between two metal plates.

When an acoustic wave strikes one or both of the plates, the metal vibrates. This vibration is transferred to the piezoelectric crystal. The crystal generates weak electric currents when it is subjected to mechanical stress. Therefore, an AC voltage develops between the two metal plates, with a waveform identical to that of the acoustic waves. When the device operates in this mode, it is an *acoustic pickup*.

If an AC signal is applied to the plates, it causes the crystal to vibrate in sync with the current. The result is that the metal plates vibrate also, producing an acoustic disturbance in the air. In this mode, the transducer is an *acoustic emitter*.

Figure 8-10
A piezoelectric transducer.

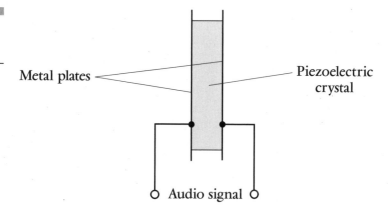

Metal plates — Piezoelectric crystal

Audio signal

A major asset of piezoelectric transducers is that they can operate at high acoustic frequencies, considerably beyond the range of human hearing. These transducers are common in ultrasonic motion detectors and intrusion alarm systems. Miniature piezoelectric transducers are relatively inexpensive, and are commonly found in consumer devices such as travelers' alarm clocks and electronic calculators. They can generate a loud noise for their size.

Electrostatic Transducer

Another type of acoustic transducer that can function as either a pickup or an emitter is the *electrostatic transducer*, a simplified diagram of which is shown in Fig. 8-11.

When an acoustic wave strikes the flexible plate, the plate vibrates. This causes rapid changes in the spacing between the flexible plate and the rigid plate, producing fluctuation in the *capacitance* between the plates. A DC voltage source is connected to the plates. As the capacitance fluctuates, the plates alternately charge and discharge. This causes a weak AC signal to be produced, whose waveform is identical to that of the acoustic disturbance. The *blocking capacitor* allows this AC signal to pass to the transducer output, while keeping the DC confined to the plates.

If an AC audio signal is applied to the terminals of the transducer, it passes through the blocking capacitor and onto the metal plates. The signal creates a fluctuating voltage between the plates. The DC source, combined with the audio signal, results in an electrostatic field of constant polarity, but of rapidly varying intensity. This exerts a fluctuating force between the plates, causing the flexible plate to vibrate and generate an acoustic wave in the air.

Figure 8-11
An electrostatic transducer.

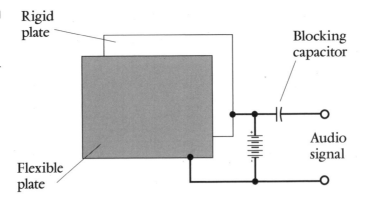

Electrostatic microphones and speakers are used in some hi-fi sound systems. Large electrostatic speakers are known for their excellent sound reproduction, but they have a limited power rating.

Dynamic Transducer

A *dynamic transducer* is a coil-and-magnet device that converts mechanical vibration into electrical currents, or vice versa. The most common examples are the *dynamic microphone* and the *dynamic loudspeaker.* But dynamic transducers can be used as sensors in a variety of applications.

Figure 8-12 shows a simplified pictorial diagram of a dynamic transducer suitable for converting sound waves into electric currents, or vice versa. A diaphragm is attached to a permanent magnet. The magnet is surrounded by a coil of wire. Sound vibrations cause the diaphragm to move back and forth; this moves the magnet, in turn causing fluctuations in the magnetic field within the coil. The result is AC output from the coil, having the same waveshape as the acoustic waves striking the diaphragm.

If an audio signal is applied to the coil of wire, it creates a magnetic field that produces forces on the permanent magnet. This causes the

Figure 8-12
A dynamic transducer.

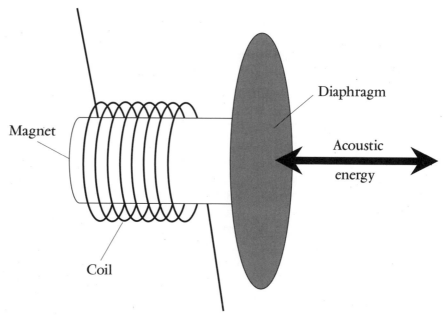

magnet to move, pushing the diaphragm back and forth, thereby creating acoustic waves in the air that exactly follow the waveform of the signal.

Many, but not all, dynamic transducers can work efficiently in both the emitter mode and the pickup mode. In some applications, such as ultrasonic intrusion detectors and *sonar,* specialized dynamic transducers are used in this two-way fashion. However, a typical hi-fi loudspeaker cannot be interchanged with a typical microphone. *Dynamic pickups* usually have small, thin diaphragms that move with the slightest disturbance; *dynamic emitters* have larger, thicker diaphragms that can displace a greater number of molecules in the transmitting medium.

Acoustic Communications

Acoustic waves can be used for wireless communication over short distances through air or underwater. This section describes the operation of a practical ultrasonic communications system that can convey computer data at moderate speeds. This system, like the modulated-light system described earlier in this chapter, can be built by a motivated hobbyist. *Please note that this information is not intended as a design guide. You must experiment to find the best components and adjustments if you want to build this type of system and make it work.*

A Simple Transmitter

The ultrasonic transmitting station consists of a computer, a modem, an oscillator, an amplitude modulator, an amplifier, and an ultrasonic transducer. A hi-fi stereo *tweeter* makes an excellent transducer for this purpose.

The oscillator should be designed to produce a near-perfect sine wave at about 30 kHz. This is a high enough frequency so you can be sure people will not be irritated by it (although animals and insects might react), but it is low enough to be within the operating range of most hi-fi amplifiers and tweeters. The amplitude modulator can be a simple passive circuit, or it can employ a bipolar or FET to provide some gain. The output of the computer modem is fed to the modulator through a capacitor or transformer to ensure that the modem is not damaged by the DC bias in the modulator. The amplifier is a high-quality, reasonably high-power stereo unit; only one channel is used. The output of the

Figure 8-13
At A, an ultrasonic
data-communications
transmitter. At B, an
ultrasonic data-com-
munications receiver.

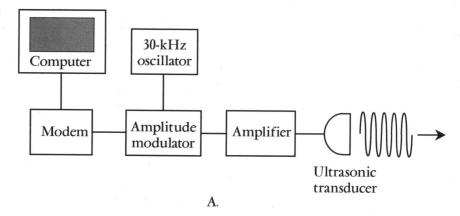

A.

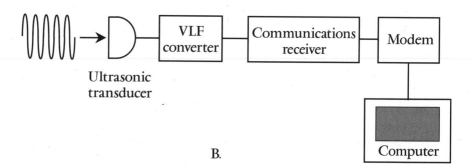

B.

amplitude modulator is fed to the tape or phono input of the hi-fi amplifier. The tweeter is connected to the amplifier's speaker output terminals. The amplifier should be set for maximum treble (high-frequency) gain and minimum bass (low-frequency) gain. A block diagram of the transmitter is shown in Fig. 8-13A.

The output of the computer modem consists of two audio tones at frequencies below 3 kHz, alternating rapidly back and forth. One tone represents the high logic state (digital 1) and the other tone represents the low logic state (digital 0). When these tones are combined in the modulator with the 30-kHz oscillator signal, *sidebands* result. These sidebands appear at frequencies between 27 and 33 kHz (up to plus or minus 3 kHz from the carrier frequency). This signal is amplified to approximately 25 W by the hi-fi amplifier. At this point, the signal would become a 30-kHz EM wave 10 kilometers (km; 6 mi) long if fed to an antenna. But instead, it is fed to the tweeter, where it is converted into acoustic waves measuring about 11 mm ($^7/_{16}$ in).

A Simple Receiver

The receiving system is diagramed in Fig. 8-13B. It consists of a transducer, a VLF converter of the type used for radio reception, a general-coverage shortwave or amateur-radio communications receiver, a modem, and a computer. A hi-fi tweeter can be used as the transducer in this system, if a more sophisticated and sensitive ultrasonic pickup cannot be found.

The VLF converter can be constructed from designs in amateur-radio or shortwave-listening periodicals and books. Alternatively, it can be purchased ready-made. The communications receiver should be capable of receiving AM signals, and should be set to operate in that mode. The ultrasonic transducer is connected to the antenna terminals of the VLF converter. The output of the converter is fed to the receiver antenna input. The modem is connected to the headphone or speaker output of the receiver. A transformer or series-connected capacitor should be used to ensure that the modem is not damaged by the audio output and/or DC voltages from the receiver. A potentiometer might be connected in series with the receiver output and modem input, to be sure that the modem does not receive an excessively strong audio signal from the receiver.

To receive the ultrasonic signal from the transmitter, the communications receiver should be tuned to a frequency either 30 kHz above or 30 kHz below the local-oscillator output of the VLF converter. That is, the receiver should be tuned in the same way as it would be if you wanted to listen to a 30-kHz VLF radio signal. When the receiver is properly adjusted, the receiving computer modem should "hear" exactly the same analog signal that it would "hear" if it were used on the telephone line. But, of course, a wireless link is employed instead of the telephone wires.

In Practice

The above-described system can provide an experimenter with plenty of fun. But there are some problems with acoustic communications of any kind. It won't take long before the experimenter realizes this.

Of course, the biggest limitation on acoustic communications is imposed by the fact that the waves travel at only 1100 feet (335 meters) per second in air. If two stations are 1 mi apart, the delay each way will be about 5 seconds. It's not likely, however, that the system described here would function over a distance that great. Background noises exist even at frequencies above the range of human hearing, and they will produce interference of various fascinating sorts. In addition, the atmosphere is not a stable medium; it is filled with vortices, wind shear, and obstructions. The result is a severe *flutter*, or rapid variation, in the strength of the received signal. In general, this effect gets worse and worse as the acoustic frequency increases, because the atmospheric disturbances become larger and larger with respect to the wavelength of the signal. The flutter also becomes more pronounced as the distance between the transmitter and receiver increases.

Underwater acoustic communication will work over a longer range than atmospheric acoustic communication. The frequency of an underwater system can be lower than that of an atmospheric system (perhaps 10 kHz), because underwater sounds do not efficiently penetrate above the surface. Acoustic waves travel much faster in water than in air, and the waves are of a nature that allows propagation for longer distances. The hi-fi tweeters in the systems of Fig. 8-13 can be directly replaced with underwater ultrasonic transducers for experimentation.

Fiberoptic Cable

A *fiberoptic cable* is a bundle of glass strands, designed to carry modulated light or IR. A ray of light or IR can, in theory, carry millions of independent signals along one thread-thin strand of clear solid. Such a strand is called a *fiber* or an *optical fiber.*

Manufacture

Optical fibers are made from glass, to which impurities have been added. The impurities do not cloud the glass, but they change its *index of refraction* (also called the *refractive index*). The refractive index is the property of glass that makes lenses magnify. The greater the refractive index of a clear solid, the more it will bend light that enters it at an angle.

An optical fiber has a core surrounded by a tubular *cladding*, which has a lower index of refraction than the core. There are two ways in which the refractive index can vary in an optical fiber. In the *step-index optical fiber*, the core has a uniform index of refraction and the cladding has a lower index, also uniform; the transition occurs at a well-defined, cylindrical boundary concentric with the outside of the fiber. In the *graded-index optical fiber*, the core has a refractive index that is greatest along the central axis and steadily decreases outward from the center. At the boundary between the core and the cladding, there is an abrupt drop in the refractive index.

Optical fibers are specially manufactured to have the lowest possible optical loss. That is, they are much clearer than ordinary window glass. As a result, visible light and IR rays can propagate within a fiber for miles before the energy must be intercepted, the signals amplified, and a new beam generated and sent along another length of fiber.

Modes of Operation

The behavior of the ray within an optical-fiber core varies somewhat, depending on whether the fiber is of the step-index type or the graded-index type. In Fig. 8-14A, showing a step-index fiber, ray *X* enters the core exactly parallel to the fiber, so it travels without striking the boundary between the core and the cladding unless there is a bend in the fiber. If there is a bend, ray *X* will veer off center with respect to the fiber and will start to behave like ray *Y*. Ray *Y* enters at an angle and hits the boundary repeatedly. Each time ray *Y* encounters the boundary, *total internal reflection* occurs, so the ray always stays in the core. The reflection takes place because of the difference in refractive index between the core and the cladding. In Fig. 8-14B, showing a graded-index fiber, ray *X* enters the core exactly parallel to the fiber, and it travels without striking the boundary between the core and the cladding. If there is a bend in the fiber, ray *X* will veer off center with respect to the fiber and will start to behave like ray *Y*. Ray *Y* enters at an angle. As ray *Y* moves farther from the center of the core, the index of refraction decreases, causing refraction (bending) of the ray back toward the center of the fiber. If ray *Y* enters at a sharp enough angle, it will reach the boundary between the core and the cladding, in which case total internal reflection will occur, as it does in a step-index fiber.

Some fiberoptic cables have only one fiber; other cables have several fibers. Optical fibers are bundled into cable in much the same way as wires are bundled. The individual fibers are protected from damage by plastic jackets. Common coverings are polyethylene and polyurethane. Along with the fibers, steel wires or other strong materials can be used to add strength to the cable. The whole bundle is encased in an outer jacket. This outer covering can be reinforced with wire or tough plastic compounds.

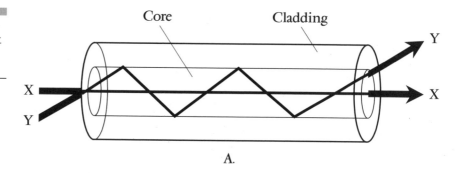

A.

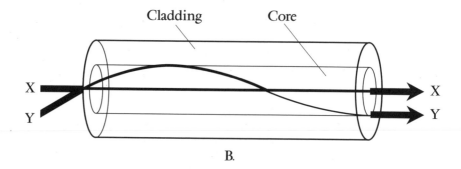

B.

Fiberoptic Data Transmission

Fiberoptic systems are used in electronic and electromechanical devices such as robots and computer-controlled cars. Fiberoptics is gradually replacing conventional cable for data communications. The scheme is ideal for networking, or interconnecting computers. Fiberoptic technology lends itself to long-distance multimedia communications. Ultimately, fiberoptic cables might completely replace older cable systems in applications ranging from telephone systems to television networks to the Internet. At the subscriber level (that is, among individual businesses and households), fiberoptic installations might someday be as common as telephone lines are today.

Assets

Fiberoptic systems have numerous advantages over wire, cable, and radio systems. Some of these assets are as follows.

Immunity to Interference   Besides allowing signals to be sent at high speeds, optical fibers do not suffer from *electromagnetic interference.*

A strong radio signal, or a lightning discharge, will not disrupt the data, as can happen with cable systems.

Zero Electromagnetic Radiation   All the signal energy stays within the optical fiber. There are no *extremely-low-frequency* fields or other EM energy that can cause interference with other systems.

Hard to Tap   You can wrap a coil of wire around a cable and intercept the data without cutting into the line. This requires some expertise in communications engineering, but it is possible. When a signal is carried by a light beam inside an optical fiber, it is impossible to tap without breaking the fiber.

Abundant Materials   The sand from which glass fibers are made is cheap and plentiful. The supply will never run short. Sand can also be "mined" with minimal environmental impact.

Durability   Fiberoptic cables can be submerged in lakes and oceans, or buried in soil, and will not corrode the way metal wires do. Therefore, fiberoptic cables last much longer than wire cables and need less maintenance.

How It Works

At the transmitting station (*source*), analog or digital data is impressed on a beam of visible light or IR by means of a modulator. Lenses feed this energy into the fiber. At the receiving station (*destination*), a lens focuses the beam onto a photosensitive device such as a photodiode. This converts the optical or IR energy into electric currents that are amplified, so the signals regain their original form. Along the fiberoptic route, *repeaters* periodically intercept weakened optical or IR energy, demodulate it, amplify the signals, and retransmit another beam at a similar wavelength.

A single beam of light or IR can carry a huge number of signals. The frequency of the rays is many times higher than that of any data signal, and the available bandwidth is practically infinite. One thin strand of glass can therefore replace a bundle of wires of much greater diameter and weight. A fiberoptic cable can contain dozens of individual fibers. This increases the versatility of the system still further.

Optical Computer Technology

One form of alternative computer technology makes use of laser beams, rather than electric currents, to perform digital computations. This is called *optical computer technology.*

Light versus Electricity

An electric current flows because *charge carriers* pass from atom to atom in a conductor or semiconductor. In a wire, these charge carriers are *electrons.* In some materials, they are electron shortages called *holes.* An electric current consisting of electrons flows about 18,000 miles per second in a wire, approximately 10 percent of the speed of light. A current consisting of holes moves somewhat more slowly than that. But light beams travel at 186,282 miles per second in free space, and only a little more slowly in optical fibers. This means that, all other things being equal, an *optical computer* might perform operations up to 10 times faster than an electronic one.

Light and IR beams have another advantage over electric currents: the ability to pass through each other without interacting. Two or three or a million laser beams can be shone so they intersect, but there is no interference among them. Electric currents do not work that way. Because of this, optical computers can perform millions of operations in parallel, something unheard of with electronic devices. The basic principle is shown in Fig. 8-15. There are eight inputs and eight outputs (represented by binary numbers 000 through 111) in this example. But there could be 16, 32, 64, or any power of 2, defining binary numbers having any number of bits. The only limit would be imposed by the number of lasers and sensors that could be packed into a given amount of physical space.

Highs and Lows

In an optical computer, logic 1 (or *high*) is usually represented by the presence of a light beam or pulse, while logic 0 (or *low*) is the absence of a light beam or pulse. However, there are other ways to encode logic into light. The color can be changed; for example, red=1 and green=0. Or the polarization might be altered; vertical=1 and horizontal=0. It is possible to modulate laser beams via a modem, putting electronic data into beams of light. Many such signals can be carried by one laser.

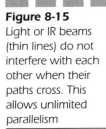

Figure 8-15

Light or IR beams (thin lines) do not interfere with each other when their paths cross. This allows unlimited parallelism

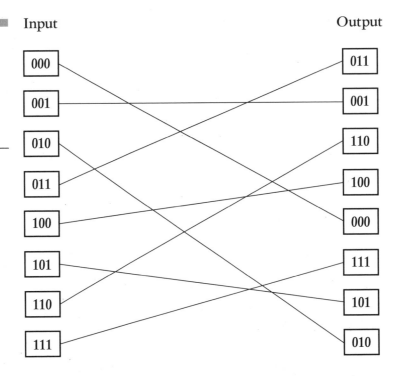

Perhaps, in time, optical computers will render electronic computers obsolete. The AS-400 computer by International Business Machines (IBM) was one of the first midrange systems to incorporate fiberoptic cables in its bus architecture. This resulted in extremely high data-transfer rates between the main system and peripherals such as the disk drives. But the evolution of computer technology to optical- and IR-based hardware will probably be gradual, and will most likely be partial, not total. Some optical integrated circuits have already been put into use. One such device takes a video picture and encodes it directly into light beams that are sent along optical fibers. Instead of silicon, the semiconductor material most commonly used to make conventional integrated circuits or "chips," optical devices have been fabricated from lithium niobate (a compound of lithium and niobium). This allows control of optical pulses via electric currents.

Optical technology can offer breakthroughs in some fields, but it is not likely that optical technology will completely replace older "hard-wired" and radio-spectrum systems. There are many things that electric currents and radio waves can do perfectly well, such as convey data to and from orbiting satellites, facilitate the use of cordless and cellular telephone systems, and provide power to common household appliances.

Digital
Technology

A signal or quantity is *digital* when it can attain only a finite number of levels or values. This is in contrast to *analog* signals or quantities that vary over a continuous range of levels or values.

Analog Quantities

An analog quantity is continuously variable; it can attain any possible instantaneous value within a specified range of values. This is what makes analog quantities different from digital quantities. Digital quantities exist at defined levels or states, for example, 0 and 1 (low and high) or 0 and +5 direct-current volts.

Signals

The sound of a voice, a musical note, a jet engine, or anything else has a certain waveform. You can look at the waveform of any sound by connecting a microphone to an amplifier, and then connecting the output of the amplifier to an oscilloscope. If you whistle into the microphone, a simple analog wave will appear on the screen (Fig. 9-1A).

The level, or *amplitude*, of the wave shown in Fig. 9-1A varies continuously from instant to instant. Engineers say that the *instantaneous amplitude* changes in an analog manner. Some sounds, such as whistling, produce simple analog waveforms. But for other sounds, the shape of the waveform is extremely complicated. The noise of a jet taking off, if

Figure 9-1

An analog sine wave (A), and a possible digital rendition of this wave (B).

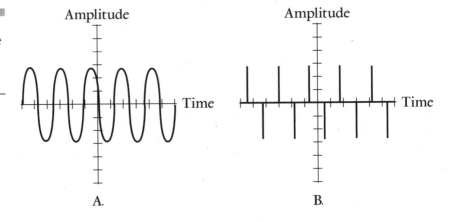

A.

B.

picked up by the microphone and fed into the oscilloscope, would create a jumble so complex that its simple components would be impossible to identify.

You might pass an analog signal through a device called an *analog-to-digital* (A/D) *converter.* Then the signal would have only a few specific levels or values, as shown in Fig. 9-1B. This is a digital approximation of the analog waveform at A. It is a very rough approximation; more sophisticated converters can create digital signals closer to the true shape of the analog wave. If you want to recover the analog signal from a digital one, you must pass the digital impulses through a *digital-to-analog* (D/A) *converter.* If the wave at B were passed through a D/A converter, the wave at A, or something similar to it, would result. A compact disk (CD) player, for example, uses D/A conversion to get music from the huge array of digital bits stored on the optical disk.

Analog Limitations

We humans live and think in an analog world; continuously varying phenomena are easy for our senses and brains to deal with. But these same analog variables, if input directly to a digital computer, will totally confuse it.

To get an idea of how hard it is for a computer to "imagine" an analog signal, suppose you read this sentence into the microphone/oscilloscope speech viewer described above. You watch the jumble of waves on the screen. Now, write down the mathematical function that represents the waveforms you see. An approximation is not good enough; it must be exact. This is practically impossible; the world's greatest mathematicians would have a terrible time with a problem like that. A digital computer would reject it as unworkable; the machine would demand some A/D conversion scheme or program to make sense out of it.

If you were to feed the voice signal through an A/D converter, a computer would store the sentence as a sequence of ones and zeros. The resulting array of data bits would be an exact rendition of the voice signal for all practical purposes. This is how computers store and transfer digital voice information. It also applies to images, music, and any other analog quantity. In wireless communications, this is important, because digital data can be transmitted in ways that make noise and interference less of a problem, thereby resulting in higher data speeds and lower error rates than would be possible if analog methods were used.

Numbering Systems

Digital electronic systems employ various methods of numbering. The main difference among these schemes is the number of digits used to represent numerical quantities. People are used to dealing with the *decimal number system,* which has 10 unique digits, but machines use schemes that have some power of 2 unique digits—most often 2 (2^1), 8 (2^3), or 16 (2^4).

Decimal

The *decimal number system* is also called *modulo 10, base 10,* or *radix 10.* All these names are mathematically imprecise; a more absolute way of stating this is that in the decimal system, the number of digits is the same as the number of fingers on a typical human being. (This is probably the reason most societies evolved to prefer the digital system.)

How It Works The digits in the decimal system are representable by the set {0, 1, 2, 3, 4, 5, 6, 7, 8, 9}. Digits are written in a certain sequence to get a unique number. Depending on the position of a digit in the sequence, its value is multiplied by some power of 10. The digit just to the left of the decimal point is multiplied by 100, or 1. The next digit to the left is multiplied by 10^1, or 10. The power of 10 increases as you move farther to the left. The first digit to the right of the decimal point is multiplied by a factor of 10^{-1}, or 1/10. The next digit to the right is multiplied by 10^{-2}, or 1/100. This continues as you go farther and farther to the right. Once the process of multiplying each digit is completed, the resulting values are added. This is what is represented when you write a decimal number. For example,

$$2704.53816 = 2\times10^3 + 7\times10^2 + 0\times10^1 + 4\times10^0 + 5\times10^{-1}$$
$$+ 3\times10^{-2} + 8\times10^{-3} + 1\times10^{-4} + 6\times10^{-5}$$

Computer Numbers For a computer, the decimal number system is an awkward and inefficient way to make use of the electronic circuits. It is easier for a computer to work with powers of 2 than with powers of 10. Even though a computer does not work directly with decimal numbers, it converts the decimal numbers the operator gives it into its own system (usually binary) and also converts its own numbers into decimal form before they appear on the display. As far as a human operator can tell, the computer understands modulo 10, although technically the operator is communicating with the machine through a two-way numerical translator.

Binary

The *binary number system* is a method of expressing numbers using only the digits 0 and 1. It is sometimes called *base 2, radix 2,* or *modulo 2.* Normally, people write numbers in decimal, or base 10, form. But digital communications equipment, and also digital computers, work most efficiently with binary numbers.

Digits   In binary notation, the digit immediately to the left of the radix point is the "ones" digit, just as in base 10. But the next digit to the left is a "twos" digit; after that comes the "fours" digit. Moving farther to the left, the digits represent 8, 16, 32, 64, 128, 256, 512, 1024, etc., doubling every time. The values increase in integral powers of 2. Any given number (other than 0 and 1) requires more digits to express in binary form than it requires in decimal notation. But for a digital system, the sheer quantity of numerals is not a problem. The important thing is that the machine be able to easily recognize, and act upon, each digit. In the binary number system, the digits are always either 0 or 1. Electronically, these states can be represented as "off" and "on," or "low" and "high," or "space" and "mark." This dichotomy is the simplest scheme for expressing numerical quantities, and provides the best possible accuracy in many communications situations.

Decimal versus Binary   Consider an example using the decimal number 94. In the decimal system, this quantity is broken down as

$$94=(4\times10^0)+(9\times10^1)$$

In the binary number system the breakdown is

$$1011110=(0\times2^0)+(1\times2^1)+(1\times2^2)+(1\times2^3)+(1\times2^4)+(0\times2^5)+(1\times2^6)$$

which is another way of saying that

$$94=2+4+8+16+64$$

One-to-One   Every decimal number has one and only one binary representation. Every binary number, likewise, has one and only one decimal equivalent. Mathematicians would say there exists a *one-to-one correspondence* between numbers in decimal and binary form. When you work with a computer or calculator, you give it a decimal number that is converted into binary form. The computer or calculator does its operations with zeros and ones. When the process is complete, the machine converts the result back into decimal form for display. In a communications system, various binary numbers represent alphanumeric characters, shades of color, frequencies of sound, and other variable quantities. In this way, a binary digital communications system can convey anything: text, a computer program, a picture, a drawing, a movie, or a sound bite. Someday, perhaps even tastes, smells, and tactile sensations will be conveyed using data encoded into digital zeros and ones.

Octal and Hexadecimal

Another scheme, sometimes used in computer programming, is the *octal number system,* so named because it has eight symbols (according to our way of thinking), or 2^3. Every digit is an element of the set {0, 1, 2, 3, 4, 5, 6, 7}. Octal digits are sometimes used by software engineers to represent three-digit groups of binary numbers. Octal numbers can also be used like decimal numbers.

Octal numbers have some interesting properties. For example, in the octal system, if a number is a power of "10" (that is, 1 with any number of zeros after it), then it can be divided in half repeatedly until the result is the number 1. Octal numbers are also "musical." In 2/4 music time, there are four measures to the octal number 10; in 4/4 time there are two measures to the octal number 10.

Yet another numbering scheme, also used in computer work, is the *hexadecimal number system,* so named because it has 16 symbols (according to our way of thinking), or 2^4. These digits are the usual 0 through 9 plus six more, represented by A through F, the first six letters of the alphabet. The digit set thus becomes {0, 1, 2, 3, 4, 5, 6, 7, 8, 9, A, B, C, D, E, F}. These hexadecimal digits are used by computer software engineers to represent four-digit groups of binary numbers. Hexadecimal numbers can also be used like decimal numbers. If you ever see something like 1C8BF depicted as a numeral, it probably represents a hexadecimal number.

If you've done much work with computer graphics, you've been exposed to hexadecimal numbers. Color combinations are sometimes specified by six-digit hexadecimal numbers. An example is 005CFF. In the *RGB color model,* the first two digits represent the red (R) intensity in 256 levels ranging from 00 to FF. The middle two digits represent the green (G) intensity, and the last two digits represent the blue (B) intensity. In this example, the color contains no red, a fair amount of green, and a maximum amount of blue.

Logic

Logic refers to methods of reasoning used by people and electronic machines. The term is also sometimes used in reference to the circuits that comprise most digital devices and systems.

Boolean Algebra

Boolean algebra, also called *propositional logic* or *sentential logic,* is a simple system of purely mathematical reasoning using the numbers 0 and 1, the binary operations AND and OR, and the unitary operation NOT. In the boolean system, AND is represented by multiplication, NOT by negation, and OR by addition. Combinations of the operations lead to other binary operations such as NAND (NOT AND) and NOR (NOT OR). Boolean functions are used in the design of digital logic circuits.

Notation and Values In boolean algebra, X AND Y is written as XY or X·Y. NOT X is written with a line or tilde over the quantity involved, or as a minus sign followed by the quantity involved, such as $-X$ or $-(XYZ)$. X OR Y is generally written as $X+Y$. Table 9-1A shows the values of these functions, where 0 indicates "falsity" and 1 indicates "truth." Some rules of conventional algebra also apply in boolean algebra. Logic equations often resemble their arithmetic counterparts. The statements on either side of the equal sign are always logically equivalent. When you say "X=Y" it means "If X, then Y, and if Y, then X." Mathematicians would say "X if and only if Y," writing "X iff Y."

Theorems Table 9-1B shows several logic equations. These are facts, or *theorems,* in boolean algebra. They have been proved true in the past, and logicians recognize them as such. Once you have proved a theorem true in a mathematical or logical system, you never have to repeat that proof. You can use the theorem to help you prove other theorems in the future. Boolean theorems can be used to analyze complicated logic

TABLE 9-1A

Boolean
Operations

X	Y	−X	X·Y	X+Y
0	0	1	0	0
0	1	1	0	1
1	0	0	0	1
1	1	0	1	1

TABLE 9-1B

Common
Theorems in
Boolean Algebra

$X+0=X$	OR identity
$X \cdot 1 = X$	AND identity
$X+1=1$	
$X \cdot 0 = 0$	
$X+X=X$	
$X \cdot X = X$	
$-(-X)=X$	Double negation
$X+(-X)=X$	
$X \cdot (-X)=0$	
$X+Y=Y+X$	Commutativity of OR
$X \cdot Y = Y \cdot X$	Commutativity of AND
$X+(X \cdot Y)=X$	
$X \cdot (-Y)+Y=X+Y$	
$X+Y+Z=(X+Y)+Z=X+(Y+Z)$	Associativity of OR
$X \cdot Y \cdot Z=(X \cdot Y) \cdot Z=X \cdot (Y \cdot Z)$	Associativity of AND
$X \cdot (Y+Z)=(X \cdot Y)+(X \cdot Z)$	Distributivity
$-(X+Y)=(-X) \cdot (-Y)$	DeMorgan's theorem
$-(X \cdot Y)=(-X)+(-Y)$	DeMorgan's theorem

functions. This is an aid to engineers who design digital electronic equipment of all kinds. Working with boolean algebra is often much easier than analyzing schematic diagrams of digital logic circuits.

Predicate Logic

Predicate logic, also called *predicate calculus,* goes a step further than boolean algebra, in that it breaks down individual sentences. The structure of a sentence can drastically affect its behavior in a system of reasoning.

In predicate calculus, a sentence is not represented as a single letter like P. Instead, it is represented as a function, such as Px. You probably learned the word "predicate" when breaking down and diagraming sentences in high school English grammar courses. Predicate calculus is not generally used in electronic systems.

Reductio ad Absurdum

It is possible to prove the truth of some statement, say P, by assuming that its opposite $-$P is true. Then, based on this premise, a contradiction, such as (Q AND $-$Q), is derived. This method of logical proof is called *reductio ad absurdum* (meaning "proof by contradiction" or "derivation of an absurdity") and is familiar to logicians and mathematicians.

A well-known theorem in propositional and predicate logic can be summed up as "From a contradiction, anything follows." This means that, if you accept a direct contradiction (Q AND $-$Q) as being true, then you must accept every statement in the whole logical system as being true. This reduces the whole system to absurdity. In boolean algebra, a contradiction is therefore assigned the value 0 (false).

Theorem-proving computer programs make use of this technique, which is based on the principle of double negation. The denial of a false statement is a true statement. *Reductio ad absurdum* has been used to prove some of the most famous theorems in all of mathematics.

Mathematical Induction

A special form of reasoning known as *mathematical induction* allows a person or machine to prove an infinite number of facts in a finite number of steps. Usually this is done for a series of "countably infinite" objects such as whole numbers.

Suppose you want to prove that something is true for all the numbers *n* in the set *N*={1, 2, 3,...}. You must first show that the fact holds for the number 1. Then you must show that if it is true for any number *n* in the set *N*, it is true for *n*+1. In this way, by means of a sort of logical domino effect, you prove the fact for all the natural numbers 1, 2, 3, and so on, without limit. Imagine knocking over the first domino in an infinitely long row of dominoes, and you can get an idea of the fun logicians have when they prove something by mathematical induction.

Computers have been used in an attempt to "prove" propositions for all the numbers in the set *N*={1, 2, 3,...} by proving validity up to some huge value such as 10^{15} (one quadrillion or 1,000,000,000,000,000). But such exercises are not mathematically rigorous proofs. However large the number a computer might reach, there is always a larger number, for example, $10^{15}+1$.

Inductive Reasoning

All the logic forms mentioned so far are examples of *deductive reasoning*. That means they are 100 percent rigorous methods of proving things. When you speak, or hear, of having proved something mathematically, deductive logic has been used.

Another kind of logic, commonly called *inductive reasoning*, does not prove things beyond any possible doubt. Instead, it derives statements that are true most of the time or that are reasonable based on the data available. This is the sort of logic that lawyers use in the courtroom when they cannot prove things deductively. "Beyond a reasonable doubt" is the phrase that gives away the fact that inductive reasoning has been used. Inductive reasoning is therefore not at all like mathematical induction, even though these terms are similar.

Computers are extremely powerful for use in inductive reasoning. This is because such reasoning is based on statistical probability, which involves taking huge numbers of samples and combining the data into meaningful form. One of the original intentions of computer engineers was to get a machine to do this kind of work, saving countless human-hours of counting and paper shuffling.

Trinary Logic

Trinary logic allows for a neutral condition, neither true nor false, in addition to the usual true/false (high/low) states of *binary logic*. These three values are representable by logic −1 (false), 0 (neutral), and +1 (true).

You might wonder how trinary logic could possibly be of any use. Aren't things always either true or false? In mathematical systems, at least, the answer is no. In 1930, a mathematician by the name of Kurt Godel rigorously proved the *incompleteness theorem*. This basically says that, in any logical system, there are statements whose truth value cannot be determined. That is, certain propositions are undecidable as to truth or falsity. The publication of the incompleteness theorem caused an upheaval in the mathematical world.

An undecidable statement might have truth value in an absolute, cosmic sense; but we humans, and our machines too, are somehow barred by fate from unraveling it. The notion of undecidability creates a made-to-order niche for trinary logic, which allows for a "don't know" state. Trinary logic can be easily represented in electronic circuits by positive, zero, and negative currents or voltages.

Fuzzy Logic

A quite different concept in logic is *fuzzy logic.* This might also be called "analog logic." It hearkens back to the days in which *analog computer technology* was more widely used than it is today.

In fuzzy logic, values cover a continuous range from "totally false," through neutral, to "totally true." Fuzzy logic reflects the inexact nature of the real world. It allows for, and in fact practically insists upon, some error, which is inevitable in experimental science. Fuzzy logic is well suited for the control of certain processes. Its use will probably become more widespread as the relationship between computers and robots matures. Fuzzy logic can be represented to some extent by digital circuits, but only in discrete steps. For a true range of values, analog circuits must be used.

Logic Gates

All digital electronic devices employ thousands, millions, or billions of switches that perform certain functions. These switches, called *logic gates,* have anywhere from one to several inputs, and one output.

Positive and Negative Logic

Usually, the binary digit 1 stands for "true" and is represented by about five volts positive (+5 V). The binary digit 0 stands for "false" and is represented by about zero volts (0 V). This is called *positive logic.* There are other logic forms, the most common of which is *negative logic* (in which the digit 1 is represented by a more negative voltage than the digit 0). The remainder of this discussion deals with positive logic.

Three Simple Gates

The simplest logic gates are the *NOT gate* or *inverter,* the *OR gate,* and the *AND gate.*

An inverter has one input and one output. It simply reverses the state of the input. Thus, if the input is 1, the output is 0; if the input is 0, the output is 1.

An OR gate can have two or more inputs (although it usually has just two). If both, or all, of the inputs are 0, then the output is 0. If any of the inputs are 1, then the output is 1. It only takes one "true" input to make the output of the OR gate "true."

An AND gate can have two or more inputs (although it usually has just two). If both, or all, of the inputs are 1, then the output is 1. Otherwise the output is 0. For the output of an AND gate to be "true," all of the inputs must be "true."

Other Gates

Sometimes an inverter and an OR gate are combined. This produces a *NOR gate.* If an inverter and an AND gate are combined, the result is a *NAND gate.*

An *exclusive OR gate,* also called an *XOR gate,* has two inputs and one output. If the two inputs are the same (either both 1 or both 0), then the output is 0. If the two inputs are different, then the output is 1.

The functions of the logic gates mentioned here are summarized in Table 9-2. Their schematic symbols are shown in Fig. 9-2.

Digital Circuits

In contrast to analog circuits such as the oscillators and amplifiers described in previous chapters, digital circuits are designed to deal with signals that attain discrete, well-defined levels. Many digital circuits are nothing more than sophisticated electronic switches.

Gate type	No. of inputs	Remarks
TABLE 9-2		
NOT	1	Changes state of input
OR	2 or more	Output high if any inputs are high
		Output low if all inputs are low
AND	2 or more	Output low if any inputs are low
		Output high if all inputs are high
NOR	2 or more	Output low if any inputs are high
		Output high if all inputs are low
NAND	2 or more	Output high if any inputs are low
		Output low if all inputs are high
XOR	2	Output high if inputs differ
		Output low if inputs are the same

Logic Gates and Their Characteristics

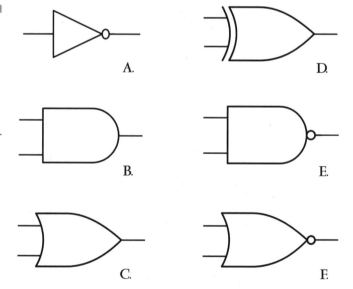

Figure 9-2
Logic gate symbols.
An inverter or NOT
gate (A), an AND
gate (B), an OR gate
(C), an XOR gate (D),
a NAND gate (E),
and a NOR gate (F).

A.

D.

B.

E.

C.

F.

Binary Data

Binary (two-level) signals are used in many kinds of communications systems. Theoretically, it is possible to convert any analog signal into a binary digital signal. Binary circuits consist of simple parts, but they can be built up into massive systems and networks. The *Internet* is the classic example of this, being the largest binary system in the world. Binary data is less susceptible to the effects of noise and other interference than analog data, or multilevel digital data.

Morse Code The *Morse code* is the oldest binary means of sending and receiving messages. It is a binary code because it has just two possible states: on (key-down) and off (key-up). There are two different Morse codes in use by English-speaking operators today. The more commonly used code is called the *International Morse Code* or the *Continental Code*. A few telegraph operators use the older *American Morse code*. The code characters vary somewhat in other languages. Communications speeds are many times slower than the speeds of digital computers and other machines. Communications devices today can function under weak-signal conditions that would frustrate a human operator. But when human operators are involved, the Morse code has always been the most reliable means of getting a message through severe interference. This is because the bandwidth of a Morse-code radio signal is extremely narrow on account of its slow speed, and it is comparatively easy for the human ear to distinguish between the background noise and a Morse signal.

Baudot The *Baudot* code is a five-unit digital code for the transmission of teleprinter data. Letters, numerals, symbols, and a few control operations are represented. Baudot was one of the first codes used with mechanical printing devices. Sometimes this code is called the *Murray code*. It is not widely used by today's digital equipment, except in amateur radio communications. There are 2^5, or 32, possible combinations of binary pulses in the Baudot code, but this number is doubled by the control operations LTRS (lowercase) and FIGS (uppercase). In the Baudot code, only capital letters are sent. Uppercase characters consist mostly of symbols, numerals, and punctuation.

ASCII The *American National Standard Code for Information Interchange*, more commonly called by its acronym ASCII, is a seven-unit digital code for the transmission of teleprinter data. Letters, numerals, symbols, and control operations are represented. ASCII is designed primarily for computer applications. Each unit is either 0 or 1. There are 2^7, or 128, possible representations. Both upper- and lowercase letters can be represented. In addition, numerous symbols can be sent and received.

Flip-Flops

A *flip-flop* is a form of *sequential logic gate*. In a sequential gate, the output state depends on both the inputs and the outputs. The term "sequential" comes from the fact that the output depends not only on the states of the circuit at the moment, but also on the states immediately preceding. A flip-flop has two states, called *set* and *reset*. Usually, the set state is logic 1 (high), and the reset state is logic 0 (low). There are several different kinds of flip-flop. In schematic diagrams, a flip-flop is usually shown as a rectangle with two or more inputs and two outputs. If the rectangle symbol is used, the letters FF are printed or written at the top of the rectangle, either inside or outside.

R-S The inputs of an *R-S flip-flop* are labeled R (reset) and S (set). The outputs are Q and −Q (Often, rather than −Q, you will see Q″, or perhaps Q with a line over it.) As their symbols imply, the two outputs are always in logically opposite states. The symbol for an R-S flip-flop is shown in Fig. 9-3A. In an R-S flip-flop, if R=0 and S=0, the output states do not change. If R=0 and S=1, then Q=1 and −Q=0. If R=1 and S=0, then Q=0 and −Q=1. That is, the Q and −Q outputs will attain these values, no matter what states they were at before. When S=1 and R=1, an R-S flip-flop becomes unpredictable. Because of this, engineers avoid letting both inputs of an R-S flip-flop get into the logical high state. Table 9-3A is a truth table for an R-S flip-flop.

Synchronous An R-S flip-flop output changes state as soon as the inputs change. If the inputs change at irregular intervals, so will the outputs. For this reason, the aforementioned circuit is sometimes called an *asynchronous flip-flop*. In contrast, a *synchronous flip-flop* changes state only at certain times. The change-of-state times are determined by a circuit called a *clock*, which puts out a continuous series

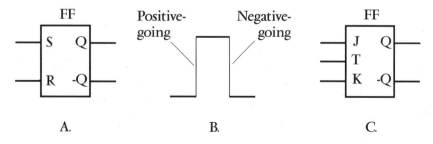

Figure 9-3
Symbol for an R-S flip-flop (A); pulse edges are either negative-going or positive-going (B); symbol for a J-K flip-flop (C).

TABLE 9-3

Flip-Flop States

A: R-S			
R	S	Q	−Q
0	0	Q	−Q
0	1	1	0
1	0	0	1
1	1	?	?

B: J-K			
J	K	Q	−Q
0	0	Q	−Q
0	1	1	0
1	0	0	1
1	1	−Q	Q

of pulses at regular intervals. There are several ways in which a synchronous flip-flop can be triggered, or made to change state. In *static triggering,* the outputs can change state only when the clock signal is either high or low. This type of circuit is sometimes called a *gated flip-flop.* In *positive-edge triggering,* the outputs change state at the instant the clock signal goes from low to high, that is, while the clock pulse is positive-going. The term *edge triggering* derives from the fact that the abrupt rise or fall of a pulse looks like the edge of a cliff (Fig. 9-3B). In *negative-edge triggering,* the outputs change state at the instant the clock signal goes from high to low, or when the pulse is negative-going.

Master/Slave When two flip-flops are triggered from the same clock, or if one of the outputs is used as an input, the input and output signals can sometimes get confused. A *master/slave (M/S) flip-flop* overcomes this problem by storing inputs before allowing the outputs to change state. An M/S flip-flop consists of two R-S flip-flops in series. The first flip-flop is called the *master,* and the second is called the *slave.* The master flip-flop functions when the clock output is high, and the slave acts during the next low portion of the clock output. This time delay prevents confusion between the input and output of the circuit.

J-K   A *J-K flip-flop* works just like an R-S circuit, except that it has a predictable output when the inputs are both 1. Table 9-3B shows the input
and output states for this type of flip-flop. The output changes only
when a triggering pulse is received. The symbol for a J-K flip-flop is
shown in Fig. 9-3C. It looks like the R-S flip-flop symbol with the addition of a third (trigger) input, labeled T.

Other Types   A *D flip-flop* operates in a delayed manner, from the
pulse immediately preceding the current pulse. An *R-S-T flip-flop* has
three inputs called R, S, and T. It operates exactly as the R-S flip-flop
works, except that a high pulse at the T input causes the circuit to
change state. A *T flip-flop* has only one input. Each time a high pulse
appears at the T input, the output state is reversed.

Clocks

In electronics, the term *clock* refers to a circuit that generates pulses at
high speed and at precise intervals. A clock is a specialized form of oscillator that sets the tempo for the operation of certain kinds of digital
devices.

In a computer, the clock acts like a metronome for the *microprocessor.*
Clock pulses occur many times every second. The clock speed is measured in megahertz (MHz) or gigahertz (GHz). If you know the clock
frequency of a computer, you might actually tune a radio receiver to
that frequency, bring it near the computer's main unit, and listen to the
signal from the clock.

In general, when all other factors are held constant, a higher clock
speed means a faster computer. But clock speed is not the only thing
that determines how many calculations a computer can do in a given
amount of time. The microprocessor and bus architectures make a big
difference. Another factor is the nature of the computer's *instruction set.*
As computer technology advances, microprocessors become able to do
more with a single clock pulse.

Counters

A digital circuit that keeps track of the number of cycles or pulses entering it is called a *counter.* A counter consists of a set of flip-flops or equivalent circuits. Each time a pulse is received, the binary number stored by

the counter increases by 1. Counters can be used to keep track of the number of times a certain event takes place. Some counters measure the number of pulses within a specific interval of time, for the purpose of accurately determining the frequency of a signal.

A *frequency counter* is a device that measures the frequency of a periodic wave by actually counting the pulses in a given interval of time. The usual frequency-counter circuit consists of a *gate*, which begins and ends each counting cycle at defined intervals. The accuracy of the frequency measurement is a direct function of the length of the gate time; the longer the time base, the better the accuracy. At higher frequencies, the accuracy is limited by the number of digits in the display. An accurate reference frequency is necessary in order to have a precision counter; quartz-crystal oscillators are used for this purpose. The reference oscillator frequency can be synchronized with a time standard by means of a *phase-locked loop* (PLL). Frequency counters are available that allow accurate measurements of signal frequencies into the gigahertz range. Frequency counters are invaluable in the test laboratory, since they are easy to use and read. Typical frequency counters have readouts that show 6 to 10 significant digits.

Digital Metering

Many metering devices, from car speedometers to radio-receiver tuning dials, are designed to provide direct numerical readouts. This scheme is called *digital metering*.

Assets

The main advantage of a digital meter is that it is easy to read, and there is no possibility of making interpolation errors. This is ideal for utility meters, clocks, and some ammeters, voltmeters, and wattmeters. It works well when the value of the measured quantity does not change frequently. A digital meter with a large numerical display can be read at a single glance, from some distance away, without the need for visual squinting and mental calculation.

In contrast to a digital meter, an *analog metering* device has a scale and an indicator, such as a pointer, that moves along a defined continuum. Such a meter allows interpolation, gives the operator a sense of the measured quantity relative to other possible values, and follows along as a parameter fluctuates. But the limitations of analog meters often out-weigh their assets. An analog meter usually has moving parts that can break if the device is subjected to physical shock. The indication can be hard for some people to read. These problems, along with a general (and sometimes overhyped) trend toward digitization in the public mind, have contributed to the popularity of digital meters.

Limitations

There are some situations in which digital meters will not work as well as analog devices. Digital meters require a certain length of time to lock into the current, voltage, power, frequency, or other quantity being measured. If this quantity never settles at any one value for a long enough time, the meter can never lock in. An example is the signal-strength indicator, called the *S meter* (as opposed to the frequency indicator) in a radio communications receiver. An S meter bounces up and down as signals fade, as the radio is tuned, or sometimes as the signal is modulated. A digital meter in this application would usually display a randomly fluctuating, meaningless set of numerals.

Another potential problem with digital meters is that the reader (operator) might mentally misplace the decimal point. If you are off by one decimal place, the error will be an order of magnitude (a factor of 10); if you are off by two decimal places, the error will be two orders of magnitude (a factor of 100). Some digital meters will display and automatically position the decimal point so such errors cannot occur. With other digital meters, you must multiply the reading by a certain power of 10 to get the true indicated value. It is important to double-check to be sure of the correct position of the decimal point in a numerical display.

Bar-Graph Meter

A *bar-graph meter* is a special form of digital meter consisting of several lamps, light-emitting diodes (LEDs), or liquid-crystal-display (LCD) elements. The components are arranged in a row, so the meter has an incrementally graduated scale.

Figure 9-4 shows a bar-graph meter (at A) and an analog meter (at B), both connected to a signal source having a strength of 75 units. The bar-graph meter has 10 LCD elements. The first seven segments of the bar graph are completely darkened (black). There are four possible states that

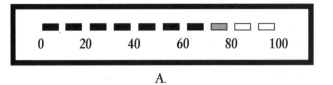

Figure 9-4
An LCD bar-graph
meter (A) and an
analog meter
(B), both indicating
75 units.

A.

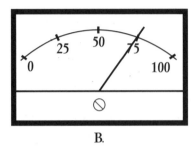

B.

the eighth segment might have, depending on the meter design: it might be partially darkened (gray), it might be flickering rapidly, it might be undarkened (white), or it might be black. The last two segments are white.

It is impossible to precisely interpolate the reading given by the meter in drawing A. The best an observer can do is tell that the reading is between 70 and 80 units. The analog meter allows much finer interpolation; from Fig. 9-4B it is easy to see that the reading is extremely close to 75 units. This demonstrates the main limitation of bar-graph meters: they are not very accurate.

Bar-graph meters have a big advantage over numeric digital meters in some applications. If the measured quantity fluctuates, the variations can be seen to some extent on a bar-graph meter. This gives the observer some sense of how the quantity rises and falls. In communications equipment, it is important that the operator have a sense of how a quantity, such as signal strength, varies.

The Digital Computer

When you hear people talking about computer technology, they're probably discussing *digital computer technology*. There are other ways in which computers can work; some of these are being explored. However, digital technology dominates.

Machine Language

Machine language, used by computers to exchange and store information, has only two possible states. These are represented by the binary digits (bits) 1 and 0. Other expressions include on/off, bright/dark, red/green, yes/no, plus/minus, true/false, or anything that can attain two clearly different conditions. All the data you put into a digital computer, no matter what its real-life form, must be converted into ones and zeros for a digital computer to make sense of it. The computer does this for you, so you can input data in a form you're comfortable with, such as typewritten text, images, or spoken words. You don't have to type expressions like "10011011 11001010 00001101" to input data to a computer.

The digital machine translates its binary language into a form you can understand before presenting you with output. Thus, you won't see a string of ones and zeros on your screen or on your printer paper. You'll see or hear words, music, and images. Or you might witness the output of a computer in mechanical form, such as the movement of a robot.

Strengths

Digital technology has prevailed in the design of computers almost from the beginning of the computer age. People have become accustomed to digital computers because these machines are good for numerical calculations, word processing, and other applications that lend themselves to digital codes.

Letters of the alphabet, numerals, and other symbols can be represented by digital signals. Digital highs and lows stand out clearly from noise on communications lines. Digital data is easy for a machine to tell apart from analog data.

Some kinds of data, such as printed matter, lend themselves better to digitization than other forms, such as music or color photographs. This is vividly illustrated by the fact that a full-page or full-screen image will consume dozens, or even hundreds, of times as much storage space as a full page or screen of text. But if the data speed is high enough, and if the storage/memory capacity is great enough, a computer can digitally encode anything.

Limitations

It is usually possible to make digital approximations almost identical to analog reality. A digital machine can nearly always follow an analog effect so closely that the difference is not worth worrying about. But there are a few exceptions.

There is an absolute limit to the speed at which a digital computer can operate. There's also a limit to how much data can be stored in a given physical space. As people demand faster, more powerful digital computers, these limits will eventually impose constraints on the technology. Impulses cannot travel faster than 186,282 miles per second (the speed of light in free space), nor is it likely that researchers will ever find a way to get a single electron of matter to represent many bits of data.

Some researchers, having anticipated the limitations of digital technology, have been looking at *alternative computer technology* for years. This is most apparent in *artificial intelligence* (AI), especially among those engineers who hope to develop a human-like electronic brain. Not surprisingly, the military is interested in all forms of computer technology, including alternative computers.

Human-likeness

Digital machines lack at least three things that humans have in abundance: emotion, empathy, and intuition. It is hard to imagine a digital computer falling in love, or having a sense of self, or acting on a hunch, although such ideas can make good science fiction.

Some alternative technologies come closer to human behavior than digital systems. But even in a purely logical field such as mathematics, digital technology can sometimes fall short. Pure mathematics, especially to those who practice it, often borders more closely on art than on hard science. Machines can prove some mathematical theorems; the approach is one of brute force, to try millions of avenues of reasoning. But some of the greatest theorems of all time were proved with intuition. The mathematician "smelled" the solution rather than deploying a massive logical dragnet. The imprecise human mind can do some things that a precise digital machine cannot. Alternative computer technology might someday prove useful in communications systems. For example, AI could aid in the development of unbreakable ciphers to maintain privacy or secrecy, or help to fill in gaps in received messages.

Binary Data

Digital data is binary when there are only two possible states, represented by 0 and 1. Binary data is the simplest and most elegant form of digital data. Binary data results in the greatest possible communications efficiency. If multilevel signaling is required, then all the levels can be represented by strings, or groups, of binary digits. The number of levels can be set at 2^n, where n is some positive whole number. Then a string of n binary digits can represent any of the 2^n levels. For example, if eight levels are required, they can be represented by 2^3 strings of three-digit binary numbers: 000, 001, 010, 011, 100, 101, 110, and 111.

Bits and Bytes

The term *bit* comes from the words *binary digit*. A bit is an elementary unit of digital data, represented by either logic 0 or logic 1. In digital computers and communications circuits, a data unit consisting of eight bits is known as a *byte*. In communications, a byte is sometimes called an *octet*. Bytes and octets, and multiples thereof, are commonly employed to express the sizes of data files and the lengths of digital transmissions.

Occasionally you will hear of data units called *kilobits, megabits,* and *gigabits*. One kilobit is equal to 1024 bits. A megabit is 1024 kilobits, or 1,048,576 bits. A gigabit is 1024 megabits, or 1,073,741,824 bits.

One byte is about the same as one character, such as a letter, numeral, punctuation mark, space, line feed command, or carriage return command. Data quantity, especially in reference to computer files, is almost always specified in bytes. But because the memory and storage capacities of modern computers are large, *kilobytes* (units of $2^{10} = 1024$ bytes), *megabytes* (units of $2^{20} = 1,048,576$ bytes), and *gigabytes* (units of $2^{30} = 1,073,741,824$ bytes) are given. The abbreviations for these units are KB, MB, and GB, respectively. Alternatively you might see them abbreviated as K, M, and G.

There are larger data units. The *terabyte* (TB) is equivalent to 2^{40} bytes, or 1024 GB. The *petabyte* (PB) is 2^{50} bytes, or 1024 TB. The *exabyte* (EB) is 2^{60} bytes, or 1024 PB.

Baud, bps, and Bandwidth

In digital communications, some machines "talk" faster than others, just as some people speak faster than others. The speed at which digital signals are transmitted can be measured in various ways. The most common method of measuring data speed is to specify the number of data bits (high and low signal states) that occur in one second. This is called *bits per second* (bps). *Baud* refers to the number of times per second that a signal changes state. Technically, bps and baud are not the same, although many people treat them as if they were. When people say that such and such a modem works at a certain "baud rate," they are usually referring to the speed in bps.

Modems When digital devices such as personal computers are linked by telephone, satellite, or fiberoptic systems, the signals are sent and received at standard speeds. The higher the signal speed, the faster the computers respond to each other and to their human operators' commands. When you access an *online service* with a home computer, you use a *modem* that can send and receive data at certain speeds. The computer at the other end of the line also has a modem, with the ability to send and receive at various speeds. Sometimes the two modems have different maximum speeds. The slower modem determines the highest speed at which the machines can communicate. There is a limit to the data speed that any communications circuit can handle. The higher the speed, the larger the *bandwidth,* or range of frequencies that a signal takes up. A telephone line can handle bandwidths of up to about 3000 cycles per second, or 3 kilohertz (kHz). This will accommodate the signals commonly used with personal computers. Table 9-4 shows common data speeds in bps and the time required to send 1, 10, and 100 pages of text at each speed. It is assumed that the text is typed at 10 characters per inch, with normal margins and double spacing. This table is not an expression of data speed limits. Modems have been tested that can work at several million bits, or *megabits per second* (Mbps).

Advanced Links Satellite, radio, and fiberoptic communications systems can handle larger signal bandwidths than telephone circuits. Therefore, much higher speeds are possible in these modes. Amateur radio operators often use systems having larger bandwidths than telephone systems, especially at microwave radio frequencies. On the so-called shortwave frequencies, however, data speeds are slower, sometimes less than 1200 bps. At these frequencies, radio-wave propagation anomalies can cause distortion that severely limits the speed at which accurate data communication can be realized. The speed of data transmission and reception is an important factor in how "smart" a computer network can become. The faster a system can "think," the higher its potential level of intelligence. However, regardless of the speed at which data is transmitted,

	Bits per second	To send 1 p.	To send 10 pp.	To send 100 pp.
TABLE 9-4	1,200	9.00 s	1 min 30 s	15 min
Time Needed to	2,400	4.50 s	45.0 s	7 min 30 s
Send Data at	4,800	2.25 s	22.5 s	3 min 45 s
Various Speeds	9,600	1.13 s	11.3 s	1 min 53 s
	14,400	0.75 s	7.5 s	1 min 15 s
	19,200	0.56 s	5.6 s	56 s
	28,800	0.38 s	3.8 s	38 s
	38,400	0.28 s	2.8 s	28 s
	57,600	0.19 s	1.9 s	19 s
	115,200	0.09 s	0.94 s	9.4 s
	230,400	0.05 s	0.47 s	4.7 s
	460,800	0.03 s	0.24 s	2.4 s

propagation delays limit the speed with which computers can exchange data in a two-way link. This is because no signals can travel faster than the speed of light, about 186,282 miles per second in free space.

A/D and D/A Conversion

Because digital modes are, in general, more efficient than analog modes, an increasing number of voice, music, and image communications systems "digitize" the analog signals at the *source* (transmitting end) and "undigitize" the signals at the *destination* (receiving end). Computer modems for use with telephone lines work in the opposite way: they "analogize" the signals at the source and "re-digitize" them at the destination.

Analog-to-Digital

Any analog, or continuously variable, signal can be converted into a string of pulses whose amplitudes have a finite number of states, usually some power of 2. This is called *analog-to-digital (A/D) conversion*.

A/D conversion is used in communications, and also in high-fidelity (hi-fi) recording. In computers, A/D converters are used to change voices, pictures, music, or test-instrument readings into signals that the machine can understand. In CD hi-fi recording systems, the music is converted from analog to digital form; the digital information appears as pits on the disk surface. Digital data has many advantages over analog data. Digital signals are less susceptible to interference than analog signals. Also, it is possible to get many more digital signals into a circuit, or onto a communications line, than is the case with analog data.

An A/D converter works by checking the instantaneous amplitude of an analog signal every so often, and outputting a series of pulses having discrete, predetermined levels, as shown in Fig. 9-5. The standard levels can be expressed as binary numbers. In the example shown, there are eight standard levels, represented by three-digit binary numbers from 000 to 111. The vertical pips represent the pulses obtained by measuring, or *sampling*, the analog waveform, and rounding the instantaneous signal amplitude off to the nearest standard level. Although this example shows a system with eight levels, there might be fewer levels, or there might be more. The number of levels is called the *sampling resolution*. In most digital signals, the sampling resolution is a positive whole-number power of 2.

Figure 9-5

An analog waveform and an eight-level digital representation.

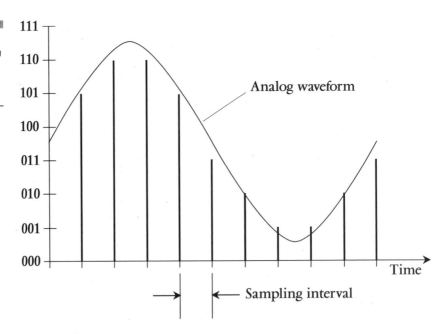

Computer data is binary to begin with, and therefore can be represented by signals having a sampling resolution of two levels. A sampling resolution of eight levels is good enough for voice transmission, and is the standard resolution for commercial digital voice circuits. A resolution of 16 levels is adequate for CDs used in advanced hi-fi systems. For video signals, such as might be used in high-speed animated computer graphics or virtual reality (VR), the sampling resolution must be higher, such as 32 or 64 levels. If represented as a string of binary numbers, a sampling resolution of 8 levels would have 3 bits; a resolution of 16 levels would have 4 bits; a resolution of 32 levels would have 5 bits; a resolution of 64 levels would have 6 bits.

The efficiency with which a signal can be digitized depends on how often a sample is taken. In general, the *sampling rate* must be at least twice the highest data frequency. For an audio signal with components as high as 3 kHz, the minimum sampling rate is 6 kHz, corresponding to a *sampling interval* of 167 microseconds (μs). The commercial voice standard is 8 kHz, or one sample every 125 μs. For hi-fi digital transmission, the standard sampling rate is 44.1 kHz, or one sample every 22.7 μs. This is based on a maximum audio frequency of 20 kHz, the approximate upper limit of the human hearing range. In animated computer graphics, games, television, robotic telepresence, or VR, the sampling rate might need to be several megahertz. The faster the sampling rate, the more realistic the images up to a certain point, beyond which a further increase in the sampling rate will not result in significant improvement.

Digital-to-Analog

The scheme for *digital-to-analog (D/A) conversion* depends on whether the signal is binary (consisting of just two logic states) or multilevel. In communications receivers, the A/D conversion process is carried out by a microprocessor that undoes what an A/D converter originally did. In computer modems, the digital impulses are converted to audio tones having different pitches.

In a binary signal, different patterns of high and low states (logic ones and zeros) represent tiny encoded fragments of signal information. The microprocessor simply decodes these logic pulses, reversing the A/D process that was done in recording or transmission. The high-pitched incoming tone is converted to one logic state; the low-pitched tone is converted to the other logic state.

Multilevel digital signals can be converted back to analog form by "smoothing out" the pulses. This can be intuitively seen by examining Fig. 9-5. Imagine the train of pulses being smoothed into the continuous curve.

Serial and Parallel Data

Binary digital data can be sent and received one bit at a time along a single communications circuit or channel. This is typical of narrowband wireless systems, and is called *serial data transmission*. Higher communications speeds can be obtained by using a wideband channel and *multiplexing* (see Chap. 7), sending independent sequences of bits along each subchannel. This mode is called *parallel data transmission*.

The term "serial" derives from the fact that in serial data transmission, logic bits are sent in a predetermined sequence, one after another. In digital communications, serial data transfer requires that each byte or word be split up into its bits, then sent over the line, and reassembled at the other end of the line in the same order as originally.

Serial data needs only one transmission line or channel. This is the main asset of serial data transfer. Amateur radio operators have been using this mode for decades in *radioteletype* and *packet communications* systems. The main limitation of serial transmission is that, all other things being equal, it can be done at only a fraction of the speed of parallel data transfer.

When binary data is split up, and the components are transmitted simultaneously along two or more lines or channels, it constitutes parallel transmission. In digital communications and computer practice, parallel data transmission is done by sending all the bits in a byte at the same time, over eight parallel lines. Bytes are generally sent one after the other. Some systems use *words* consisting of 16 bits, 32 bits, or 64 bits. These systems need 16, 32, and 64 parallel lines or channels, respectively.

Parallel data transfer has the advantage of being more rapid than serial data transfer. The extent of the improvement depends on the number of lines, which is usually equal to some positive whole-number power of 2, and is the same as the number of bits per word. Suppose a certain data file requires 256 seconds (s) to be transmitted serially. Using *8-bit parallel transmission,* the time is cut to 32 s. Using *16-bit parallel transmission,* it is cut down to 16 s. With *32-bit parallel transmission* it takes 8 s to send the file; using *64-bit parallel transmission* it takes 4 s.

The advantage of parallel transfer is that it can be largely offset (or more than offset) by the common denominator all engineers must face: cost! Parallel data transfer is expensive, and the cost goes up according to the number of lines required. Some engineers would argue that the cost increases according to the square, or even to the cube, of the number of lines.

Parallel-to-serial (P/S) conversion gathers the bits from multiple lines or channels, and transmits them out one at a time, at a regular rate and in a defined sequence, along a single line or channel. The output of the P/S converter must, over a period of time, keep up with the input. If the output is slower than the input, bits will accumulate in the converter. A *buffer memory* stores the bits from the parallel lines or channels while they are awaiting transmission along the serial line or channel. *Serial-to-parallel (S/P) conversion* gathers bits up in groups from a serial line or channel and sends them in parallel along several lines or channels. The output of an S/P converter cannot go any faster than the input, but the circuit is useful when it is necessary to interface between a serial-data device and a parallel-data device.

Figure 9-6 illustrates a circuit in which a P/S converter is used at the source and an S/P converter is used at the destination. This arrangement might be used for data transfer over a long-distance, narrowband radio channel. In this example, the words are 8-bit bytes. However, the words could have more bits, say 16 or 32, or even 64.

Data Compression

Data compression is a way of maximizing the amount of digital information that can be stored in a given amount of data space or sent within a certain period of time. Data compression can speed up data communication by minimizing the number of bits in a transmitted file.

Text files can be compressed by replacing often-used words and phrases with symbols such as =, #, &, @, etc., as long as none of these symbols occurs in the uncompressed file. For example, in this section, the string of characters [data compression] might be replaced by a less-than symbol [<], and the character string [computer] could be replaced by an asterisk ['] (after deleting this sentence to avoid confusing the system). The word [the] could be replaced with the symbol [^]. When the compressed data is received, it is uncompressed by substituting the original words and phrases in place of the symbols. This type of data

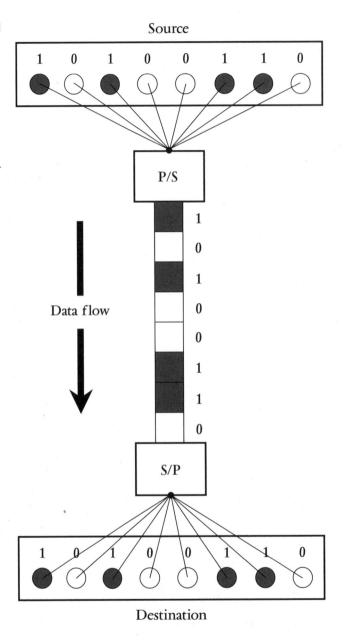

Figure 9-6
A communications circuit employing parallel-to-serial (P/S) conversion at the source, and serial-to-parallel (S/P) conversion at the destination.

compression is a simple form of *encryption;* the reverse process is a form of *decryption.*

The compression of digitized images is more complex and sophisticated than the compression of text. There are several different schemes in common use, all of which can be categorized as either *lossless compres-*

sion or *lossy compression.* In lossless compression, the image detail is not sacrificed at all, even though the size of a file or the length of a transmission is reduced. Truly redundant data alone is eliminated. In lossy compression, some of the image detail in the file is lost, although it cannot usually be noticed by a person who sees the image at the destination (receiving end).

Suppose you want to send someone, via wireless, the entire contents of a CD-ROM. That will take a long time even at high data speeds. But if the data on the CD-ROM is compressed, the actual rate of transmitted information is higher than when the data is not compressed. This is true even though the literal speed in bps, as sent and received by the modems, remains constant.

Data can be compressed as part of the transmission process if it has not been compressed beforehand. For example, you might have a large random-access memory (RAM) filled with data. Suppose you want to send it to a friend via wireless. The data can be compressed at your computer, the source, as it is sent via your modem, and then decompressed (expanded) at the destination as it is received by the other modem.

Packet

Anyone who has a computer can connect it to the telephone lines or to a radio transceiver. Then the computer can exchange data with other computers or communications terminals. *Packet communications,* also called simply *packet,* is the most common way in which computers send and receive signals to and from each other.

Packet Networks

A packet is a block of digital data sent from a computer at a source station to another computer at a destination station. It is not necessary for the operator at the destination station to be physically present for a packet to be received; the computer can store it. For this reason, packet communications is sometimes spoken of as being a form of *time-shifting communications.*

Wireless packet networks are getting larger, easier to access, and more sophisticated all the time. The data speed is high, allowing long messages to be sent in short signal bursts. Packet communications is self-correcting. The destination equipment detects discrepancies in received data, and compensates by having the source retransmit packets that do not "look right."

Reliability and ease of use are important features of packet networks. In an ideal packet network, you can address a message to any destination, anywhere in the world, and be confident that the network will automatically get it there within minutes or, at most, a few hours. As the Internet evolves, coordinating radio, satellite, microwave, and fiberoptic data links, speed and reliability of packet communications are improving.

Protocol

In order to make a packet network function, all the computers must use a *protocol* (data format) that adheres to a standard model. The universal standard is called the *open systems interconnection reference model,* abbreviated *OSI-RM.*

OSI-RM is arranged into seven levels of activity, called *layers,* ranging from the *physical layer* to the *application layer.* Most amateur-radio packet networks do not make use of the full capabilities of OSI-RM, but advanced government systems and those proposed by well-known figures in the computer world employ OSI-RM to its full extent. This allows for the exchange of software, as well as text and image data, via wireless links, and for remote control of computers and peripheral equipment.

When you send a packet, the computers do all of the routing work, once the source computer knows the *local node* of the desired destination station. The operator only needs to key in the destination-station information in order to establish the route. The protocol keeps the connection intact by means of *rerouting,* in case there is some disruption along a given route.

Packet Wireless

Amateur-radio packet communications is called *packet radio;* in general, packet done via radio is called *packet wireless.* Instead of connecting the computer to the telephone lines, you interface the computer with a radio transceiver using a *terminal node controller* (TNC).

A simple amateur packet-radio station is shown in Fig. 9-7A. The personal computer is equipped with a *telephone modem* as well as a TNC, so messages can be sent and received via conventional online services as

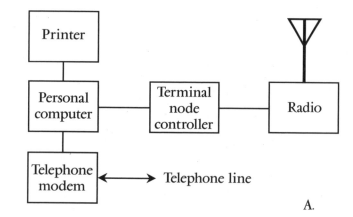

Figure 9-7
A typical amateur packet-radio station (A). Passage of a packet through nodes in a wireless communications circuit (B). Black dots represent end users.

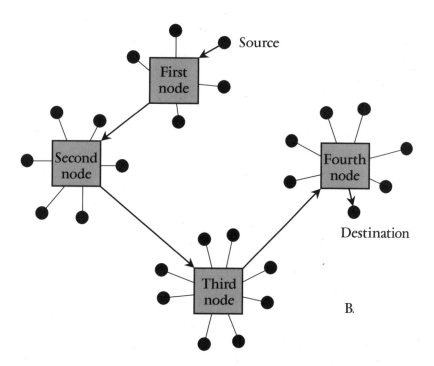

well as via the radio transceiver. With this arrangement, a packet can be received from the radio, stored in the computer, and sent to someone at an address in, say, America Online. A message might be received from the Internet, stored in the computer, and sent to an amateur radio operator via packet radio.

Figure 9-7B is a functional depiction of how a packet-wireless message gets routed. The black dots represent *subscribers* (end users) with packet-wireless capability. The four rectangles are local nodes, probably located in four different cities. Each local node serves several stations in its immediate vicinity via short-range links at very high frequencies (VHF), ultra-high frequencies (UHF), or microwave frequencies. The nodes are interconnected by terrestrial VHF, UHF, or microwave links if the cities are relatively near each other. If the cities are widely separated, satellite or high-frequency (HF) links are used. The source station needs to know the local node of the destination station, as well as a specific address for that station, to send the message. The transfer then takes place automatically, in much the same way as data is sent in online computer networks. The main difference is that the entire link is via radio; none of it takes place over wire or cable.

Packet wireless is evolving toward the ultimate goal of a worldwide system. A major landmark in the evolution of amateur packet radio was the so-called NET/ROM firmware developed by Mike Busch, W6IXU and Ron Raikes, WA8DED. NET/ROM is protected by copyright, as is all commercial software and firmware. For comprehensive information about packet-radio protocols, amateur-radio magazines and clubs are recommended as information sources. The American Radio Relay League (ARRL), 225 Main Street, Newington, CT 06111 (www.arrl.org) publishes books that contain in-depth information about all aspects of packet radio.

Bulletin-Board Systems

A *bulletin-board system* (BBS) is a set of stored messages, accessible by means of a computer. Most computer users are familiar with BBSs. They allow you to use a modem to connect the computer to telephone lines, via which the BBS is accessed. There might be a telephone toll charge associated with the use of a BBS. Wireless makes it possible to communicate over long distances without telephone lines, and therefore without tolls. This is a *packet-wireless bulletin-board system* (abbreviated PBBS). The amateur-radio counterpart is the *packet-radio bulletin-board system* (also abbreviated PBBS). With a TNC connecting a computer to a radio transceiver, you can access PBBSs. Messages can be left for other computer users, or retrieved, in a short time.

The PBBS is a form of time-shifting communications. It is a means of sending and receiving messages, even when the destination operators are not physically at their stations. Because of this, there is often some delay in transferring the message, simply because the operator of the destination station must check the PBBS before he or she can get the message (hence the term "time shifting," which sounds better than "time delayed"). A delay might also occur if the network has difficulty routing the packets. This might happen because the network is overloaded with users, or because part of it has been compromised. Any PBBS is a public storage system; messages for a large number of people are left on the bulletin board.

Sometimes, you might want to leave a message for someone who has a packet radio station, but not do it on the general PBBS. A quasi-private way to leave messages for an individual is *electronic mail,* also known as *e-mail.* If you must have an absolute guarantee of privacy, however, no form of electronic media should be used; this includes voice mail and telephone conversations. The lack of true privacy is one of the less pleasant aspects of modern data communications. The Postal Service ("snail mail") is still privacy-optimal for the average citizen, when face-to-face visits are impractical or impossible. Of course, face-to-face meetings are always an option when the people involved happen to live near each other. Such meetings can take place in out-of-the-way places. Hardcopy documents sent by certified "snail mail" are still preferred by many business executives when legal authenticity is required.

The Gray-Scale System

The *gray-scale system* is a method of creating and displaying a digital video image. The image is made up of *pixels.* One pixel is a single picture (pix) element. This acronym is sometimes shortened to *pel.*

Degree of Detail

Suppose an image has an array of at least 256 by 256 pixels, or $2^8 \times 2^8 = 216 = 65,536$ elements. This number, in digital lingo, is 64K. An image with moderate resolution has an array of at least 512 by 512 pixels, or $2^9 \times 2^9 = 2^{18} = 262,144 = 256K$ elements. (These values are intended only for illustrative purposes. Actual video display standards vary.)

The pixels are little squares, each with a shade of gray that is assigned a digital code. Table 9-5 shows a hypothetical 16-level gray scale. Four binary digits, or bits, are needed to represent each level of brightness from black=0000 to white=1111. This is a coarse gray scale; finer scales might use 64 or even 256 different brightness levels. These can be expressed in binary codes of 6 or 8 bits, respectively.

If an image is 256 pixels square, with a 16-level gray scale, then the memory needed to represent a single nonmoving frame is

TABLE 9-5

Hypothetical
16-Shade
Gray-Scale Codes

Code	Relative shade	Percent brightness
0000	Black	0.00
0001		6.67
0010	Very dark gray	13.33
0011		20.00
0100	Dark gray	26.67
0101		33.33
0110	Medium-dark gray	40.00
0111		46.67
1000	Medium gray	53.33
1001		60.00
1010	Medium-light gray	66.67
1011		73.33
1100	Light gray	80.00
1101		86.67
1110	Off-white	93.33
1111	White	100.00

$256 \times 256 \times 16 = 1,048,576$ bits. A byte consists of 8 bits; this coarse-gray, low-resolution image would therefore need 128 KB. If the image is 512 pixels square with a 256-level gray scale, then the memory required for one frame is $512 \times 512 \times 256 = 67,108,864$ bits. That is 8 MB.

Analog versus Digital

The above examples show that analog technology still enjoys a few advantages over digital technology. These memory/storage figures represent a single image sent once. For a fast-scan television picture, at least 16 images are required per second. Yet even these images would represent only gray-scale, and not color, pictures. Lengthy digital color television programs cannot, at the present time, be effectively stored on magnetic disks. The required storage space is simply too great. But long television programs are easily recorded on analog tapes of manageable size, and played on analog equipment of reasonable cost. This equipment is, of course, the familiar *videocassette recorder* (VCR).

Eventually, computers, televisions, and telephones might merge into single appliances, or at least be made compatible. Satellite links will provide people with the wireless option, so it will be possible to be in full two-way communication with the world no matter where you are. You might be on a mountain-climbing expedition, in an orbiting space station, or on a ship in the middle of the ocean, and you'll be able to link up to the Internet as if you were at home or in your office.

The RGB Color Model

All visible colors can be obtained by combining red, green, and blue light in varying proportions. The *RGB color model* is a scheme for digital video imaging that takes advantage of this fact.

Hue, Saturation, Brightness

Visible light consists of *electromagnetic (EM) waves.* The color you see is a function of the wavelength. There are often many wavelengths combined. When light energy is concentrated mostly at one wavelength, you see a vivid *hue.* A rainbow has some energy at all visible wavelengths, so you see a continuum of all possible hues.

When visible-light energy is confined to a single wavelength (as is the case with a laser), the hue is intense. More often, the hue is less concentrated; sometimes it is diffuse and hard to ascertain. The relative vividness of a hue is called the *saturation.*

The *brightness* of a color is a function of how much total energy the light contains. A ruby laser has high brightness; a small LED has low brightness. In a television receiver or computer monitor, there is a control for adjusting the brightness, also called *brilliance.*

3D Color

You can get an excellent full-color palette by combining three pure colors in various degrees of brightness. For radiant light, these colors are

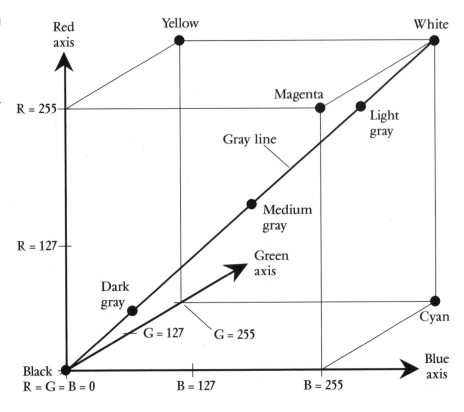

red, green, and blue. For pigments in printing they are magenta (pinkish red), yellow, and cyan (greenish blue).

Color computer monitors and color television sets have thousands of tiny dots arranged in an interlaced pattern. One-third of the dots are red; one-third are green; one-third are blue. When you look at the screen from a distance, the dots blend together to form a continuous color image.

Imagine each of the three primary hues in its purest possible form, that is, with maximum saturation. The brightness of each can be varied independently, from zero to some arbitrary maximum. Assign each hue an axis in cartesian three-space, as shown in Fig. 9-8. Call the axes R (for red), G (for green), and B (for blue). The values of R, G, and B can range from 0 (no output) to 255 (maximum output). These levels can be digitized as 256 different values ranging from 00000000 to 11111111. The result is a cube, within which there are $256 \times 256 \times 256 = 16{,}777{,}216$ points. Any point, say (R,G,B)=(45,55,75), represents a unique color. There are 16,777,216 possible colors ("millions of colors") that can be represented in this cube. This is a geometric representation of the RGB color model. In the drawing, several points are labeled, in addition to the "gray line" representing the set of points for which R=G=B.

Some RGB systems use fewer than 256 possible levels for each color. An alternative is to use only 16 levels. This results in 4096 possible colors ("thousands of colors").

Error Correction

In digital systems, errors occur as misinterpreted bits. A machine might see a logic 0 when it should see a logic 1, or vice versa. It is also possible for bogus data bits to be inserted or legitimate ones to be dropped. This almost never happens in a computer or within a piece of digital communications equipment. In digital communications networks (wire, fiberoptic, or radio), errors occur more often than they do within an individual machine. Wireless systems are especially susceptible to errors. But computers are not tolerant of errors, especially when data contains software. A single misplaced bit can make the difference between a program that runs well and one that does not run at all. The error tolerance is lower for programs than it is for plain text or image data.

With *digital signal processing* (DSP), most digital errors can be corrected as data is sent over a communications circuit. This cannot be done with analog data, and that is one of the reasons why digital communications are so much more accurate and reliable, in general, than analog communications. In the case of digital data, the logic 0 and 1 states (lows and highs) can be distinguished by a circuit designed especially for that purpose, even when they contain some noise. The noise can be removed, and the original "clean" lows and highs restored. This allows transmission of digital signals over great distances through many repeaters or nodes. For analog data, the signals can be converted to digital impulses via A/D conversion, then subjected to DSP, and then converted back to analog form by means of D/A conversion. This is done in some analog communications networks, particularly among computers in online services. Additional details about DSP can be found in Chap. 7.

CHAPTER **10**

Satellite
and Space
Communications

Until the 1960s, long-distance wireless communication and broadcasting depended on atmospheric and ionospheric effects. During the 1960s and 1970s, artificial-satellite links became commonplace. In the 1980s and 1990s, people began to talk about the "overcrowding of space" because there were so many satellites in existence, and because so many new satellites (and fleets of satellites) were being planned. Yet some engineers believe satellite communications is in its infancy compared to systems that might someday exist. The extent to which the potential is realized will depend on public demand and on the willingness of people and corporations to pay for the systems envisioned by leading engineers.

Overview

Satellite data transmission involves the use of orbiting, wideband *repeaters,* more commonly called *transponders,* to get information between sources and destinations on the earth's surface. Electromagnetic (EM) signals, usually at very high frequencies (VHF), ultra-high frequencies (UHF), or microwave frequencies, are transmitted to a satellite, where they are received and retransmitted on another frequency at the same time by a transponder (Fig. 10-1).

Schemes

Most communications links employ either *geostationary satellites* or *low-earth-orbit (LEO) satellites.* Their orbits are nearly perfect circles. A few satellites revolve around the earth in elongated, elliptical orbits.

A geostationary-satellite link employs one or two satellites, each strategically placed so that it stays over the same spot on the earth's surface at all times. Such a satellite must be directly over the equator, and it must orbit at an altitude such that it revolves around the earth precisely in step with the planet's rotation on its axis. Continuous contact can then be maintained between any two stations that are both within range of a satellite. From any point on the surface, a geostationary satellite appears to hover in a fixed spot.

A LEO-satellite link employs a fleet of satellites, each in a fairly low-altitude orbit that takes it over, or close to, the earth's north and south geographic poles. The number of satellites can vary. In general, the more satellites there are in the fleet, the more reliable will be the communica-

Figure 10-1
Simplified block diagram of a satellite communications link.

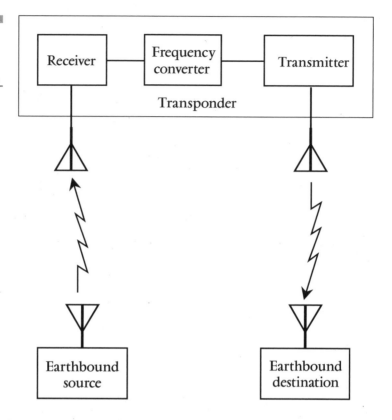

tions. Ideally, there should be enough satellites in the fleet so every point on the surface is always within range of at least one. The LEO satellite system works like a *cellular telephone network,* except the transponders are moving, not fixed, and they are in space, not on the ground.

Uses

Satellite data links do not require long-distance cables. It is as easy to access a satellite system from a remote island, mountain, boat, or aircraft as it is to make the connection from a large city.

To access a satellite network, you need a *transceiver* (transmitter/receiver) and a suitable antenna. If you want to use the network for data or packet communications, you also need a computer and a *modem* or *terminal node controller* (TNC). A basic system works in *half duplex,* allowing the workstation to operate as both a signal source and a signal destination, but not at the same time. A more sophisticated satellite data link allows *full duplex,* allowing the workstation to operate as a source and destination concurrently.

A set of satellites called the *global positioning system* (GPS) allows a computer to figure out, to within a few feet, its location on the surface of the earth. This is an invaluable navigational aid for those who own or rent boats or aircraft. Some cars and trucks are equipped with GPS links, so their drivers can determine exactly where they are when driving in unfamiliar territory. A computer display will show the location on a detailed road map. The driver can "zoom" to see the vehicle's position within a country, state, or city.

Amateur radio operators, also called "radio hams," have been exchanging computer data by radio for decades. This is known as *packet radio.* Radio hams have pooled their resources and placed communications transponders on satellites. If you are technically inclined, you might want to study electronics and take the test to get an amateur radio license. You can then send and receive Internet e-mail via satellite from practically anywhere, at no cost (except a few dollars for club memberships in some cases).

Challenges

One of the biggest problems with long-distance two-way data communications networks is the fact that it takes time for signals to get between sources and destinations. This is true no matter how the data is sent. Nothing can travel faster than the speed of light (186,282 miles, or 299,792 kilometers, per second).

A satellite might hover in a geostationary orbit, at some fixed spot 22,300 miles (mi) above the earth's equator. When such a satellite is used, the total path length is always at least 44,600 mi (up and back), and usually a little more. The shortest possible one-way signal delay is therefore 44,600/186,282=0.24 second (s). A signal's round trip, typical of a two-way data conversation involving queries, responses, and counterresponses, requires twice this length of time, or about $^1/_2$ s. High-speed, two-way data communication and processing is impossible with a delay that long. This problem can be overcome to some extent by placing satellites in lower orbits. For a link to be continuously maintained, a fleet of LEO satellites can be used.

One of the greatest challenges facing communications and computer scientists is figuring out how to provide high-speed links between computers or terminals that are far from each other. There is no way to overcome the fact that the speed of light is, in computer terms, slow if the computers are not located in close proximity. This is not a problem for one-way data transfer, but it hinders the linkage of computers worldwide to form supercomputers. The speed of such a system, as a whole, is slowed by signal propagation delays, no matter how fast the individual microprocessors might be.

Orbital Characteristics

There are, in theory, infinitely many possible orbits a satellite can follow around the earth. An orbit might be as low as approximately 100 mi above the earth's surface, or many thousands of miles up. Some orbits are essentially perfect circles, while others are greatly elongated. Some orbits take a satellite over the geographic poles, while other orbits stay near the equator. There are, nevertheless, some general characteristics that apply to all orbits.

Eccentricity

Any earth-orbiting satellite follows either of two types of path through space. A *circular orbit* has a defined radius as measured from the center of the earth, and this radius never changes. When a satellite follows an *elliptical orbit,* the center of the earth is at one focus of the ellipse. The extent to which the orbit differs from a circle is called the *eccentricity* of the orbit. Whenever a satellite is placed into orbit around the earth, or around any other large celestial body, there is almost always some deviation from a perfectly circular path. In geostationary and LEO satellite systems, the objective is to minimize this deviation. This requires precise launch speeds and trajectories.

A circular orbit is said to have eccentricity zero. As the orbit becomes elliptical and deviates more and more from a perfect circle, the eccentricity increases. Some satellites have orbits that are greatly elongated ellipses. As the orbit becomes more and more elongated, the eccentricity approaches a value of 1.

There is a limit to how elongated an orbit can become before the satellite ventures so far out into space that the earth's gravitational field loses control of it. If a spacecraft is launched with sufficient speed (more than about 25,000 miles per hour), it will go into a *parabolic trajectory* with an eccentricity equal to 1, or into a *hyperbolic trajectory* with an eccentricity greater than 1. In either of these cases the spacecraft will escape the earth's gravitation and will fall into an elliptical orbit around the sun. In the extreme case, if a spacecraft is given a tremendous boost, it might escape the gravitational influence of the solar system and wander into interstellar space. The *Voyager* probes are examples of such spacecraft.

Perigee and Apogee

The point at which a satellite attains its lowest altitude is called *perigee*. It occurs once for every complete orbit. At perigee, the satellite travels faster than at any other point in the orbit. The point of maximum altitude is called *apogee*. Like perigee, apogee occurs once for every complete orbit. At apogee, the satellite travels more slowly than at any other point in the orbit.

The motion of a satellite in an elliptical orbit obeys *Kepler's second law.* This rule was originally formulated to describe the motions of planets, asteroids, and comets around the sun. However, the rule applies equally well to satellites orbiting the earth. The law, as applied to earth satellites, states that a line between the satellite and the center of the earth will sweep out equal areas in equal amounts of time. Thus, in a given period of time, the line, called the *radius vector,* will rotate through a large angle near perigee and a comparatively small angle near apogee (Fig. 10-2).

Active satellite communications is most difficult when the satellite is at or near perigee, if it has a very elongated orbit. This is because the satellite moves fast then, and its position in the sky changes rapidly. The antenna at the ground station must be constantly turned to fol-

Figure 10-2

The radius vector of a satellite moves fastest near perigee and most slowly near apogee.

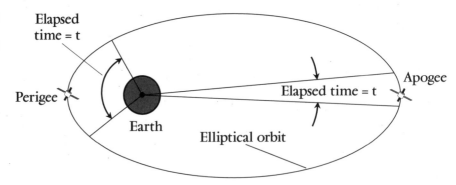

Elapsed time = t

Perigee

Earth

Elapsed time = t

Apogee

Elliptical orbit

low the satellite during this time. Conversely, satellite communication is the most convenient when the spacecraft is at or near apogee, because the satellite moves slowly then, and its position in the sky does not change much for awhile. The antenna at the ground station does not have to be turned to follow the satellite during this time.

Inclination

With the exception of geostationary satellites, earth-orbiting satellites usually orbit at a slant with respect to the equator. The *inclination* is a measure of this slant.

When a satellite in a normal orbit (going in the same direction that the earth rotates) crosses the equator going from south to north, it does so at an angle of up to 90 degrees. This is the *inclination angle*. It is approximately the same as the highest latitude reached by the satellite in the northern hemisphere. A satellite with an inclination of, or close to, 90 degrees is said to be in a *polar orbit*, because it passes over the north and south geographic poles of the earth.

It is possible for a satellite to have an inclination angle of more than 90 degrees, up to 180 degrees. Such angles indicate that the satellite is in a *retrograde orbit*, going in the opposite direction from the rotation of the earth.

Groundtrack

At any time, there is one, and only one, point on the earth's surface from which an orbiting satellite or spacecraft appears directly overhead. This point moves as the satellite orbits the earth. As the point moves, it follows a path over the surface, whose position and direction depend on the orbit of the satellite. This path is called the *groundtrack*. All satellite groundtracks are *great circles* around the earth. For most satellites, the groundtrack shifts toward the west for each succeeding orbit because the earth rotates eastward underneath the satellite.

For geostationary satellites, the groundtrack is a single point that never changes. For satellites in equatorial orbits, the groundtrack lies exactly along the equator.

Groundtrack information is predicted for nongeostationary satellites, so its users know when and where the satellite will appear. This allows a satellite user to program a computer to aim the system antenna in the proper direction and to be sure the antenna tracks the satellite correctly as it moves across the sky.

Nodes and Passes

A satellite usually has a groundtrack that crosses the equator twice for each orbit. The only exceptions are geostationary satellites and satellites whose orbits are exactly over the equator. The points where, and times when, the groundtrack is exactly over the equator are called *nodes*.

For every orbit, there is one node when the groundtrack moves from the southern hemisphere to the northern, and one node when the groundtrack moves from the northern hemisphere to the southern. The first of these, going south to north, is called the *ascending node*. The other is called the *descending node*. Ascending nodes are commonly used as reference points for locating satellite positions at future times. The position of the node is given in degrees and minutes of longitude.

When a satellite has an orbit that is slanted relative to the equator, its groundtrack is moving generally northward half of the time. This period starts when the satellite attains the southernmost latitude in its orbit, and lasts until it reaches its northernmost latitude. For any given earthbound location, an *ascending pass* is the time during which the satellite is accessible, while it is moving generally northward. The pass time depends on the altitude of the satellite, and on how close its groundtrack comes to the earth-based station. A *descending pass* is the time during which the satellite is accessible and is moving generally toward the south.

Satellite Power Supplies

An active satellite requires a reliable source of power to remain operational under all conditions. Unlike earthbound repeaters and transponders, satellites are extremely inconvenient (if not impossible) to repair.

Most satellites, regardless of the type and shape of their orbits, spend some time in the earth's shadow, and some time in direct sunlight. In general, the higher the average orbital altitude of a satellite, the greater the proportion of time it spends in sunlight. Averaged over long periods, the sun is eclipsed by the earth at most 50 percent of the time, although a single eclipse period might last for several hours. In space, the temperature difference between "day" and "night" is far greater than anything encountered on the earth's surface. This necessitates the design of power supplies that can keep functioning despite excursions between extreme heat and extreme cold.

Electrically, the power supply on a typical satellite is similar to stand-alone solar power plants that some people have installed in their homes and businesses. When a satellite is in sunlight, *solar panels* provide all the power needed for operation, and also some surplus power, which is used to charge batteries. When the satellite is in the earth's shadow, the batteries must supply all the power. The batteries must have sufficient energy-storage capacity to meet the needs of the satellite during the entire eclipse period. Otherwise the satellite might "go to sleep" if it is heavily used during a long eclipse. A greatly simplified diagram of a satellite power supply is shown in Fig. 10-3. The diodes allow the current to flow from the solar panels to the batteries and to the electronics in the satellite, but prevent battery current from passing through the solar panels.

The amount of power consumed by a satellite depends, in part, on the total received signal power. This in turn depends on how many people are trying to access the satellite and also on how strong the incoming

Figure 10-3

Simplified diagram of a satellite power supply.

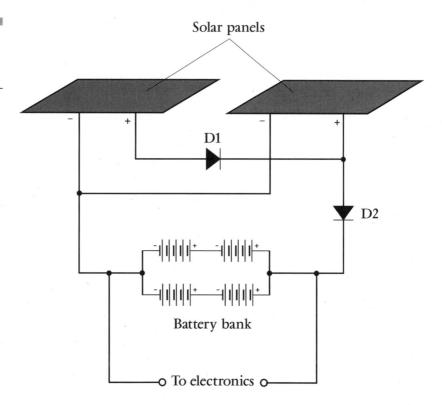

Solar panels

D1

D2

Battery bank

To electronics

signals are. If too many people attempt to communicate through a satellite, and/or some of them use transmitters that radiate too much power, the satellite's battery might be depleted during an especially long eclipse period. For this reason, it is advisable for earth-based stations to limit the transmitter output power to a level sufficient to ensure reliable communications, and not to exceed that limit.

Transponders

In the simplest type of active communications satellite, there is a broadband *repeater* that receives the signals sent up from the earth, converts them to another frequency, and retransmits them back to the earth. This device is a *transponder.* Some satellites have two or more transponders, each working on different sets of frequency bands. More sophisticated satellites have transponders that employ multiplexing, data storage and retrieval systems, and other schemes to maximize the amount and variety of data they can handle.

Operation

A transponder is something like a repeater, because it retransmits the signals at the same time they are received. But a transponder converts a whole band of frequencies at once, whereas a repeater works with only one input frequency and one output frequency. A transponder can carry many contacts at a time. Also, full duplex is possible using a transponder.

Figure 10-4 is a block diagram of a transponder. The *mixer* can convert a band of frequencies, containing many signals, to another band. Depending on whether the sum or difference output of the mixer is used for retransmitting, the signals in the transponder output might be right-side-up (the same as the way they came in) or upside-down (inverted in frequency from the way they came in). The first kind of device is called a *noninverting transponder.* The second kind is called an *inverting transponder.* If a satellite has an inverting transponder, continuous-wave signals will not be affected. But the mark and space in frequency-shift keying will be reversed, and a single-sideband (SSB) signal will be transformed to the opposite sideband. This does not cause a loss of signal

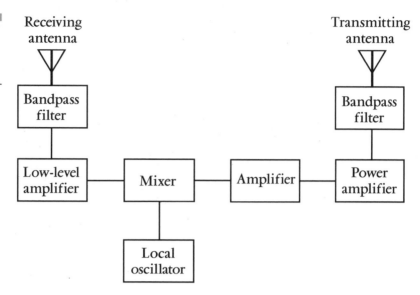

Figure 10-4
Simplified block diagram of a basic satellite transponder.

information, but it must be taken into account if the transponder is of the inverting type.

In a noninverting transponder, the output band has the same sense as the input, or band. That is, the lowest-frequency signal going in is also the lowest-frequency signal coming out. In an inverting transponder, however, the entire band is flipped upside-down, so that the lowest-frequency signal going in becomes the highest-frequency signal coming out.

Uplink and Downlink

The term *uplink* is used to refer to the frequency or band on which an active communications satellite receives its signals from earth-based stations. The uplink frequency or band is different from the *downlink* frequency or band, on which the satellite transmits signals back to the earth. Many satellites receive and retransmit signals at the same time. For this to be possible, the uplink and downlink frequency bands must be substantially different. Also, the uplink band should not be harmonically related to the downlink band. The receiver must be well designed so it is relatively immune to *desensitization* and *intermodulation* effects. The receiving and transmitting antennas should be aligned in such a way that they are electromagnetically coupled to the least possible extent.

At a ground-based satellite-communications station, the uplink antenna can sometimes be nondirectional if high transmitter power is used, or if the satellite is in a low orbit or near perigee. But for more distant satellites, and especially for satellites in geostationary orbits, directional antennas are preferable. These maximize the gain in both the transmitting and receiving modes, and also minimize the susceptibility of the ground-based receiver to interference from nonsatellite signals and local noise sources. If full-duplex operation is contemplated, the ground station, like its satellite counterpart, must be designed so that the transmitter does not interfere with the receiver. This means that separate transmitting and receiving antennas must be used, and bandpass filters employed (especially in the receiver system).

A transponder can only produce a certain maximum power output, distributed among all the signals it sends back to earth. Therefore, when a transponder is heavily used, the individual signals tend to be weaker than they are when the transponder is not dealing with many signals. A very strong uplink signal can cause a transponder to use up almost all of its available power to retransmit that signal. If there are many contacts in progress via a satellite, and an excessively powerful uplink signal is received, all the other signals will be attenuated. This effect can be mistaken for receiver desensitization, although its cause is different. It does not take a fantastic amount of uplink power to produce this effect in a small satellite (such as an amateur-radio satellite).

Geostationary Satellites

A *geostationary satellite* is an artificial satellite that follows a specialized orbit called a *geostationary orbit*. Such satellites are extensively used for television (TV) broadcasting. These satellites are also employed for communications, for gathering weather and environmental data, for *radiolocation,* and for *radionavigation.*

All satellite TV signals are broadcast via geostationary satellites. When you set up a dish antenna for receiving these signals, you must aim the dish at a geostationary satellite. Once you have adjusted the position of the dish, you need not change it unless you move to a different part of the country or the world, or you want to view the signals from a different satellite.

If you use online computer services, or if you often make international telephone calls, you probably work through geostationary satellites on occasion. Some Internet service providers employ direct downlinks from geostationary satellites that subscribers can pick up with small dish antennas similar to those used for digital satellite TV. The data from the end user's computer, called the *upstream data link,* is sent out via the telephone lines or TV cable system; the *downstream data link* comes from a geostationary satellite straight into the end user's workstation. This allows for high-speed downloading of data from the Internet to individual subscribers. It also bypasses some of the bottlenecks that plague the land-based Internet hardware system.

Fixed in the Sky

For a satellite in a circular orbit, the revolution period gets longer as the altitude increases. At an altitude of about 22,300 mi, a satellite in a circular orbit takes exactly 24 hours (one solar day) to complete an orbit. If a satellite is placed in such an orbit over the equator, and if it revolves in the same direction as the earth rotates, the satellite will follow the earth as the planet spins and will stay above the same place at all times. Figure 10-5A is a view of this situation from far above the earth's north geographic pole.

From the vantage point of someone on the surface of the earth, a geostationary satellite always stays in the same spot in the sky. One such "bird" covers about 40 percent of the earth's surface. A satellite over Quito, Ecuador, for example, can link most cities in North America and South America. Three satellites in geostationary orbits spaced 120 degrees apart (one-third of a circle) provide coverage over the entire world, with the exception of the polar regions. A geostationary satellite cannot provide coverage for the immediate north and south geographic polar zones, because as "seen" from those locations, the satellite is below the horizon.

In geostationary-satellite networks, earth-based stations can communicate via a single "bird" only when they are both on a line of sight with it. If two stations are nearly on opposite sides of the globe, they must operate through two satellites to obtain a link (Fig. 10-5B). In this situation, signals are relayed between the two satellites, as well as between either satellite and its respective ground-based station.

Figure 10-5
A geostationary
satellite follows the
rotation of the earth
(A); a data link using
two satellites (B).

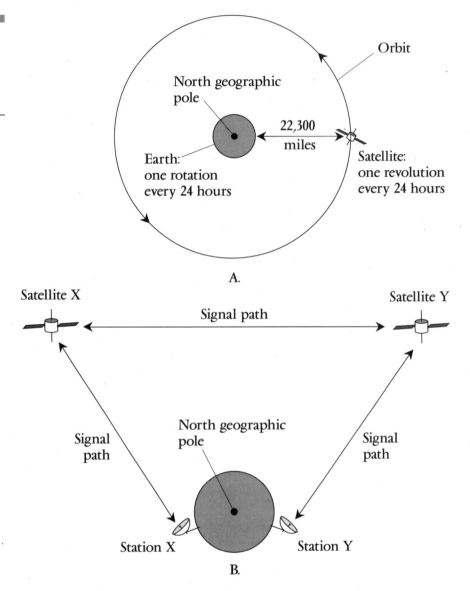

A.

B.

Long Signal Path

The main limitation of geostationary-satellite communication is the fact
that the signal path is long: at least 22,300 mi up to the satellite, and at
least 22,300 mi back down to the earth. If two satellites are used in the
circuit, the path is substantially longer. The signals are delayed because it
takes time for them to span such a distance, even though the waves travel

at the speed of light (186,282 miles per second). The delay is about $1/4$ to $1/3$ s. This does not cause problems in TV broadcasting or in one-way data transfers, but it slows things down when computers are linked with the intention of combining their processing power. It is also noticeable in telephone conversations.

When you use a long-distance telephone network or an online computer service and geostationary satellites are involved, the telephone companies provide all the hardware, including the satellite equipment. You must dial certain numbers or type certain commands to get a connection, and that is all you need be concerned about. It is another matter to set up an independent satellite data communications station. Amateur radio operators have been communicating through satellites for decades; some Internet subscribers do it as well. The main difficulty with this scheme is the fact that it is comparatively expensive.

Low-Earth-Orbit Systems

The earliest communications satellites orbited only a few hundred miles above the earth. They were *LEO satellites*. Because of their low orbits, LEO satellites took only about 90 minutes to complete one revolution around the earth. This made communications inconvenient, because a satellite was in range of a ground station for only a few minutes during each pass, and there were usually only two good passes every day. This is one of the reasons why geostationary satellites gained predominance. In recent years, however, the LEO satellite concept has been revitalized.

There are some problems with geostationary-satellite data links. The orbit requires constant adjustment; the slightest change in altitude will cause the satellite to get out of sync with the earth's rotation. Geostationary satellites are expensive to launch and maintain. When communicating through them, there is always a noticeable delay because of the path length. It takes rather high transmitter power, and a sophisticated, precisely aimed antenna, to communicate reliably.

Of course, geostationary satellites have many advantages, and they will continue to be used in communications systems of all kinds. But the limitations of geostationary satellites have brought about a revival in the LEO concept. Instead of a lone LEO satellite, the new concept is to have a large fleet of them.

Imagine dozens of LEO satellites in orbits such that, for any point on the earth, there is always at least one satellite in range. Further suppose that the satellites can relay messages throughout the fleet. Then any two points on the surface will always be able to make contact through the satellites. This is a form of *cellular communications network* in which the repeaters are in motion. The scheme employs satellites in polar orbits strategically spaced around the globe. Because of the large number of satellites, this system is called a *big LEO satellite network.*

Big LEOs can make wireless communications easier than ever before. Amateur radio operators are especially interested in the systems. A big-LEO satellite communications link is easier to access and use than a geostationary-satellite link. A small, simple antenna will suffice, and it does not have to be aimed in any particular direction. The transmitter can reach the network using only a few watts of power. The propagation delay is much shorter than is the case with a geostationary link, generally less than 0.1 s. Using big LEOs, individual Internet subscribers can connect computers to simple radio transceivers and "log on" from anywhere in the world. A big LEO network will work in the Himalayas or Antarctica, on a boat or a plane. The universal coverage makes big LEOs appealing to developing countries.

Amateur Satellites

Satellites are used by amateur radio operators, mostly at VHF and UHF. All modern amateur satellites are of the active type.

Early Amateur Satellites

The first amateur satellites, called by the acronym OSCAR (short for "orbiting satellite carrying amateur radio"), were placed in low, nearly circular orbits. Contacts had to be carried out within a few minutes via these satellites, because they remained above the horizon for only a short time during each pass. It was necessary to keep changing the azimuth and elevation bearings of the antenna in order to keep it pointed at the satellite, or else to use low-gain, omnidirectional antennas at the surface. It was also necessary to keep constant track of orbital decay effects, because any satellite in low earth orbit is subject to atmospheric drag. Eventually, it will fall into the atmosphere and burn up.

More recent amateur satellites have been placed in elliptical orbits. This causes them to swing out far away from the earth, and during this time, they move much more slowly than a satellite in low orbit. It is possible to access the satellite for periods of hours, instead of just minutes, when the satellite is at apogee. Also, a directional antenna can be left alone for some time, not needing constant readjustment. The effective range of the satellite is much greater, because it "sees" a larger part of the earth's surface from its higher altitude. But more gain is generally required for antennas at the surface.

Future Amateur Satellites

In the future, many amateur satellites will probably be placed in geostationary orbits. This allows for a constant azimuth/elevation setting for surface-station antennas. An antenna can be mounted in a fixed position, aimed at the satellite, and then left alone without the need for tracking or for azimuth/elevation rotators. Three geostationary amateur satellites, each spaced equally around the world at angles of 120 degrees with respect to each other, can provide for communications among radio amateurs everywhere in the civilized world.

LEO systems offer some opportunities for amateur radio use. The advantage of these systems is the fact that directional antennas are not needed and antenna-aiming problems are eliminated. The main limitation of LEO systems for amateur use is the fact that, in order to offer 100 percent coverage and reliability, there must be many satellites. Such a system is far too expensive for amateur radio organizations to launch for their exclusive use. The physical and spectrum space for an amateur transponder on a commercial satellite must be bought or rented. This is now done with individual satellites; it is affordable in these cases because only one satellite and one transponder is involved for each communications system. Placing a transponder on board every one of a fleet of several hundred "birds" in a big LEO system might be too expensive for amateur organizations to afford.

There is another problem as well, which began in relation to satellites but which could become an ongoing challenge to the existence of amateur radio as an institution in the United States. In recent years, there has been a tendency to regard EM spectrum space as a commodity that can be bought and sold, like real estate. Commercial enterprises with vast financial resources can easily outbid amateur organizations for space in big LEO systems. In fact, some commercial interests have proposed that certain existing amateur frequencies be used for business LEO communications networks, creating concern among radio amateurs that their EM spectrum allocations might be bought out from under them. This situation has given rise to some bizarre and sarcastic end-point questions posed by frustrated amateur radio operators. For example, what is a reasonable price for the yellow and green portions of the visible-light spectrum? There will be much debate on this issue before, or if, it is ever resolved.

Future amateur satellites can be expected to operate over larger bandwidths, and probably also on higher frequency bands, than current amateur satellites. It will probably be possible to communicate using two satellites if necessary. In this way, almost every amateur station in the world will have immediate, continuous access to almost every other amateur station, 24 hours a day, every day.

Operation

The transponder in an amateur satellite converts signals from one band to signals in another band. It is, in this way, like a transmitting converter or a repeater.

The input band and output band are of the same width, but the output is often "upside-down" relative to the input, because of the conversion process. Therefore, if you increase the frequency of your transmitter signal, the output frequency from the satellite might go down, rather than up. The whole band is thus inverted (Fig. 10-6). In this case, upper-sideband uplink signals will be lower-sideband in the downlink, and vice versa. In radioteletype, if this mode is allowed through the satellite, the mark and space signals will be reversed when the transponder is of the inverting type. There is no effect on CW signals. An advantage of inverting transponders is that *Doppler effect* is minimized. This effect is described in more detail later in this chapter.

Figure 10-6
An inverting transponder flips a band "upside-down." In this linear graph, frequency increases from left to right.

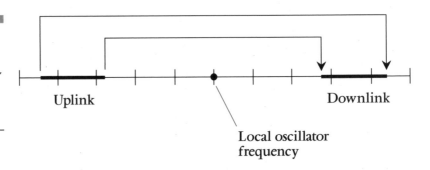

Uplink

Downlink

Local oscillator frequency

The transponder can handle numerous signals all at once. This makes full duplex operation possible; other parties can interrupt you while you talk to them. You can even listen to your own downlink signal.

It is a fundamental rule in amateur satellite work that you never use more power than you need. This is a good rule (and actually carries the force of law) for all amateur communications, but with satellites, there is an added importance to it. If a single station uses far more power than necessary to access, or to communicate through, the satellite, the transponder will pay a disproportionate amount of attention to that one station. The result will be that the other stations' signals are greatly attenuated while the strong station is transmitting.

Normally, only about 4 watts (W) *effective radiated power* (ERP) is needed at the earth-based transmitting station to access low-orbiting satellites, and about 30 W ERP is needed to access high-altitude satellites. If significantly more power is used than this, the station might "hog" satellite transponder power.

Antenna Systems

In satellite communications and broadcasting, antenna systems usually employ *circular polarization*. This is because the relative orientation of the satellite changes with respect to the earth. Some satellites use directional antennas; others use nondirectional antennas.

Antennas on Geostationary Satellites

From the viewpoint of a geostationary satellite, orbiting 22,300 mi up, the earth's disk subtends only 18 degrees of arc, or one-twentieth of a celestial circle. For this reason, directional antenna systems are always preferable in geostationary satellites; otherwise most of the transponder's output energy would go off into space, and the transponder's receiver would be subject to interference from solar and cosmic noise.

Dish antennas are commonly used on geostationary satellites. *Horn antennas* are also used. In some cases, *phased arrays* provide coverage for specific parts of the globe, for example, North America or Africa. The antennas must be precisely aimed. The slightest perturbation in the satellite's orbit, or the slightest change in its rotational speed, can misalign the antenna and cause communications failure.

With circular polarization, it becomes possible to use all the frequencies in a band in effect twice. One set of signals can be sent and received using circular polarization in the clockwise sense; the other set of signals can be sent and received in the counterclockwise sense. If antennas with extremely narrow beamwidth are employed, bands can also be reused for different geographical regions. For example, a geostationary satellite orbiting over Quito, Ecuador, might have two dishes with 5-degree beamwidths aimed at the center of North America and two dishes with 5-degree beamwidths aimed at the center of South America. One of the North America dishes might use clockwise circular polarization while the other uses counterclockwise polarization; the same would be true for the South America dishes. This would allow a frequency band to be used in effect four times.

Antennas on LEO Satellites

In contrast to geostationary-satellite antennas, the antennas on LEO satellites are generally omnidirectional. As a satellite passes over the surface, each individual ground station changes position relative to the satellite. The earth's disk subtends almost half the celestial sphere. From the vantage point of any single LEO satellite, only a small percentage of the earth is covered at any given time, and individual ground stations can be separated by angles ranging from zero to almost 180 degrees. The satellite must be able to send and receive signals to and from other satellites, as well as to and from ground stations. The antenna system on a LEO satellite therefore need not, and in fact cannot, be aimed.

Antennas on LEO satellites, being *isotropic radiators,* have no gain. However, this is offset by the fact that the distance between the satellite and an earth-based station is never very great. Earth-based transmitters using omnidirectional antennas can easily access a LEO satellite. The transmitter on board the satellite can employ a low-power transmitter, using a relatively small power supply. This greatly reduces the cost of a LEO satellite compared with a geostationary satellite. Of course, an effective LEO system requires dozens of satellites. Instead of finesse, which is requisite in geostationary satellites, a LEO system relies on sheer force of numbers to maintain reliable communications links. This translates into a high initial cost, because many satellites must be launched.

Earth-Based Antennas

For communication through geostationary satellites, earth-based antennas must have reasonably high gain and directivity. Ideally, the earth-based antenna system is similar or identical to that on board the satellite. The antenna must be carefully aimed. For a given geographical location, once the antenna is set in position, it need not be reoriented unless communication is wanted via a different satellite.

For single, nongeostationary satellites, antennas with moderate or high gain are generally used. If a directive antenna is used, the satellite must be tracked; that is, the orientation of the earth-based antenna must be constantly adjusted to keep it pointing toward the satellite. This is done using two antenna rotators, one for the *azimuth* (compass bearing) and the other for the *elevation* (angle with respect to the horizon). A computer can be programmed, using satellite tracking data, to turn the rotators at the correct rates to keep the antenna pointing at the satellite.

A simple antenna system for an earth-based station, intended for use with individual nongeostationary satellites, consists of two pairs of *Yagi antennas*. Each pair of antennas is set on a common boom. Each pair's members are oriented at right angles and are fed **90** degrees out of phase, producing circular polarization. One pair of antennas is cut for the uplink band, and the other is cut for the downlink band. The antenna booms are parallel, so both antenna pairs point in the same direction. A simple illustration of this scheme is shown in Fig. 10-7. In this example, the uplink frequencies are higher than the downlink frequencies. Thus

Figure 10-7
An antenna system for a ground-based station, for use with a nongeostationary satellite. This is a top view with the system aimed horizontally. The mast is perpendicular to the page and is not shown.

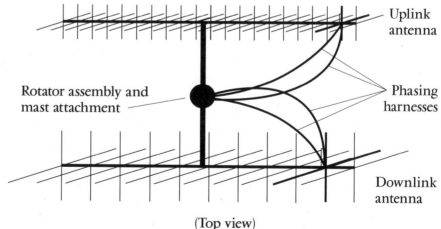

(Top view)
(Antennas aimed at horizon)

the uplink antenna has shorter elements than the downlink antenna, but higher gain because there are more elements. If the satellite employs circular polarization, the senses of both the uplink and downlink antennas must agree at the earth-based station and the satellite. The antenna system shown in this drawing is commonly used by radio amateurs.

For communication via LEO satellite systems, nondirectional antennas can be used at the ground stations. The ideal antenna is an isotropic radiator. Such an antenna can be placed on a house rooftop, car roof, boat mast, or even on a hiking backpack. The antenna orientation is not critical. However, it will work best when it is clear of surrounding obstructions such as buildings, which can cut off links to satellites close to the horizon.

Other Considerations

In some ways, satellites represent the ultimate communications solution, especially for media such as the Internet. But they aren't perfect.

Cost to Subscribers

One big problem with satellite communication is the fact that the equipment is comparatively expensive and is beyond the budgets of many individuals. This is especially true of earth-based stations designed for true wireless two-way communications via geostationary satellites. These installations require specialized dish antennas and moderate-power UHF or microwave radio transmitters. The monthly service fees can also be rather steep. Even amateur radio operators, who pay small fees or none at all, sometimes find themselves nearly broke as a result of having spent money on communications gear for use with satellites. These systems have a definite appeal to *technophiles*, people who love to fiddle around with electronic gadgets. Such high-tech "toys" can drain a hobbyist's bank account in a hurry.

One Internet system, available at the time of this writing at a fairly reasonable cost, makes use of a direct satellite downlink and employs the telephone lines for the uplink. This eliminates the need for a sophisticated transmitting installation at the subscriber site. The workstation requires a small dish (about 2 ft across) and a converter for receiving high-speed, broadband data from a geostationary satellite and feeding it to a computer. A conventional telephone modem is used to send data from the computer into the Internet. The *upstream data* (subscriber-to-Internet) is transmitted at a far lower speed than the *downstream data* (Internet-to-subscriber). Most of the data quantity, in terms of bytes, comes downstream; upstream data consists mostly of requests for websites, electronic mail (e-mail) messages, and other items that do not contain many bytes. For most individuals, especially those who like to "surf the Web," this arrangement works almost as well as a true two-way satellite link, at a fraction of the cost.

Atmospheric Effects

Active communications and broadcast satellites operate at VHF, UHF, and microwave frequencies. While EM waves at these frequencies are not affected by the ionosphere to the extent that waves at lower frequencies are affected, some fluctuation in polarization can take place as the energy passes through the ionized layers. One common effect is called *Faraday rotation*; it primarily affects systems that use *linear polarization*. It does not affect circularly polarized signals. Faraday rotation tends to occur to a lesser extent, and less often, as the frequency increases. At frequencies higher than about 10 gigahertz (GHz; wavelengths shorter than 3 centimeters), little or no Faraday rotation occurs.

Suppose there is a satellite that has a simple dipole antenna for transmitting. Imagine you have access to a powerful telescope and can look at the satellite and see that the dipole appears to be horizontally oriented relative to your location. If there is no Faraday rotation, then the signals you receive from the satellite will arrive horizontally polarized. However, if Faraday rotation takes place, the incoming EM field polarization will change. It might arrive with horizontal polarization at a certain instant in time; a few seconds later it might arrive with vertical polarization. Although the polarization of the incoming signal will be linear at any given instant in time, the orientation of the electric-field component of the EM wave will constantly change. This will result in severe fading if a linearly polarized receiving antenna is used at the ground station. A circularly polarized receiving antenna, however, will eliminate this problem.

In some cases, the *troposphere*, or the atmosphere within about 10 mi of the earth's surface, can affect the propagation of signals to and from satellites. In general, this effect tends to decrease as the frequency increases. It also decreases as the satellite's elevation angle increases.

Figure 10-8
The troposphere has
the least effect when
a satellite is high in
the sky (path X), and
the most effect when
a satellite is low in
the sky (path Y).

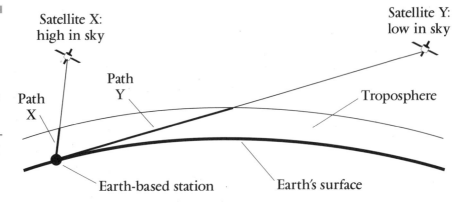

When a satellite is low in the sky, *tropospheric bending* can, at times, cause greatly increased path loss and fading. On rare occasions, *duct effect* can cut off a communications path completely. *Tropospheric scatter* increases path loss when satellites are low in the sky, all other factors being equal. This is because, at such times, the signals must pass through more air than is the case when the elevation angle is high (Fig. 10-8).

Natural and Artificial Obstructions

In order to effectively transmit and receive signals to and from a satellite, the ground station and the "bird" must be on a direct line of sight. Major obstructions, such as hills, mountains, and steel-frame buildings, will cut off the link. At microwave frequencies, even trees and houses can interfere considerably.

If you happen to live in a valley between mountain ranges, or downtown in a large city, it does not necessarily mean you cannot use a direct satellite link. If you happen to be lucky and can mount your antenna system somewhere that has a "clear shot" at a geostationary satellite, you can enjoy a reliable link no matter how atrocious the situation in general, because the satellite will always be in the same position in the sky. But if you are in a position from which no geostationary satellites lie on a line of sight, you cannot communicate via this mode unless you can find a place for your antenna that will allow communications, or else work through a repeater that is favorably situated and is designed to facilitate communications through a geostationary satellite.

Nongeostationary individual satellites cross the sky in various paths depending on a large number of factors. Therefore, even people in poor locations can usually communicate via such satellites on occasion. The fewer obstructions that exist, and the smaller the proportion of the sky that they obscure, the more often it will be possible to work through these satellites, and the longer the "openings" will last on the average.

In the case of LEO satellite systems, the more severe the obstruction problem, the less reliable the communications will be. The ideal LEO system is designed so that, if you are in an ideal location (say, the open sea or a farm in central Nebraska), you will always be able to establish a connection. In hilly country, such as is found in much of California or New England, you might find that communications can be maintained for 80 to 90 percent of the time. In deep mountain valleys, this percentage might drop to 60 or 70 percent. In the downtown area of a large city, it might diminish to practically zero in the worst locations. These figures are only rough estimates. The larger the number of satellites in the LEO system, the higher the proportion of the time that contact can be maintained in marginal locations.

Earth-Moon-Earth Communications

The moon is near enough to return signals from transmitters affordable to private citizens. *Earth-moon-earth* (EME) communications, also called *moonbounce,* is routinely done by amateur radio operators at VHF and UHF. The moon's orbit around the earth is an elliptical orbit. The distance between the earth and the moon varies between about 225,000 and 253,000 mi. The eccentricity of the moon's orbit is not very great, but it is enough to affect EME communications.

Equipment

A sensitive receiver, using a low-noise *preamplifier* and having narrow bandwidth, is mandatory if EME is contemplated. Some radio amateurs use high-frequency transceivers with converters. In recent years, good VHF and UHF gear has become available. These radios will receive moonbounce if used along with low-noise preamplifiers.

A high-gain antenna is always needed. Stacked Yagi arrays, or even matrixes (bays) of four or more Yagis, are common. A dish antenna is a good choice at frequencies above approximately 1 GHz. The transmitter output power should be at least several hundred watts. Radio amateurs almost always use the maximum legal limit of 1500 W.

Digital modes are required for reliable communication via EME. The signal-to-noise ratio is not usually good enough to allow communications by analog means such as SSB. Severe and rapid fading is common, and this will almost always introduce intolerable distortion into an analog signal when digital signals would be easily readable. Synchronized digital modes offer the best EME communication of all.

Technique

The main difficulty with moonbounce is signal path loss. Any signal received via EME is weak. Patience and a "good ear" are absolute requirements.

Another problem is that the high-gain antennas must always be kept aimed at the moon. This can be done using a *clock drive* and an *equatorial mount* for the array or dish, but the moon does not exactly follow the stars in the sky, so correction is needed to compensate for the moon's orbital motion around the earth.

Still another problem is *solar noise*. Moonbounce is most difficult near the time of the new moon, when the moon is near the sun in the sky. The sun is a broadband generator of EM energy. Problems can also occur when the moon passes near other noisy regions in the radio sky. The constellation Sagittarius lies in the direction of the center of the Milky Way galaxy, and reception is somewhat degraded when the moon is in this region.

The moon keeps the same face more or less toward the earth at all times, but there is some back-and-forth wobbling. This motion produces rapid, deep fluctuations in signal strength, known as *libration fading*. This effect becomes more pronounced as the operating frequency increases, but it is dramatic even at the lowest commonly used EME frequencies.

Path loss increases with increasing frequency. But this effect is offset by the more manageable size of high-gain antennas as the wavelength decreases.

Doppler Effect

Any wave effect, such as EM energy, has observable frequency and wavelength, both of which depend on the relative motion of the wave source and the observer. When a wave source moves radially with respect to an observer, there is an apparent change in the frequency and wavelength of energy from the source, compared to the situation when the source remains at the same distance from the observer. This phenomenon is known as the *Doppler effect*.

In Theory

If a wave source is getting closer to an observer, the apparent frequency rises, and the apparent wavelength shortens, compared with the case when there is no relative radial motion. If a wave source is getting farther away from an observer, the apparent frequency falls, and the apparent wavelength increases. The extent of frequency/wavelength change depends on the relative radial speed between the source and the observer, and also on the speed with which the wave disturbance propagates. For a given source of wave energy, various observers will witness emission frequencies and wavelengths that depend on the relative radial motion of each observer with respect to the source.

The theory of Doppler effect is shown in Fig. 10-9. In A, a source of EM waves and an observer are both stationary; there is no Doppler

Figure 10-9
Doppler effect.
Source stationary
relative to observer
(A); source receding
from observer (B);
source approaching
observer (C).

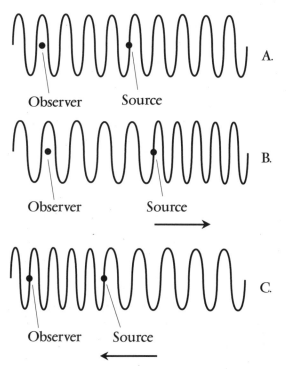

Observer   Source   A.

Observer   Source   B.

Observer   Source   C.

effect. B shows what happens when the source moves away from the observer; the wavefronts become "stretched out," and the observed frequency drops. C shows what happens when the source moves toward the observer; the wavefronts are "compressed," and the observed frequency rises. The extent of the frequency/wavelength change depends on the source speed relative to the speed of light.

Only the radial component of motion produces Doppler effect. If a source of wave energy is revolving in a circle around an observer who is at the center of the circle, for example, there is no relative radial motion, and thus there is no Doppler effect.

In Practice

Doppler effect is important to users of communications satellites other than the geostationary type. When a low-orbiting satellite makes a pass within range of a ground station, the received frequencies of all the downlink signals seem to constantly drop. When the satellite first comes over the horizon, it has an approaching radial speed vector. The observed downlink signal frequency is therefore higher than that actually emitted by the satellite. This speed vector decreases as the satellite comes nearer, reaching zero when the satellite is most nearly overhead; at that point, the observed downlink frequency is the same as the actual satellite downlink frequency. When the satellite has passed the point where it is most nearly overhead, it acquires a receding radial speed vector, causing the observed downlink frequency to fall below the actual emitted downlink frequency. This effect increases until the satellite disappears below the horizon. The higher the communication frequency, the more noticeable is the Doppler effect on the downlink signals. Also, the more nearly overhead the satellite passes, the more dramatic is the shift in frequency during the pass.

Communications in Deep Space

In the case of earth-orbiting spacecraft, conventional EM transmitters and receivers can be used for communications and data transfer. Even during the Apollo missions to the moon, the EM equipment was of the same sort, and communications was of basically the same nature, as with earth-based and earth-orbiting systems. But in deep space, should human beings venture there, the situation will be vastly different. In this section, two major obstacles to interstellar and intergalactic communications are discussed.

Path Loss

In earth-based and near-earth communications, *path loss* is overcome by means of high-gain antennas, high-power transmitters, and sensitive receivers. Engineers built equipment that could send signals to, and receive data from, the *Voyager* space probes even when the craft were in the outer reaches of the solar system. But communications with spacecraft traveling among the stars and galaxies will require transmitter power, receiver sensitivity, and antenna designs of a sort that humanity has not yet developed.

The well-known *inverse-square law* governs the path loss of an EM signal, in terms of power per unit surface area. Basically, the law is as follows: The incident EM power from a point source, striking a surface having a given area, and impinging on that surface at a right angle, diminishes according to the square of the distance between the source and the destination. For example, doubling the distance cuts the power field strength to one-fourth; multiplying the distance by 10 cuts the power field strength to one-hundredth.

Suppose you shine a flashlight at a wall 100 ft away and the resulting intensity of the beam is x microwatts per square meter ($\mu W/m^2$) on the wall. Then if you move to a position 200 ft from the wall, the beam will have an intensity of $x/4$ $\mu W/m^2$. The amount of power in the beam remains the same, but it shines over an area four times as great, so its power density is "diluted" by a factor of 4. This principle is shown in Fig. 10-10A. If you back off until you are 1000 ft from the wall, the power striking the wall will amount to only $x/100$ $\mu W/m^2$. Now imagine that you travel to a star 1000 *light-years* away and shine the flashlight at the wall. (A light-year is approximately 6,000,000,000,000, or 6×10^{12}, miles.) The incident power will be so tiny as to be undetectable by any devices now known to science. Of course, the star's light will wash out the flashlight beam anyway.

The intensity of an EM field at radio wavelengths is commonly specified in *microvolts per meter* ($\mu V/m$). This quantity diminishes according to the simple inverse of the distance. For example, doubling the distance cuts the voltage field strength in half; the electric-field component of

Figure 10-10
Inverse-square law
for EM field strength
in terms of power
(A); inverse law for
field strength in terms
of voltage (B).

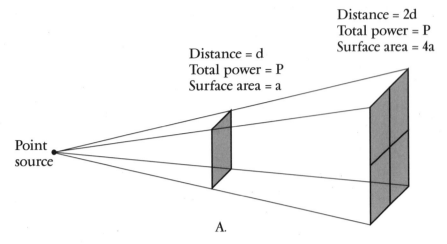

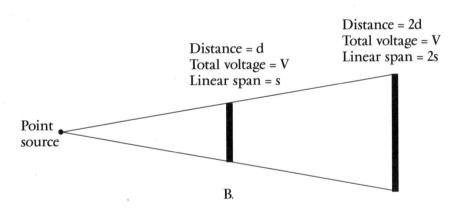

the wavefront is spread out over a span twice as long (Fig. 10-10B), so it is "diluted" by a factor of 2. Multiplying the distance by 10 cuts the voltage field strength to 1/10. At first it is tempting to think that this "inverse first-power law" is nowhere near as severe as the inverse-square law relating to path loss for power. But remember that power is proportional to the square of the voltage, assuming a constant impedance (the receiver input impedance). Thus the microvolts-per-meter decrement represents exactly the same state of affairs as does the microwatts-per-square-meter decrement.

Suppose two spacecraft are 6000 mi apart and one of them transmits a signal to the other, resulting in a 10-μV signal at the antenna input of the receiver. If the same two vessels are 1000 light-years apart, the field strength will be 0.00000000001 (10^{-11}) μV—far less than anything detectable by conventional radio receivers. Even if a receiver were built that would detect such a signal, that receiver would be subject to interference from cosmic noise sources, which might drown out the intended signal anyway. Moreover, a span of 1000 light-years is not a great distance on an interstellar or intergalactic scale. The Milky Way, our own galaxy, is believed to be approximately 100,000 light-years in diameter; the nearer galaxies are millions (1,000,000s) of light-years away, and some galaxies are billions (1,000,000,000s) of light-years away.

The foregoing discussions assume that outer space is a lossless medium. This would be true if space were a perfect vacuum, but it is not. Interstellar dust, gas, and "dark matter" result in losses in addition to simple divergence loss. The extent of this additional loss varies with the EM wavelength, but it is always present to some extent.

Propagation Delays

The speed of EM wave propagation in free space is a constant 186,282 miles (299,792 kilometers) per second. This is, for many practical purposes, an infinite speed for local wireless communications and networking. But in space communications, the speed of light is anything but infinite. When distances become vast, the speed of light becomes slow. On an interstellar and intergalactic scale, EM waves propagate so sluggishly that the mode is of little or no practical use.

The limitations of the "speed of light" begin to be noticed when geostationary satellites or EME modes are used. People who watched the TV coverage of the Apollo moon missions remarked about the long pauses in conversation that always followed questions and comments by earth-based personnel. These delays occurred because of the moon's distance from the earth; signals took 1.35 s to get from the earth to the moon and another 1.35 s to return, creating a total delay of 2.7 s (in addition to the pauses typical of human conversations). The moon astronauts noticed similar delays in the responses from earth.

When working with interplanetary probes, scientists contend with lags of minutes, and in some cases hours, between the time a control signal is transmitted and the time the results can be observed. The same delays will occur in communications among space travelers within the solar system. A round-trip signal delay of a few seconds can be tolerated (with difficulty) in two-way conversations. But when the time lag stretches to minutes, communications resemble exchanges of *electronic mail* (e-mail) or *voice mail* more than real conversations. Text and image data will probably be the preferred method of communications (rather than voice signaling) when human astronauts venture beyond the earth-moon system. When the time lag becomes a matter of hours, as will be the case if astronauts travel to the outer planets, true two-way communications will be impossible. Long e-mail messages will be the rule.

The nearest stars are several light-years from the earth. More distant stars are hundreds, and even thousands, of light-years away. On a galactic scale, astronauts need not travel far in order to get beyond communications range with the earth. A star 50 light-years from the solar system is sufficiently distant that, were a message sent to earth from its vicinity, the astronauts could not expect to receive a reply within their lifetimes. The Andromeda galaxy, one of the closest neighboring galaxies to our Milky Way, is about 2,000,000 light-years away. Astronauts traveling there will be riding on a one-way ticket. If they send an EM message to earth, chances are that by the time the signal arrives here, civilization will have forgotten all about them (if civilization as we know it still exists). Anyone familiar with *special relativity* knows that while it is theoretically possible to journey to the Andromeda galaxy, *time dilation* will place the travelers (and their signals) irretrievably into the future by 2,000,000 years. At present, there is no known way that this can be circumvented.

A technological breakthrough will be necessary if interstellar or intergalactic travelers are to have any contact with their fellow adventurers, or with people back home on the earth. Perhaps something akin to the "subspace" media of science fiction, wherein signals are transmitted through "hyperspace," thereby reducing or eliminating divergence loss and time lag, will be discovered. Most scientists today would feel silly spending time and money in search of such a scheme, but if humanity attempts to travel to other star systems or galaxies, attitudes will probably change.

Wireless on Other Planets

The nature of wireless communications on other planets will depend on the needs of the users and on the environment of the planet in question.

A planet with little or no ionosphere, such as the moon, Mars, or Mercury, will not allow "skip" of the sort shortwave listeners and amateur radio operators are familiar with. A planet with little or no atmosphere will allow no tropospheric bending, ducting, or scattering. Planets with powerful magnetic fields, such as Jupiter, might exhibit propagation phenomena never observed on the earth, for example, pronounced and constant auroral reflection. Planets with dense atmospheres will probably allow bending and ducting to a degree, and over a range of frequencies, far beyond anything observed on our planet.

As an example of a planet on which long-distance communications might be contemplated in the foreseeable future, consider the moon. It has essentially no atmosphere. There are no ionized layers. Thus EM waves always travel in essentially straight lines, regardless of frequency, with the possible exception of surface-wave propagation. Even at low radiation angles (takeoff angles close to the horizon), EM waves will not be bent or reflected back toward the surface unless an artificial reflector is used for that purpose. The moon might be ringed by passive satellites designed to reflect EM waves above a certain cutoff frequency. If there were a sufficient number of these satellites, they could simulate the effects of an ionospheric layer. Such satellites could be large metallic spheres, perhaps inflatable balloons similar to the Echo satellites used for earth communications during the 1960s.

It is not known whether surface-wave propagation will work at very low frequencies and low frequencies on the moon, in the way this mode operates on the earth. The conductivity of the moon's surface is probably not as good as that of the earth in general. However, it should be possible to use lower frequencies on the moon than can be used on the earth, because the lack of ionized layers eliminates the low-frequency cutoff caused by the waveguide effect. Because surface-wave propagation improves as the frequency decreases, it is not unreasonable to suppose that long-distance communications could be had on the moon at frequencies on the order of 1 kHz or less. This would limit the speed at which data could be transferred, but the tradeoff might be worthwhile for simple text messages.

Conditions on Mars are believed to be similar to those on the moon. While Mars has a thin atmosphere, it is probably not dense enough to cause appreciable bending. Some atmospheric scatter might take place. Surface-wave propagation might work fairly well on Mars, because it is thought to contain metallic oxides and small quantities of water.

The planet Mercury is not the sort of place astronauts would be likely to spend much time; propagation conditions there probably resemble those on the moon. The main problem with Mercury is the extreme heat, which can reach approximately 1000° Celsius.

There are planets in the solar system that might well support long-distance EM communications. Venus presents interesting possibilities for tropospheric propagation. It has a dense atmosphere that will probably cause significant bending and scattering. Jupiter has a strong magnetic field and a deep atmosphere, and this might support ionospheric and tropospheric propagation even better than does the earth. Saturn, Uranus, and Neptune all have atmospheres as well. The main problem with these planets is that their environments are probably too hostile to allow space travelers to land on them. Jupiter might not even have a surface as we know it. Jupiter, Saturn, and Uranus all have numerous moons, however, and some of these are likely candidates for space expeditions.

On some planets, communications might be had via subterranean currents. These systems would work in a manner similar to the Omega scheme used to communicate with submarines in the earth's oceans. But active LEO or planet-stationary satellite networks will probably be the media of choice for long-distance communications among humans who venture to other planets.

Location and Navigation Systems

372

Chapter 11

Wireless devices of all kinds are employed to pinpoint objects, people, and animals on the surface of the earth, under water, or in space. In addition, wireless devices can provide an indication of whether or not a moving vessel or vehicle is "on course." Radio-frequency (RF) location and navigation systems are the oldest; more recently developed devices function at infrared (IR) and visible wavelengths. Acoustic location and navigation systems also exist.

RF Systems

Radiolocation is a process whereby the position of a vehicle, aircraft, or ocean-going vessel is determined. *Radionavigation* is the use of radio apparatus, by personnel aboard moving vessels, for the purpose of plotting and maintaining a course.

Radiolocation

A radiolocation system can operate in various ways. The simplest method is the directional method. Two or three fixed receiving stations, separated by considerable distance, are used. *Radio direction-finding* (RDF) equipment is used at each of these stations, in conjunction with an omnidirectional transmitter aboard the vessel, to establish the bearings of the vessel with respect to each station. The vessel location corresponds to the intersection point of great circles drawn outward from the receiver points in the appropriate directions. This scheme requires that the vessel be equipped with a transmitter and that the captain wants to be located.

If the captain of a vessel does not want to be located, or if the radio equipment aboard the vessel is not working, other means of radiolocation are necessary. *Radar* can be used to locate such vessels. In wartime, enemy craft can sometimes be located by visual or IR apparatus. In recent years, satellites have been developed that can locate enemy ships and missiles, in some cases with an error smaller than the length of the vessel itself. One such satellite network is the *global positioning system* (GPS).

Radionavigation

There are several means of radionavigation. The intersecting-line method is probably the simplest. Two or three land-based transmitters are needed. Their locations must be accurately known. An RDF device on the vessel is used to determine the bearings of each of the transmitters. After the bearings have been determined, *great circles* are drawn on a map outward from each transmitter, so that the lines all intersect at a common point. That point represents the position of the vessel.

An alternative intersecting-line radiolocation method requires a transmitter aboard the vessel. The vessel's position is determined by land-based personnel with the aid of RDF equipment. The land stations, in communication with each other and with the captain of the vessel, can inform the captain of his or her position.

Regardless of the method used to locate the position of a vessel, radionavigation requires that frequent readings be taken and that the positions be compared in each case with the anticipated or desired position. Computers can make this comparison and provide a display for the captain of the vessel. In some cases a computer can even initiate necessary course changes.

Aircraft radionavigation can be performed by radar, or with RDF apparatus at the very high frequencies (VHF) or ultra-high frequencies (UHF). Air-traffic controllers constantly monitor the skies via radar. An airline pilot is automatically told if his or her plane is off course. The most sophisticated radionavigation techniques employ the GPS. The GPS has been adapted for use in automobiles, although the precision is not as great as is the case in military applications. An automobile equipped with a *cellular telephone* and GPS equipment can provide its driver with a measure of safety when driving in sparsely populated areas.

Loran

The term *loran* is an acronym derived from the words "long-range navigation." Loran is used by ships and aircraft for determination of position and for navigation.

There is one system of loran in common use today. The system, called *loran C,* uses pulse transmission at a rate of about 20 pulses per second. Two transmitters are used, and they are spaced at a distance of 300 miles (mi). The operating frequency is 100 kilohertz (kHz), which corresponds to a wavelength of 3000 meters (m). Location is found by comparing the time delay in the arrival of the signal from the more distant transmitter, relative to the arrival time of the signal from the nearer transmitter.

An older system of loran, which has now been discontinued, was *loran A.* This system operated in the range 1.85 to 1.95 megahertz (MHz), or a wavelength of about 160 m. Loran C is favored because the ground-wave propagation range is greater, resulting in better reliability.

Radar

The term *radar* is an acronym derived from the words "radio detection and ranging." Electromagnetic (EM) waves having certain frequencies reflect from various objects, especially if those objects contain metals or other good electrical conductors. By ascertaining the direction(s) from which radio signals are returned, and by measuring the time it takes for an EM pulse to travel from the transmitter location to a target and back again, it is possible to locate flying objects and to evaluate some weather phenomena. During World War II, this property of radio waves was put to use for the purpose of locating aircraft.

In the years following the war, it was discovered that radar can be useful in a variety of applications, such as measurement of automobile speed (by the police), weather forecasting (rain and snow reflect radar signals), and even the mapping of the moon and the planet Venus. Radar is extensively used in aviation, both commercial and military. In recent years, radar has found some use in robot guidance systems.

A complete radar set consists of a transmitter, a highly directional antenna, a receiver, and an indicator or display. The transmitter produces intense pulses of radio microwaves at short intervals. The pulses are propagated outward in a narrow beam from the antenna, and they strike objects at various distances. The reflected signals, or echoes, are picked up by the antenna shortly after the pulse is transmitted. The farther away the reflecting object, or target, the longer the time before the echo is received. The transmitting antenna is rotated so that all azimuth bearings (compass directions) can be observed.

A typical circular *radar display* consists of a *cathode-ray tube* (CRT). The basic display configuration is shown in Fig. 11-1. The observing station is at the center of the display. *Azimuth* bearings are indicated in degrees clockwise from true north and are marked around the perimeter of the screen. The distance, or *range*, is indicated by the radial displacement of the echo; the farther away the target, the farther from the display center the echo or blip. The radar display is therefore a set of *polar coordinates*. In this illustration, a target is shown at an azimuth of about 125 (east-southeast). Its range is near the maximum for the display.

Figure 11-1
Simplified rendition of
a typical radar display,
showing the station
and one target.

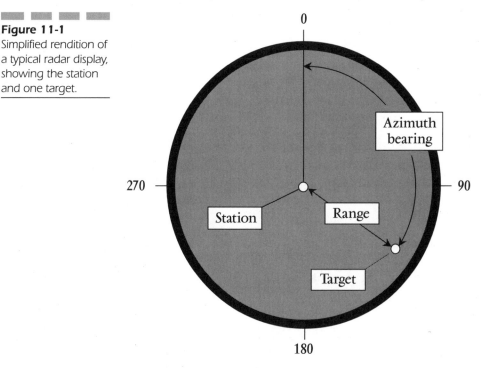

As the radar antenna rotates and echoes are received from various directions, the electron beam in the CRT sweeps in circles. The position of the sweep coincides exactly with the position of the antenna. The period of rotation can vary, but it is typically between 1 and 10 seconds (s). The transmitted pulse frequency must be high enough so that targets will not be missed as the antenna rotates.

When an echo is received, it appears as a spot on the screen, depending on the shape and size of the target. The CRT has a *persistence phosphor*, so that a blip remains visible for about one antenna rotation. This makes it easy to see the echoes on the screen, and it also facilitates comparison of the relative locations of targets in a group.

The maximum range of a radar system depends on several factors: the altitude of the antenna above the earth's surface, the nature of the terrain in the area, the transmitter output power and antenna gain, the receiver sensitivity, and the weather conditions in the vicinity. Airborne long-range radar can detect echoes from several hundred miles away under ideal conditions. A low-power radar system, with the antenna at a low height above the ground, might receive echoes from only 30 to 50 mi.

The fact that precipitation reflects radar echoes is a nuisance to aviation personnel, but it is invaluable for weather forecasting and observation.

Radar has made it possible to detect and track severe thunderstorms and hurricanes. A double-vortex thunderstorm, likely to produce tornadoes, causes a hook-shaped echo on radar. The eye of a hurricane shows up clearly on a radar display. Some radar sets can detect changes in the frequency of the returned pulse, thereby allowing measurement of wind speeds in hurricanes and tornadoes. This is called *Doppler radar.* This type of radar is also employed to measure the speeds of approaching or receding targets.

Sonar

The term *sonar* is an acronym based on the words "sound detection and ranging." Sonar can be used for determining location, plotting a course, determining distance, and for short-range *proximity sensing.* The basic principle is simple: Bounce acoustic waves off of objects, and measure the time it takes for the echoes to return.

An elementary sonar system consists of an acoustic pulse generator, an acoustic transmitting transducer (speaker), an acoustic receiving transducer (microphone), a receiver, a delay timer, and an indicating device such as a numeric display, CRT, or pen recorder. The transmitter sends out acoustic waves through the medium, which is usually water or air. These waves are reflected by objects, and the echoes are picked up by the receiver. The distance to an object is determined on the basis of the echo delay, provided the speed of the acoustic waves in the medium is known.

Acoustic waves travel faster in water than in air. The amount of salt in water makes a difference in the propagation speed when sonar is used on boats, for example, in depth finding. The density of water can vary because of temperature differences as well. If the true speed of the waves is not accurately known, false readings will result. In fresh water the speed of sound is about 4600 feet (770 fathoms) per second; in salt water it is about 4900 feet (820 fathoms) per second. In air, sound travels at approximately 1100 feet per second.

In the atmosphere, sonar can make use of audible sound waves, but ultrasound is often used instead. Ultrasound has a frequency too high to hear, ranging from about 20 to more than 100 kHz. One advantage of ultrasound is that the signals will not be heard by people working around machines equipped with sonar. Another advantage of ultrasound is the fact that it is less likely to be fooled by people talking, machinery operating, and other noises. At frequencies higher than the

Block diagram of
short-range sonar
system (A); a simple
sonar can be fooled
by long echo
delays (B).

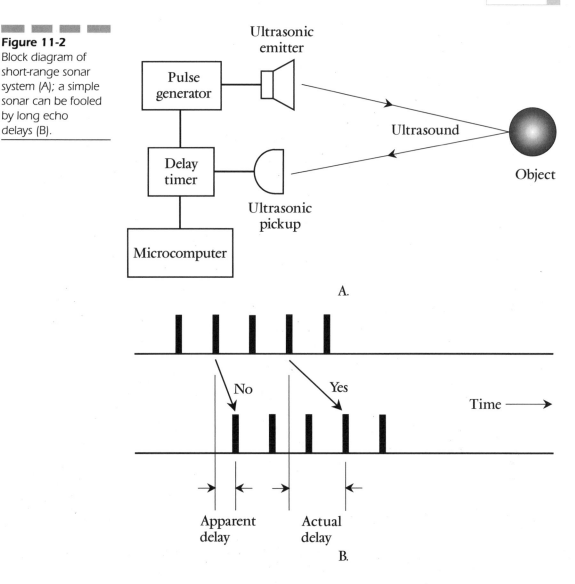

range of human hearing, acoustical disturbances do not normally occur
as often, or with as much intensity, as they do within the hearing range.

A simple robot sonar system is diagramed in Fig. 11-2A. An ultrasonic
pulse generator sends bursts of alternating current (AC) to a transducer.
This converts the currents into ultrasound, which is sent out in a narrow beam. This beam is reflected from objects in the nearby environment, and returns to a second transducer, which converts the ultrasound
back into pulses of AC. These pulses are delayed with respect to those

that were sent out. The length of the delay is measured, and the data fed to a microcomputer that determines the distance to the object in question. This system can provide a picture of the nearby environment if pulses are sent out in various directions, and a computer is incorporated to analyze the echoes. A *computer map* can then be generated on the basis of sounds returned from various directions in two or three dimensions.

One problem with this simple system is that it can be fooled if the echo delay is equal to or longer than the time interval between pulses. Suppose that the time between pulses is 0.1 s. Sound waves travel at about 1100 feet per second in air. If an object is farther than 55 feet (ft) away—say 75 ft—a simple sonar will not be able to tell which echo corresponds to the original pulse (Fig. 11-2B). In such a case a robot might think the object is 20 ft away (75−55=25). To get around this problem, several different frequencies can be used for the pulses. The computer can instruct the generator to send pulses of various frequencies in a *pseudorandom sequence*. In this way the computer can keep track of which echo corresponds to which pulse, even if the delay is several times the interval between pulses.

In its most advanced forms, sonar can rival *machine vision* systems as a means of mapping the environment. A living example of this is provided by bats, whose "vision" is actually sonar. Bats can navigate via their sonar as if they had keen eyesight, and in fact better in some sense, because illumination is not necessary. The bat's brain analyzes the echoes as efficiently as a bird's brain analyzes visual data.

In any advanced robot sonar system, some sort of *artificial intelligence* (AI) is mandatory. The robot controller must analyze the incoming pulses in terms of their phase, the distortion at the leading and trailing edges, and whether or not the returned echoes are *bogeys* (illusions). Another feature of advanced robot sonar is the ability to differentiate among objects in its vicinity. This is called *resolution*. For good resolution, the beam must be narrow, and it must be swept around in two or three dimensions. With excellent *direction resolution* and *distance resolution*, sonar can render a highly detailed map of the nearby environment.

Direction Finding

Wireless *direction finding* can be categorized broadly into two forms. The first form involves figuring out where you are, what direction you are going, and how fast you are moving. The second scheme involves determining the direction in which a vessel or machine lies, relative to your location. Direction finding is important to ship captains, airline pilots, hikers, and recreational boaters. It is also used in robotic systems.

Signal Comparison

A machine or vessel can find its position by comparing the signals from two fixed stations whose positions are known, as shown in Fig. 11-3A. By adding 180 degrees to the bearings of the sources X and Y, the machine or vessel (rectangle) obtains its bearings as "seen" from the sources (dots). The machine or vessel can determine its direction and speed by taking two readings separated by a certain amount of time. Computers can assist in precisely determining, and displaying, the position and the velocity vector.

Figure 11-3B is a block diagram of an *acoustic direction finder* such as can be used by a mobile robot. The receiver has a signal-strength indicator and a *servo* that turns a directional ultrasonic transducer. The transducer is equipped with a parabolic reflector, and is highly sensitive. There are two signal sources at different frequencies. When the transducer is rotated so the signal from one source is maximum, a bearing is obtained by comparing the orientation of the transducer with some known standard such as a magnetic compass. The same is done for the

Figure 11-3
A simple direction-finding scheme (A) and an ultrasonic direction finder (B).

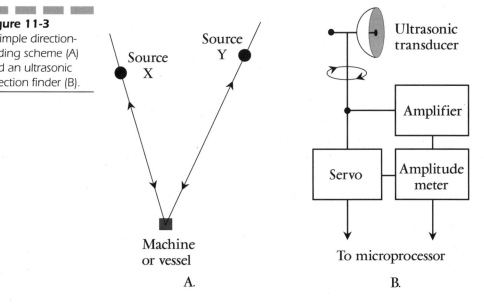

Source X

Source Y

Machine or vessel

A.

Ultrasonic transducer

Amplifier

Servo

Amplitude meter

To microprocessor

B.

other source. This information is sent to a microprocessor in the robot; the microprocessor uses triangulation to figure out the precise location of the robot.

Direction resolution is the ability of a direction-finding system to distinguish between two signal sources that lie in almost the same direction. Sometimes this is called *azimuth resolution*. It is given in degrees of arc. Two objects might be so nearly in the same direction that the direction-finding equipment perceives them as one and the same source. But if they are at different radial distances, the system can tell them apart by distance measurement. Also known as *ranging*, this can allow determination of position when triangulation is not possible.

RDF

A radio receiver, equipped with a signal-strength indicator and connected to a rotatable, directional antenna, can be used to determine the azimuth bearing (direction) from which signals are coming. *Radio direction-finding* equipment aboard a mobile vehicle facilitates determining the location of a transmitter. RDF equipment can be used to find one's own position with respect to two or more transmitters operating on different frequencies.

The receiver in an RDF system need not be sophisticated. A loop antenna is generally used. It is shielded against the electric component of radio waves, so it picks up only the magnetic flux. This minimizes inaccuracies that would otherwise be caused by the loop's tendency to operate as a simple mass of metal. The circumference of the loop is less than 0.1 wavelength at the operating frequency. Such an antenna displays a sharp null along a line passing through its center and perpendicular to the plane of the loop.

The loop is rotated until a dramatic dip occurs in the received signal strength. When the null is found, the axis of the loop lies along a line toward the transmitter. This is an ambiguous bearing, because it indicates two directions 180 degrees opposite each other. But when RDF readings are taken from two or more locations separated by a great enough distance, the transmitter can be pinpointed by finding the intersection point of the azimuth bearing lines as they are drawn on a map.

At longer wavelengths, RDF readings can be confused by the presence of steel-frame buildings, utility wires, and other objects that reflect and/or diffract EM waves. This problem, if it occurs, can usually be overcome by taking many readings from many locations.

At very short wavelengths, the physical size of an 0.1-wavelength-circumference loop is exceedingly small, and a loop antenna might not pick up enough signal energy to be useful. At ultra-high and microwave frequencies (above about 300 MHz), a directional transmitting/receiving antenna, such as a Yagi, quad, dish, or helical type, often gives better results than a small loop. When such an antenna is employed for RDF, the azimuth bearing is indicated by a signal maximum rather than by a null. This peak is less well defined than a loop null, but it is sharp enough for RDF if the antenna has high *forward gain* and therefore a narrow *main lobe*.

Ranging

Ranging is a method for a machine or vessel such as a ship, automobile, or aircraft to navigate in its environment. It also allows a central computer to keep track of the whereabouts of mobile robots. There are several ways for a machine to measure the distance between itself and some object.

Distance Resolution

Distance resolution is one important measure of the precision of a ranging system. It is the ability of the system to differentiate between two objects that are almost, but not quite, the same distance away. Distance resolution can be measured in feet, inches, meters, millimeters, or even smaller units. It can also be given as a percentage of the distance to an object.

When two objects are very close to each other, a distance-measuring system will sense them as a single object. But as the objects get farther apart, they become distinguishable. The minimum radial separation of objects necessary for a machine to tell them apart is the distance resolution for the ranging system. In Fig. 11-4, the two distant objects (*Y* and *Z*) are at almost exactly the same radial range from the machine or vessel. The delays, both long, are nevertheless almost identical. Depending on the precision of the system, the machine might or might not be able to distinguish between objects *Y* and *Z* on the basis of distance measurement alone. A direction indicator, however, would have no problem telling objects *Y* and *Z* apart.

Distance resolution sometimes depends on how far away the objects actually are. Often, nearby things can be resolved better than

Figure 11-4
Distance is proportional to echo delay time.

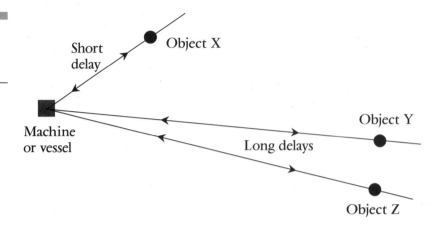

very distant ones. Suppose two objects are separated radially by 1 ft. If their mean (average) distance is 10 ft, their separation is 1/10 (10 percent) of the mean distance. But if their mean distance is 10,000 feet, their separation is 1/10,000 (0.01 percent) of the mean distance. If the distance resolution is 0.1 percent of the mean distance, then the machine can tell the nearer pair of objects apart, but not the more distant pair. Distance resolution depends on the type of ranging system used. The most sensitive methods compare the phases of laser beams arriving from, or reflected by, beacons.

Range Sensing and Plotting

Range sensing is used by vehicles and robots to measure distances in the environment in a single dimension. *Range plotting* is the creation of a graph of the distance (range) to objects, as a function of the direction in two or three dimensions.

To do one-dimensional (1D) range sensing, a signal is sent out, and the machine measures the time it takes for the echo to come back. This signal can be an acoustic wave, in which case the device is *sonar.* Or it can be a radio wave; this is *radar.* Laser beams can also be used. Localized 1D range sensing is called *proximity sensing.*

Two-dimensional (2D) range plotting involves mapping the distance to various objects, as a function of their direction. The accuracy of the rendition depends on how many readings are taken around a 360-degree circle. If a reading is taken every 5 degrees of arc, the resolution will be better than if a reading is taken every 10 degrees. Even better resolution can be obtained by taking readings every 1 degree or even every 0.5 degree. But no matter how fine the direction resolution, a 2D range plot can show things in only one plane, such as the floor level or some horizontal plane above the floor. To obtain a complete three-dimensional mapping, 360-degree maps would have to be made at numerous levels above the floor, and the result compiled by a computer in a manner similar to the way a *computerized axial tomography* scanner works in medical imaging.

Three-dimensional (3D) range plotting is usually done in spherical coordinates. The distance is measured for a large number of directions at all orientations. A 3D range plot would show ceiling fixtures, things on the floor, objects on top of the desk, and other details not visible with a 2D plot. To obtain reasonable resolution, thousands of readings are necessary. This requires considerable processing power and memory capacity on the part of the robot controller.

Eclipsing

Even the best 2D or 3D ranging systems have limitations. One major problem is the fact that nearby objects tend to eclipse more distant ones.

Imagine a robotic space vessel equipped with a range plotting radar, navigating its way through a dense swarm of asteroids. Each asteroid that happens to be on a direct line of sight will produce an echo, warning the vessel of its presence. However, many of the more distant asteroids will be eclipsed by nearer ones. This problem will exist for any type of range plotting system, be it sonar, radar, or laser, if the system depends on a single point of reference and a single plot. In fact, it will even exist for a human observer.

To obtain a true map of all the asteroids in the vicinity of the vessel, at least one (and preferably two or three) auxiliary vessels can travel along parallel paths with the main vessel. These vessels can each be equipped with 3D range plotting apparatus. The combined plots from the vessels can provide a complete 3D map of space in the vicinity. The higher the speed of the vessels relative to the asteroids, the larger the minimum separation between vessels in order to ensure safe passage. Plots must be taken at reasonably frequent time intervals.

Another way to get a fairly complete 3D map is to process range plots generated at closely spaced points in time. Asteroids eclipsed at a given time point will probably not be eclipsed at other time points. A continuous sequence of range plots can be processed by a computer, and the positions of most asteroids in the vicinity thereby inferred on a continuous basis. The greater the speed of the spacecraft relative to the asteroids, the shorter the maximum safe time interval between plots for safe passage. This scheme will not work quite as well as the auxiliary-spacecraft method, but it will be less expensive. Cost has been a consideration in electronic systems ever since the first telephone was put into use, and cost will always be a factor, even if someday we humans roam freely among the stars, navigating by radar and communicating via laser beams.

The Global Positioning System

The GPS is a network of radiolocation and radionavigation apparatus that operates on a worldwide basis. The system employs several satellites and allows determination of latitude, longitude, and altitude. The accuracy depends on the class of user (government/industry or civilian).

All GPS satellites transmit signals in the UHF part of the radio spectrum. The signals are modulated with special codes. These codes contain timing information used by the receiving apparatus to make measurements. A GPS receiver determines its location by measuring the distances to four or more different satellites. This is done by precisely timing the signals as they travel between the satellites and the receiver. The process is similar to the triangulation used in conventional radiolocation, except that with the GPS, it is done in three dimensions (in space) rather than in two dimensions (on the surface of the earth). A GPS receiver uses a computer to process the information received from the satellites. From this information it can give the user an indication of position down to a few meters (for government/industry subscribers) or a few hundred meters (for civilian subscribers). For either class of subscriber, the accuracy of the positioning readout is affected by the relative positions of the satellites with respect to the user. The larger the number of satellites involved, the more accurate the readout.

Radio signals propagate at about 186,282 miles, or 299,792 kilometers, per second through space and through the earth's atmosphere. There is some reduction in wave-propagation speed in the ionosphere. The extent of this reduction depends on the signal frequency. The GPS employs dual-frequency transmission to compensate for this effect. The delay difference is input to a computer, which generates a correction factor to cancel out the error.

An increasing number of new automobiles, trucks, and pleasure boats have GPS receivers preinstalled. If you are driving in a remote area and you get lost, you can use GPS to locate your position. Using a *cellular telephone, citizens band* (CB) radio, or *amateur radio* transceiver, you can call for help and inform authorities of your location. Arguably, every motor vehicle and boat should be equipped with GPS and cellular communications equipment to enhance the safety of passengers.

Epipolar and Log-Polar Navigation

With the exception of range plotting, location and navigation generally require observations from two or more reference points to obtain data. But if the geometry of the environment is known, the precise position and course of a vehicle can be determined on the basis of observations or readings taken from a single point of reference, usually from the vehicle itself.

Epipolar Navigation

Epipolar navigation is a means by which a machine can locate objects and/or find its way in 3D space. The scheme works by evaluating the way an image changes as viewed from a moving perspective. Your eyes and brain do this in an inexact manner without your having to consciously think about it. Machine vision systems, with the help of computers, can do it with extreme precision.

Imagine you're piloting an aircraft over the ocean. The only land you see is a small island. You have an excellent map that shows the location, the size, and the exact shape of this island. For instrumentation, you have a good video camera and a powerful computer. You let the computer work with the image of the island. As you fly along, the island seems to move underneath you, although you know it is stationary on the ocean surface. You aim the camera at the island and keep it there. The computer "sees" an image that constantly changes shape. Figure 11-5 shows three sample sighting positions (A, B, C) and the size/shape of the island as seen by the computer vision system in each case. The computer has the map data, so it knows the true size, shape, and location of the island. The computer compares the shape/size of the image it sees at each point in time, from the vantage point of the aircraft, with the actual shape/size of the island

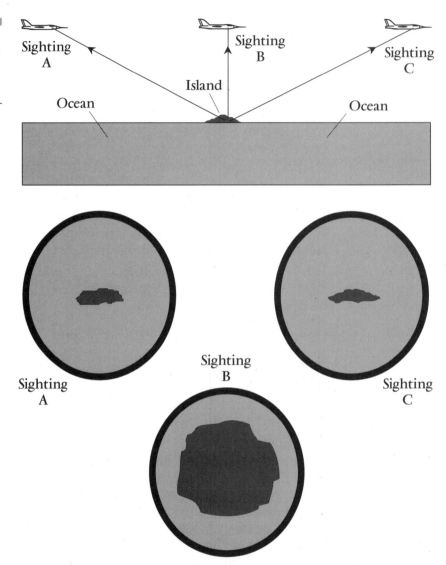

Figure 11-5
Epipolar navigation is
the electronic analog
of human spatial
perception.

that it knows from the map data. From this alone, it can tell you your altitude, your speed relative to the surface, your exact latitude, and your exact longitude. There exists a one-to-one correspondence between all points within sight of the island, and the size/shape of the island's image. The correspondence is far too complex for you to memorize exactly, although, if you have flown in that area many times, you will have an idea of your position and speed based on what the island looks like. But the computer has it all in storage. For the computer, matching the image it sees with a particular point in space is merely a series of mathematical calculations.

Epipolar navigation can function on any scale, at any speed. It can take relativistic effects into account for spacecraft traveling at speeds that are a significant fraction of the speed of light. Epipolar navigation is also a method by which robots can find their way without triangulation, direction finding, beacons, sonar, or radar. It is only necessary that the robot have a *computer map* of its environment.

Log-Polar Navigation

When mapping an image for use in a robotic vision system, it sometimes helps to transform the image from one type of coordinate system to another. In *log-polar navigation,* a computer converts an image in *polar coordinates* to an image in *rectangular coordinates.*

The principle of log-polar image processing is shown in Fig. 11-6. The polar system, with two object paths plotted, is shown at A. The rectangular equivalent, with the same paths shown, is at B. The polar radius (A) is mapped onto the vertical rectangular axis (B); the polar angle (A) is mapped onto the horizontal rectangular axis (B).

Radial coordinates are unevenly spaced in the polar map (A) but are uniform in the rectangular map (B). During the transformation, the logarithm of the radius is taken. This results in peripheral "squashing" of the image. The resolution is degraded for distant objects but is improved nearby. In robotic navigation, close-in objects are usually more important than distant ones, so this is a good tradeoff.

A log-polar transform greatly distorts the way a scene looks to people. A computer does not care about that, if each point in the image corresponds to a unique point in real space—that is, if the point mapping is a one-to-one correspondence. Machine vision systems employ television (TV) cameras that scan in rectangular coordinates, but events in real space are of a nature better represented by polar coordinates. The transform changes real-life motions and perceptions into images that can be efficiently dealt with by a TV scanning system.

Wireless Sensors

Wireless sensors are extensively used in robotics for determining the presence and evaluating the characteristics of nearby objects. The following are some examples.

Figure 11-6
Polar coordinates in
real space (A), and
coordinates after log-
polar transform (B).
Lines X and Y show
hypothetical paths in
each system.

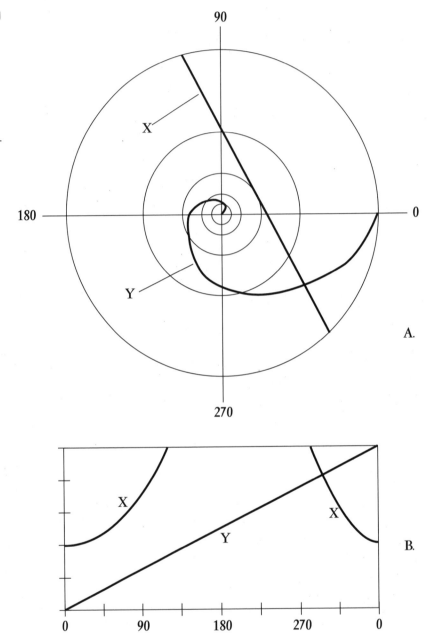

Capacitive Proximity Sensing

One of the simplest devices used for proximity sensing takes advantage of capacitive coupling. You have probably noticed the effects of capacitance when using portable TV sets or portable headphone radios. With a TV set using "rabbit ears" (an antenna mounted directly on the set), bringing your hand near either of the rods will often change the reception, especially if the station you are watching is weak. Or, if you are jogging on the beach or track with a portable headphone radio, you have almost certainly noticed how stations tend to fade in and out (especially during your favorite tunes). Part of this fading is caused by the effect of the earphone wire, which serves as an antenna, moving with respect to your body.

A *capacitive proximity sensor* uses an RF oscillator, a frequency detector, and a metal plate connected into the oscillator circuit, as shown in Fig. 11-7. The oscillator is designed so that a change in the capacitance of the plate, with respect to the environment, will cause the oscillator frequency to change. This change is sensed by the frequency detector, which sends a signal to the apparatus to be controlled. In this way, for example, a robot can avoid bumping into things.

Objects that conduct electricity to some extent, such as house wiring, people, cars, or refrigerators, are sensed more easily by capacitive transducers than are things that do not conduct, like wood-frame beds and desks. Therefore, other kinds of proximity sensors are often needed for a robot to navigate well in a complex environment, such as a household.

Figure 11-7
A capacitive proximity sensor.

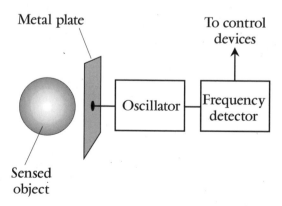

Photoelectric Proximity Sensing

Reflected light can provide a way for a robot to tell if it is approaching something. A *photoelectric proximity sensor* uses a light-beam generator, a *photodetector,* a special amplifier, and a microprocessor.

The light beam reflects from the object and is picked up by the photocell. The light beam is modulated at a certain frequency, say 1000 Hz, and the detector has a frequency-sensitive amplifier that responds only to light modulated at that frequency. This prevents false imaging that might otherwise be caused by lamps or sunlight. If the robot is approaching an object, its microprocessor senses that the reflected beam is getting stronger. The robot can then steer clear of the object.

This method of proximity sensing will not work for black objects, or for things like windows or mirrors approached at a sharp angle. These generally fool this system, because the light beam is not reflected back toward the photodetector.

Texture Sensing

A device similar to a photoelectric proximity sensor can be used to determine the smoothness or roughness of a surface. Primitive *texture sensing* can be done using a laser and several light-sensitive sensors.

Figure 11-8 shows how a laser L can be used to tell the difference between a shiny surface (at A) and a rough or matte surface (at B). The

Figure 11-8

Lasers (L) and sensors (S) analyze surfaces. A shiny surface (A) and a rough or matte surface (B).

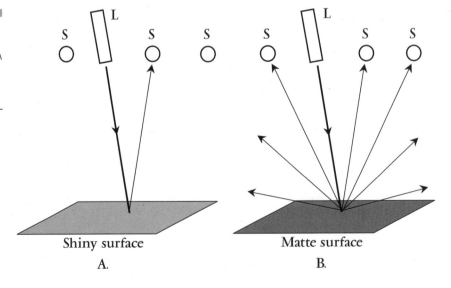

Shiny surface

A.

Matte surface

B.

shiny surface, say the polished hood of a car, tends to reflect light according to the familiar rule: The angle of reflection equals the angle of incidence. But the matte surface, say the surface of a sheet of drawing paper, will scatter the light. The shiny surface will reflect the beam back almost entirely to one of the sensors S positioned in the path of the beam whose reflection angle equals its incidence angle. The matte surface will reflect the beam back more or less equally to all of the sensors (and in many other directions).

This type of texture sensor cannot give an indication of the extent of roughness of a surface. It can only determine that a surface is either shiny or not shiny. A piece of drawing paper would reflect the light in much the same way as a sandy beach or a new-fallen layer of snow.

Robot Guidance

There are several methods via which robots can use wireless devices to navigate in their surroundings. The following are some of the more common schemes.

Beacons

A *beacon* can be categorized as either passive or active. *Passive beacons* operate without the need for a source of power; they are usually simple reflecting objects. *Active beacons* require power because they generate or retransmit signals.

A mirror is a good example of a passive beacon. It does not produce a signal of its own; it merely reflects rays of visible or IR energy that strike it. The robot requires a transmitter, such as a flashing lamp or laser beam, and a receiver, such as a photocell. The distance to each mirror can be determined by the time required for the flash to travel to the mirror and return to the robot. Because this delay is an extremely short interval of time, high-speed measuring apparatus is needed. A specialized mirror, called a *tricorner reflector*, always reflects a ray of light or IR toward its source, regardless of the direction from which the ray arrives. This passive device consists of three square, flat mirrors, attached to each other at mutual right angles, in the same way two adjacent walls of a room meet the ceiling.

An example of an active beacon is a radio transmitter. Several transmitters can be put in various places, and their signals synchronized so that they are all exactly in phase. As the robot moves around, the relative phases of the signals will change. Using an internal computer, the robot can determine its position by comparing the phases of the signals from the beacons. The system works like a small-scale version of the GPS. With active beacons, the robot does not need a transmitter, but the beacons must have a source of power and be properly aligned. In some active-beacon systems, the beacons are repeaters that receive signals from the robot, process them in some way, and then retransmit them back to the robot.

Edge Detection

The term *edge detection* refers to the ability of a robotic vision system to locate boundaries. It also refers to the robot's knowledge of what to do with respect to those boundaries.

A *robot car,* for example, uses edge detection to see the edges of a road, and uses the data to keep itself on the road. But it also needs to stay a certain distance from the center line, so that it does not cross over into the lane of oncoming traffic (Fig. 11-9). It must stay off the road shoulder. Thus it must tell the difference between pavement and other surfaces, such as gravel, grass, sand, and snow. The robot car could use beacons for this purpose, but this would require the installation of the guidance system beforehand. That would limit the robot car to roads that are equipped with such navigation aids.

A *personal robot,* of the type people often imagine doing work similar to that of a maid or butler, would need to see the edges of a door before going through the door. Otherwise it might run into walls, or into closed doors, or into windows. It will need to be able to detect the presence of a stairwell also, especially if the stairs descend.

Embedded Path

An *embedded path* is a means of guiding a robot along a specific route. One common type of embedded path is a buried, current-carrying wire. The current in the wire produces a magnetic field that the robot can follow. This method of guidance has been suggested as a way to keep a car on a highway, even if the driver is not paying attention. The wire needs a constant supply of electricity for this guidance method

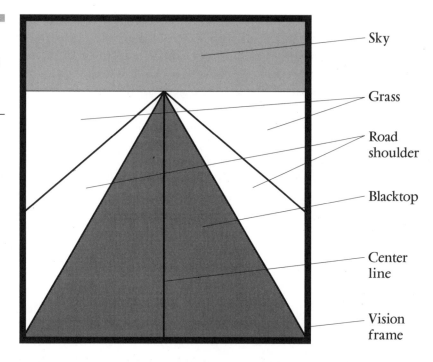

Figure 11-9
Edge detection.
Boundaries between
different objects or
substances show up
as well-defined lines.

Sky

Grass

Road
shoulder

Blacktop

Center
line

Vision
frame

to work. If this current is interrupted for any reason, the robot will lose its way.

Alternatives to wires, such as colored paints or tapes, do not need a supply of power, and this gives them an advantage. Tape is easy to remove and put somewhere else; this is difficult to do with paint, and practically impossible with wires embedded in concrete. Robots follow these devices using edge detection; the brightly colored or reflective media can assist the machine in keeping track of the boundary.

A large number of passive beacons, placed at regular intervals along a boundary, might serve as a hybrid beacon/edge/path system. The little plastic "dots," designed to reflect light back toward the source and commonly found on highways in the southern United States, could serve this purpose well. The "dots" would have to be kept in good condition and would not be reliable in places where snow falls or sand and dust accumulates.

Eye-in-Hand System

For a robot *end effector* ("hand") to find its way, a camera can be placed in the gripper mechanism. The camera must be equipped for work at close range, from several feet down to a fraction of an inch. The positioning error must be as small as possible. To be sure that the camera gets a good image, a lamp is included in the gripper along with the camera. This *eye-in-hand system* can be used to precisely measure how close the gripper is to whatever object it is seeking. It can also make positive identification of the object, so that the gripper does not go after the wrong thing.

The eye-in-hand system uses a *servo* that detects errors in the positioning and sends *feedback* signals to the robot controller. The controller processes these signals and tells the end effector to correct its position. The controller can tell when the end effector is "on course" because the error signal decreases to zero. The principle is the same as that used in automatic RDFs.

Biased Search

A *biased search* is a method of finding a destination or target, by first looking off to one side and then "zeroing in." Instead of making the most precise possible estimate of the destination (the most obvious scheme), the robot controller deliberately chooses a course that is inaccurate. The controller chooses its course on the basis of its general knowledge of the environment in which the robot operates. If the chosen course is to the left of the destination, for example, the controller "knows" this, having deliberately sent the robot on this course. Once a known landmark in the environment is spotted, the controller can use its database to calculate the position of the destination. The course is then set directly for the intended destination.

Figure 11-10 shows the technique of biased searching as a boater might apply it when approaching the dock in a dense fog. At some distance from the shoreline, the boater cannot see the dock but has a reasonably good idea of where it is. Therefore, an approach is deliberately made well off to one side (in this case, to the left) of the dock. When the shore comes into view, the boater turns to the right and follows the shoreline at a safe distance until the dock is found. For this scheme to work, the sense of the error (to the left or right) must be correct. If the captain in the example of Fig. 11-10 were mistaken, and the dock were actually to the left of the initial course, the boat would end up going the wrong direction. For this reason, the initial error must be large, so its sense is known with virtual certainty.

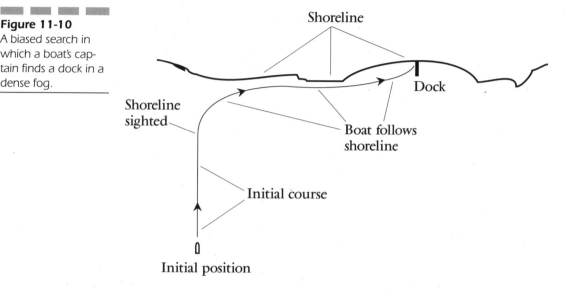

Figure 11-10
A biased search in
which a boat's cap-
tain finds a dock in a
dense fog.

Fluxgate Magnetometer

A *fluxgate magnetometer* is a magnetic compass for robot guidance. The device uses coils to sense changes in the earth's magnetic field or in an artificially generated reference field. The output of the fluxgate magnetometer is sent to the robot controller.

Navigation within a defined area can be done by checking the orientation of magnetic lines of flux in fields generated by electromagnets in the walls of the room. For each point in the room, the magnetic flux has a unique orientation and intensity. There exists a one-to-one correspondence between magnetic flux conditions and the points inside the room. The robot controller can be programmed with this information. The data allows the robot to pinpoint its position in the room, based only on the magnetic-field orientation and intensity.

Gyroscope

A *gyroscope* or *gyro* can be useful in robot navigation. It is a form of *inertial guidance system*, operating from the fact that a rotating, heavy disk tends to maintain its orientation in space. Gyroscopes are used when it is inconvenient or impossible to obtain bearings in any other way, such as radar, compass-based navigation, or visual navigation. The disk is mounted in a gimbal or set of bearings that allows it to turn up and down or from side to side. The disk is usually driven by a motor.

A common use for a gyro is to keep track of the direction of travel (bearing or heading) of a robot or vessel without having to rely on any external signals, fields, or points of reference. A gyroscope must be realigned at regular intervals because it will tend to slowly change its orientation with time.

Binaural Hearing

Even with your eyes closed, you can usually tell from which direction a sound is coming. This is because human beings have *binaural hearing*. Sound arrives at your left ear with a different intensity, and in a different phase, than it arrives at your right ear. Your brain processes this information, allowing you to locate the source of the sound, with certain limitations. If you are confused, you can turn your head until the sound direction becomes apparent to you.

Robots can be equipped with binaural hearing. Two acoustic transducers are positioned on either side of the robot. The robot controller compares the relative phase and intensity of the signals from the two transducers. This lets the robot "know," with certain limitations, the direction from which the sound is coming. If the robot is confused, it can turn until the confusion is eliminated and a meaningful bearing is obtained. If the positions of various sound sources are known, the robot can navigate its way via this directional sense of machine hearing. For this scheme to work, the robot controller must be programmed with information concerning the locations of the sound sources (beacons).

Machine Vision

One of the most advanced specialties in electronics is the field of *vision systems*, also called *machine vision*. There are several different schemes. The optimum form of machine vision depends on the intended use. Vision systems are especially important in high-level computer and robotics applications.

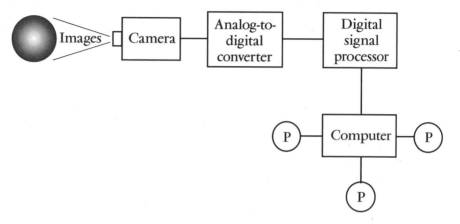

Figure 11-11
Components of a visi-
ble-light machine-
vision system. Circles
labeled P are com-
puter peripherals.

Components

A visible-light vision system must have a device for receiving incoming images. This is usually a *vidicon* camera tube or a *charge-coupled device*. In bright light an *image orthicon* can be used.

The camera produces an analog video signal. For best results, this must be processed into digital form. This is done by *analog-to-digital conversion*. The digital signal is clarified by *digital signal processing*. The resulting data goes to the computer or robot controller. A block diagram of this scheme is shown in Fig. 11-11.

The moving image, received from the camera and processed by the circuitry, contains an enormous amount of information. It is easy to present a computer with a detailed moving image. Getting the machine to "know" what's happening is a more complex affair. Processing an image, and extracting meaning from it, is a major challenge for vision-systems engineers.

Bar Coding

Bar coding is a method of labeling objects for identification. You have seen bar-code labels or tags in stores; they are used for pricing merchandise. A bar-code tag has a characteristic appearance, with parallel lines of varying width and spacing. A laser-equipped device scans the tag, thereby retrieving the identifying data. The reading device does not have to be brought right up to the tag; it can work from some distance away.

Bar-code tags are one method by which objects can be labeled so that a robot can identify them. This greatly simplifies the recognition process. A whole tool set can be tagged using bar-code stickers with a different code for each tool. When a robot's program tells it that it needs a certain tool, the robot can seek out the appropriate tag and carry out the movements according to the program subroutine for that tool. Even if the tool gets misplaced, as long as it is within the robot's range of motion, it can easily be found again when it is needed.

Passive Transponders

A *passive transponder* is a device that allows a robot to identify an object. Bar coding is one example. Magnetic labels, such as those on credit cards, automatic-teller-machine (ATM) bank cards, and store merchandise, are another example.

All passive transponders require the use of a specialized sensor in the robot. The sensor decodes the information in the transponder. The data can be quite detailed. The transponder can be tiny, and its information can be sensed from several feet away.

Suppose a robot needs to choose a drill bit for a certain application. There might be 150 bits in a tray, each containing a magnetic label with information about its diameter, hardness, recommended operating speeds, and its position in the tray. The robot would quickly select the best bit, install it, and use it. When the robot was done, the bit would be put back in its proper place.

Optical Character Recognition

Computers can translate printed matter, such as the text on road signs, into digital data. The data can then be used in the same way as if it had been entered into the computer using a keyboard. This is done by means of *optical character recognition* (OCR), also called *optical scanning*.

In OCR of printed matter, such as this page, a thin laser beam moves across the page. White paper reflects light; black ink does not. The laser beam moves in the same way as the electron beam in a television camera or picture tube. The reflected beam is intensity-modulated. This modulation is translated by OCR software into digital code for use by the computer. In this way, a computer can "read" a magazine or book.

Smart robots can incorporate OCR technology into their vision systems, enabling them to read labels and signs. The technology exists, for example, to build a robot equipped with OCR that could get in your car and drive it anywhere, reading road signs for guidance (assuming the routes were adequately posted with signs). Perhaps someday this will be commonly done. You might hand your robot a shopping list and say, "Please go get these things at the supermarket," and the robot will come back an hour later with three bags full of groceries.

For a robot to read something at a distance, such as a road sign, the image is observed with a TV camera, rather than by reflecting a scanned laser beam off the surface. This video image is then translated by OCR software into digital data.

Color Sensing

Robot vision systems often function only in gray-scale, like simple TV. But *color sensing* can be added, in a manner similar to the way it is incorporated into TV systems.

Color sensing can help a smart robot determine what an object is. Is that horizontal surface a floor inside a building, or is it a grassy yard? If it is green, it's probably a yard. Sometimes, objects have regions of different colors that have identical brightness as seen by a gray-scale system. These objects can be seen in more detail with a color-sensing system.

Figure 11-12 is a functional block diagram of a color-sensing system. Three gray-scale ("black-and-white") cameras are used. Each camera has a color filter in front of its lens. One filter is red, another is green, and another is blue. These are the three primary colors. All possible hues, brightnesses, and saturations are comprised of these three colors in various ratios. The signals from the three cameras are processed by a microcomputer, and the result is fed to the robot's AI center.

Object Recognition

Object recognition refers to any method that a robot uses to pick something out from among other things. An example is getting a tumbler from a cupboard. It might require that the robot choose a specific object, such as "Jane's tumbler."

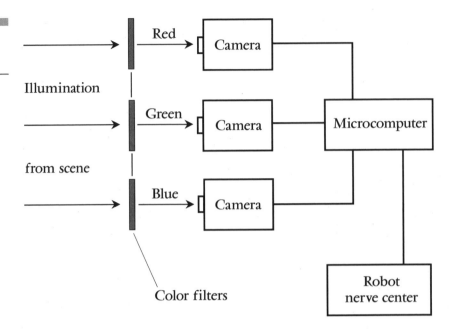

Suppose you ask your personal robot to go to the kitchen and get you a tumbler full of orange juice. The first thing the robot must do is find the kitchen. Then it must locate the cupboard containing the tumblers. One way for the robot to find a tumbler is with a vision system to identify it by shape. Another method is *tactile sensing*. The robot could double-check, after grabbing an object it thinks is a tumbler, whether it is cylindrical. If all the tumblers in your cupboard weigh the same, and if this weight is different from that of the plates or bowls, the robot can use weight to double-check that it has the right object.

If a particular tumbler is required, then it will be necessary to have them marked in some way, such as with bar-coding labels or passive transponders.

Pattern Recognition

In a robot vision system, one way to identify an object or decode data is by shape. The machine recognizes combinations of shapes and deduces their meanings using a microcomputer.

Imagine a personal robot that you keep around the house. It might identify you because of combinations of features, such as your height, hair color, eye color, voice inflections, and voice accent. Perhaps, with sophisticated enough technology, your personal robot could instantly recognize your face, just as your friends do. But this would take up a huge amount of memory, and the cost would be high. There are simpler means of identifying people.

Suppose your robot is programmed to get the attention of and look into the eyes of anyone who enters the house. In this way, the robot gets the *iris print* of the person. It has a set of authorized iris prints stored in its computer. If the robot does not recognize the iris pattern of a person entering the house, the robot can actuate a silent alarm.

Local Feature Focus

In a robotic vision system, it is rarely necessary to use the whole image to perform a function. Often, only a single feature or a small region within the image, is needed. To minimize memory space and to optimize speed, *local feature focus* can be used.

Suppose a robot needs to get a pair of pliers from a toolbox. This tool has a characteristic shape that is stored in memory. Several different images might be stored, representing the pliers as seen from various angles. The vision system quickly scans the toolbox until it finds an image that matches one of its images for pliers. This saves time, compared with trial and error, in which the robot picks up tool after tool.

Your mind/eye system uses local focus without thinking. If you are driving in a wilderness area and you see a sign that says "Watch for deer," you will be on the lookout for animals on or near the roadway. A tractor parked on the shoulder won't arouse you, but a horse probably will.

Vision and AI

There are subtle things about an image that a machine will not notice unless it has an advanced level of AI. How, for example, is a robot to know whether an object presents a threat? Is that four-legged thing a dog or a tiger? How is a machine to know the intentions (if any) of an object? Is that biped object a human being or a mannequin? Why is it carrying a stick? You can determine immediately if a person is carrying a tire iron to help you replace a flat tire or if the person intends to smash your windshield. It is important for a police robot or a security robot to know what constitutes a threat and what does not.

The variables in an image are much like those in a human voice. A vision system, to get the full meaning of an image, must be at least as sophisticated as a speech-recognition system. AI technology has not yet reached the level required for human-like machine vision and image processing.

Sensitivity and Resolution

Two important specifications in any vision system are the *sensitivity* and the *resolution.*

Sensitivity is the ability of a machine to see in dim light or to detect weak impulses at invisible wavelengths. In some environments, high sensitivity is necessary. In others, it is not needed and might not be wanted. A machine that works in bright sunlight does not need to see well in a dark cave. A robot designed for working in mines or in pipes or in caverns must be able to see in dim light, using a system that might be blinded by ordinary daylight.

Resolution is the extent to which a machine can differentiate between objects. The better the resolution, the keener the vision. Human eyes have excellent resolution, but machines can be designed with greater resolution. In general, the better the resolution, the more confined the field of vision must be. To understand why this is true, think of a telescope. The higher the magnification, the better the resolution (up to a certain point). But increasing the magnification reduces the angle, or field, of vision. Zeroing in on one object or zone is done at the expense of other objects or zones.

Sensitivity and resolution depend somewhat on each other. Usually, better sensitivity means a sacrifice in resolution. Also, the better the resolution, the less well the vision system will function in dim light.

"Invisible" and Passive Systems

Machines have an advantage over people when it comes to vision. Machines can see at wavelengths to which we humans are blind.

Human eyes are sensitive to EM energy whose wavelength ranges from about 390 to 750 nanometers (nm). The nanometer is 0.000000001 (10^{-9}) of a meter. The longest visible wavelengths look red. As the wavelength gets shorter, the color changes through orange, yellow, green, blue, and indigo. The shortest waves look violet. Energy at wavelengths somewhat longer than 750 nm is *infrared* (IR); energy at wavelengths somewhat shorter than 390 nm is *ultraviolet* (UV).

Some nonhuman creatures and machines can see energy at wavelengths to which human eyes do not respond. Insects can see UV that people cannot, and are blind to red and orange light that people can see. (Maybe you have used orange "bug lights" when camping, or those UV things that attract bugs and then zap them.) A robot might be designed to see IR and/or UV, as well as or instead of visible light. Video cameras can be sensitive to a range of wavelengths much wider than the range we humans can see.

Machines can be made to see in an environment that is dark and cold, and that radiates too little energy to be detected at any EM wavelength. In these cases the machine provides its own illumination. This can be a simple lamp, a laser, an IR device, or a UV device. Or the machine might emit radio waves and detect the echoes; this is radar. Some robots can navigate via acoustic (sound) echoes, like bats; this is sonar.

Computer Mapping

In order to function, an autonomous robot must have a controller that knows, at all times, where the machine is located relative to its surroundings. It must make a *computer map* of its environment. A computer map resembles a blueprint.

An example of a computer map is shown in Fig. 11-13. This happens to be a dining room. The robot's assignment is to set the table before the people sit down. The robot can envision, relative to the table and chairs, where to put the plates. Then it can follow its programming and place the napkins, forks, knives, spoons, and glasses in their proper places.

Note that one chair has been moved into the corner, and is not at the table. This might confuse the robot and result in its setting an extra place at the table and/or dropping utensils on the carpet in the corner. Whether or not these problems occur will depend on the quality of the map-interpretation program. The robot must be able to act appropriately under a variety of conditions.

The robot also needs a map of the kitchen, where all these utensils are to be found. It will need to recognize the items, besides knowing in which drawers or cupboards they are. Human beings can solve these "problems" with minimal conscious thought, but in terms of numbers of digital data bits, such tasks are formidable.

Figure 11-13
A computer map, in this case of a dining room.

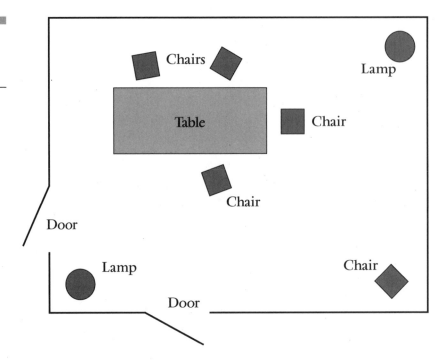

Seeing-Eye Robot

There are as many possible applications for robots as for people and animals combined. One intriguing idea is to use robots as "seeing-eye dogs" for the blind.

A *seeing-eye robot* would need excellent machine vision. It would have to guide the human user as effectively as a trained seeing-eye dog would. The machine would need AI equivalent to that of a dog. It would also need to be highly mobile, doing such diverse things as crossing a street, going through a crowded room, or climbing stairs.

A number of years ago, the Japanese, with their enthusiasm for robots that resemble living creatures, designed a seeing-eye robot called *Meldog*. It was about the size of a live seeing-eye dog. It was primitive, however, because of the limited power of its controller. Even if seeing-eye robots become affordable, some blind people would rather have a real dog. There's nothing like being awakened in the morning by a friendly lick on the face. No robot will ever be able do that as well as a live dog.

Computers, Robots, and the Internet

Wireless technology is an integral part of the so-called digital revolution. From computer accessories to mobile robots, electromagnetic (EM) links are taking the place of wires, cables, and even fiberoptics.

Computer Conveniences

In *personal computer* systems, accessories and peripherals lend themselves to wireless technology. The problems with cords and cables are well known. They eventually wear out or develop intermittent electrical failure, and they place a limit on the distance between the computer's main unit and the accessory or peripheral.

Mouse

A typical *cordless mouse* employs an infrared (IR), very-high-frequency (VHF), or ultra-high-frequency (UHF) transmitter and receiver. The transmitter is inside the mouse, and the receiver is contained in a small box attached to the computer's main unit by a cord. The box can be placed somewhere out of the way, for example, at the back of the desk. Then the mouse can be freely moved around. The link is effective at distances of up to 20 or 30 feet.

Dry cells are necessary to power the transmitter inside the mouse. These cells, which are usually size AA or AAA, must be periodically replaced. If the cells fail and you don't have a set of replacements handy, you'll have to go out and buy new cells. The cells are easy to find; almost any convenience store has them (but it won't be convenient if you have to interrupt a critical work session to replace the cells).

Keyboard

Cordless keyboards are somewhat less common than cordless mice in desktop computer systems, but you might want one if you plan to give presentations in a large room or auditorium using a projection system. You can then place the computer next to the display projector (so the cable will reach), and use the keyboard anywhere you want, for example, at a desk in front of your audience. You'll want a cordless mouse in such a situation, as well.

A cordless keyboard can be used in a "megamedia" computer worksta-tion where you use a projection device to throw the image on an entire wall, and recline in a beanbag chair. This sort of workstation is not for everyone; it can cost upward of $15,000. But if you're into technology and are a bit of a megalomaniac, it can provide unique satisfaction. If you're into computer games, three-dimensional (3D) graphics, or virtual reality (VR), a workstation that surrounds you, rather than sitting on a desk in front of you, might be worth the price.

Cordless keyboards are often used with so-called Web TV boxes. You can sit in an easy chair and surf the Internet using your television (TV) set, just as you channel surf using the TV control box. These keyboards usually employ IR links, just as do TV control boxes.

Laptop Link

If you have a laptop (notebook) computer and also a desktop computer, data transfer between the two machines can be tedious and clumsy. Files can be transferred in various ways. Diskettes are adequate only when the volume of data is small. Devices with larger storage capacity, such as the popular Iomega "Zip" drive, can be used, but unless there are two drives—one for the laptop and one for the desktop—it is necessary to attach and detach the connecting cable between the drive and the computer. It is possible to transfer data directly between the hard drives of the two computers using a cable and special software, but again, this involves an awkward arrangement because the cable normally connects to the paral-lel ports of the computers. Parallel ports are used with printers, so the printer must be disconnected from the desktop machine in order to attach the data-transfer cable; after the transfer is complete, the cables must be exchanged again.

There is a wireless way around the mechanical problems associated with data transfer. An IR link can be used rather than a cable. If both the desktop and laptop units have the necessary IR transceiver installed, then each computer can in effect work as a disk drive in the other computer. To transfer data from the desktop hard drive to the laptop hard drive, you command the desktop machine to "copy" data from its hard drive (usually drive C) to the drive letter (for example, drive G) representing the laptop hard drive. There is no limit to the volume of data that can be trans-ferred "in one swipe" via this method, except as is imposed by the avail-able space on the laptop hard drive. As long as the laptop and notebook computers are positioned so that the IR transceivers can "see" each other, there is no need to connect or disconnect any cables. Data can be transferred from the laptop to the desktop computer in the same way.

Interference

Any device that uses radio-frequency (RF) fields is subject to *electromagnetic interference* (EMI) from other RF sources. Although it is unlikely, there is a possibility that a cordless phone, cell phone, microwave oven, or radio transceiver might interfere with the operation of a cordless mouse or keyboard. If this happens, the problem can usually be remedied by making sure the offending device is not operated right alongside the computer. In the case of radio communications equipment, the radio installation should be well grounded, and the antenna should be placed as far from the computer as possible. In rare instances, it might be necessary to contact the manufacturer of the computer for technical support. Changing the frequency of the RF link between the mouse or keyboard and the computer might eliminate the problem. The ultimate solution is to convert to IR links, because IR devices are not susceptible to EMI from RF equipment. However, IR links can be affected by obstructions, placing limitations on the placement of the control units.

Networks

A *local area network* (LAN) is a group of computers linked together within a small geographical region. All the computers are near one another, so transfer delays are minimal. In any LAN, the combination of computers is far more powerful than any computer by itself. A *wide area network* (WAN) is a group of computers that are all linked together and that are separated by long distances. The interconnections are made via telephone lines or radio, or both. Because individual workstations (nodes) can be thousands of miles apart, the link delays are significant in terms of computer speed.

LANs

The way in which a LAN is arranged is called the *LAN topology*. There are two major topologies: the *client-server LAN* and the *peer-to-peer LAN*.

In the system shown in Fig. 12-1A, there is one large, powerful, central computer called a *file server*, to which all the smaller workstation personal

Figure 12-1
Client-server network
(A), and peer-to-peer
networks (B, C, D).
Personal computers
are circles labeled PC.

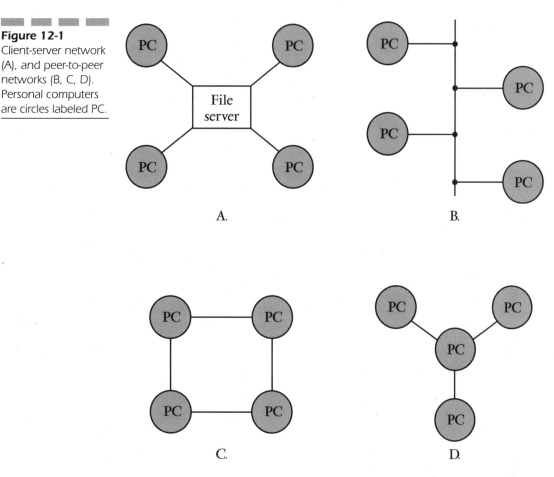

Figure 12-1
Client-server network (A), and peer-to-peer networks (B, C, D). Personal computers are circles labeled PC.

computers (labeled PC) are linked. This is a form of client-server LAN. This topology is used in medium-sized and large corporations, schools, and government offices. In small companies, schools, and organizations, LANs are often assembled without file servers; these are peer-to-peer LANs. There are several different ways in which a peer-to-peer system can be arranged. These are "subtopologies" within the general peer-to-peer scheme. Three examples are the *bus network* (Fig. 12-1B), the *ring network* (Fig. 12-1C), and the *star network* (Fig. 12-1D). In each example shown, there are four computers, a number chosen only for illustrative purposes. Any of these LANs could have as few as two, or more than a dozen, workstations.

A client-server network is more powerful than most small companies or organizations need. It is also more expensive, harder to install, and costlier to maintain than a peer-to-peer LAN. However, there are some definite advantages to the client-server arrangement, including:

1. High speed
2. Large memory and storage capacity
3. Ability to do many different tasks
4. Ability to accommodate many workstations
5. Provision for using dumb terminals as well as computers
6. Provisions for security

If you have a small or medium-sized company that is rapidly growing, and if you expect your computing needs to multiply apace, you might want to consider a client-server LAN. But it is a big investment, and you should get expert advice before making a final decision.

A peer-to-peer LAN is less expensive, in general, than a client-server system. This is one of its chief advantages for small organizations. Some other features are:

1. Comparative ease of installation
2. Moderate maintenance costs
3. Ease of use
4. Expandability
5. Choice of "subtopologies"
6. Full computing capability at each workstation

A peer-to-peer LAN will almost always meet the needs of a small, stable corporation with modest growth plans.

There are some hidden costs involved in the installation and maintenance of any LAN, however small or large. Also, training workers to use the computers in a new peer-to-peer LAN can prove to be more of a challenge than the entrepreneur at first suspects. Before purchasing a LAN, the small businessperson must decide whether or not it is really needed. There's no sense in buying a network when one or two individual computers will suffice.

Radio transceivers can be used with computers, doing away with the cables characteristic of conventional LANs. This makes it possible to move the individual workstations around freely within the network region. In the case of a LAN, the transceivers use low transmitter power, and their communications range is therefore quite limited. The transceivers operate at VHF or UHF. The antennas are generally small and unobtrusive, similar to those found on cordless telephone handsets.

WANs

When radio is used for a WAN, satellites are the preferred mode, although conventional radio or microwave links can be employed. Amateur radio operators have set up specialized WANs that use packet communications for long-distance computer communication via radio. This is known as *packet radio*. A WAN can encompass hundreds or even thousands of computers located all over the world. In effect, the files of all the computers join forces, resulting in a system with enormous data-storage capacity.

Suppose a WAN has 10,000 computers, with an average of 50 gigabytes (GB) of available space on each hard disk. Then, in theory, the WAN has $50 \times 10,000$ GB=500 terabytes (TB) of available storage. But, in practice, the power of the WAN is not equivalent to that of a single gigantic supercomputer with a 500-TB hard disk, because there are delays in data transfer among the nodes of the WAN. The delays occur because the data takes time to propagate through space, cables, or optical fibers from node to node, and also because the computers each take some time to process data. This limits the speed at which the computers can function together. While data storage capacity adds up arithmetically among the computers in a WAN, processing power does not.

Figure 12-2 depicts a simple WAN that makes use of telephone lines and satellite radio links to bring PCs together, even though the machines are located on different continents. Although only one ground station and one PC is shown for each continent, there might be dozens of ground stations and thousands of computers. The links between the ground stations and the satellite are entirely wireless. The links between the end users and the ground stations might be partly or entirely wireless. One user might have a computer connected to a cellular telephone; another might be connected via a long-distance circuit that uses a microwave link over part of the route; still another might be connected to the ground station by a traditional wire line.

Bridges

A *bridge* is a communications path between or among two or more networks. This allows the computer users in any network to obtain data from or send data to any other network, in effect creating a network of networks. The Internet contains thousands of such bridges.

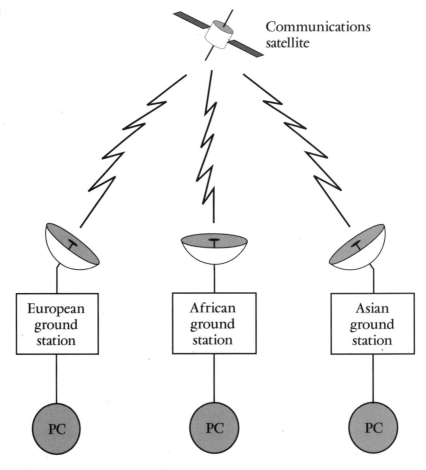

LAN-to-LAN   Bridges are commonly employed among LANs. Suppose,
for example, that a university has campuses in four towns, scattered
around a large state. Imagine a hypothetical "Pan-California University," with one campus in Los Angeles, another in San Francisco, another
in San Diego, and a main campus in Sacramento. Each campus administration has a LAN of its own. Between each LAN, there is a bridge. The
links might be via the telephone lines, but faster data speeds would be
obtained by using high-bandwidth wireless links. The circuits could
be provided by a geostationary satellite, by a low-earth-orbit system of
satellites, or by microwave links.

WAN-to-WAN   Bridges can be used to connect among WANs. For
example, Pan-California University might want to get data from and

send data to the University of Minnesota, the University of Iowa, Harvard, Princeton, and others. This would result in a "super WAN."

Long-distance bridges are used in online services, such as the Internet's Gopher, the Wide Area Information System, Archie, Veronica, and the World Wide Web, to form the so-called *information superhighway.* Via bridges of this sort, people can do things like look at library card catalogs in other states or countries, conduct research, and exchange information with colleagues. While much of the data travels over "landline" (the telephone system), some of the circuits are wireless. Satellites are preferred for worldwide communications.

Broadband Data

When computers communicate, they do so at speeds much faster than humans speak or write. Perhaps you have marveled at how a long document or program can be sent from one place to another in a short length of time.

A computer "thinks" in *digital* terms. These digital signals are converted into *analog* form for long-distance transmission from one place to another. Then, when the analog data arrives at the destination, it is converted back to digital form. The changes from digital to analog, and vice versa, are made by a device called a *modem.* The more data sent and received per unit time, the more *bandwidth* it needs. That is to say, higher data speeds require broader communications channels. This is a fundamental law of communications, and it applies to all kinds of signals, analog and digital. It is true for computer data, television images, speech, and text. *Broadband data* refers to analog computer signals that have broad bandwidth because of their high speed.

It is possible for a single communications line, or link, to carry many conversations at once. The conversations can be between people, or between a person and a computer, or between computers. Wireless links generally can accomplish this better than wire links. Radio waves or light beams can be modulated with the signals from thousands of computers simultaneously. Each signal has its own code and method of timing. The signals are all mixed up together at the transmitting end of the link, and sorted out again at the receiving end. The mixing together of signals is called *multiplexing*; the sorting out is *demultiplexing.* There are several different schemes for doing this. They all involve the use of broadband data.

There is a limit to the signal bandwidth that can be sent over common wire telephone circuits. This is about 2.7 kilohertz (kHz), the bandwidth needed to clearly convey a human voice. When telephone systems were first developed, no one took into account the possibility that the circuits might someday be used for purposes other than the transmission of voices. A home telephone line can handle one computer signal at speeds up to several tens of thousands of bits per second (bps). Higher speeds increase the bandwidth past the capacity of local telephone circuits. But computer networks can be set up with links more sophisticated than a common phone line. Two examples are microwave systems and satellite systems. Using these modes, huge bandwidths are possible, and information can be sent many times faster than over a common phone line.

Remote Control

Suppose you are on an airline flight. You have a portable computer and need some data that is stored in your home computer. You pick up the telephone on the seat in front of you, connect your portable computer to the telephone via the internal modem, and dial your home computer's phone number. You get the data you need, and in fact, you have access to most other features of your home computer. This is an example of *wireless remote control.* A simplified diagram of this particular scenario is shown in Fig. 12-3.

What You Need

When you remotely operate a computer, the machine you actually use is called the *remote computer.* The one you control or access is the *host computer.* In the above-described situation, the portable computer on the airplane is the remote unit, and the home computer is the host. Obviously, for remote control to be possible, there must be at least one remote unit, and at least one host unit.

You need a *modem* to connect the remote unit to the telephone line or other communications medium. There must also be a modem at the host computer site. The modems should have the highest operating speed possible. The speed with which you can carry out remote-control functions will be limited by the slower modem. You also need remote-control software at both sites. This allows you to look at directories (also called folders) and access data files in from the host computer. You can work in applications that do not require much processing speed and

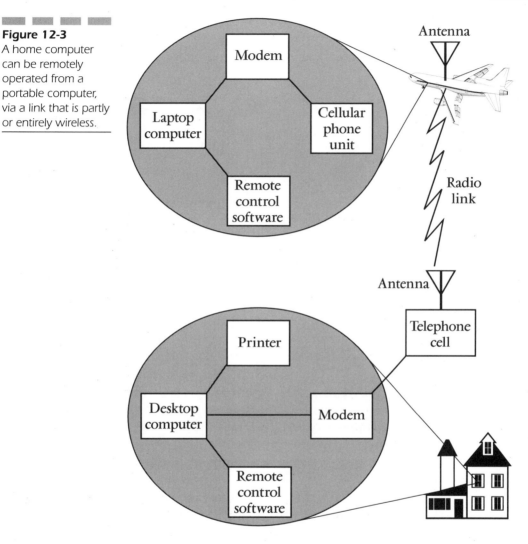

power, such as word-processed documents, small databases, small spread-sheets, and low-end graphics. You can store work on the host machine or the remote machine, or both.

Features

Various remote-control programs are available, some with basic features only and others with sophisticated capabilities. You'll want to see several programs in action before choosing one for your system. A few common features are as follows.

Operating System Flexibility   Remote-control programs are available for all versions of Microsoft Windows, Macintosh, UNIX, and other operating systems. You'll want to be sure the software you buy is compatible with your operating system(s).

Networking   Remote control is a big asset when running on networks. In fact, you can set up a WAN with one large host computer and numerous remote machines. You might want to have several remote units in operation at once with a single host.

Multiple Hosts   If you need to access more than one host computer, the software must be compatible among all the machines being used. High-end remote-control programs can let you see and work with several screens, in rotating sequence.

Modem Specifications   You must determine the range of modem speeds and the types of modems with which your system will work. Performance is directly proportional to the modem speed in bps. When two computers are in communication, however, the slower modem determines the speed at which the connection takes place.

Unattended Host   You'll want to power-up, operate, and power-down a host computer even when there is no one attending it. This will be the case, for example, if you're on a trip and need to work with your home computer. In a network situation, there might be a person at the host computer; then this feature would not be necessary.

Chat Mode   This allows you to use online services or communicate with someone attending the host computer. It can also allow the users of remote computers to communicate with each other through the host. This capability has the potential for abuse, however; thus there might be a time limitation to keep cellular and/or long-distance telephone tolls from soaring.

Timing Out   Some remote-control systems, especially those with multiple remotes, allow for a time limitation. After a certain preprogrammed length of time, the connection will terminate. This can be useful in networks when telephone tolls are involved. The time limit can be programmed in advance. In a network, this can help ensure that the remote users make efficient use of host computer time.

Printing   You might want to have the host computer print data. For this to be possible, automatic paper feed is necessary. The overwhelming consideration here is that the paper-feed system be reliable. This means that there must be an adequate supply of paper in the printer, and the printer must be in good repair so it won't be likely to jam.

Call Logging   With this feature, every time the host computer is accessed from a remote unit, data is recorded concerning the call. The information includes the date and time, and the duration of the call. If there is more than one remote unit in the system, the identity of each unit is also logged.

Hierarchical Password Protection   Also called *multilevel password protection,* this is a security feature that prevents unauthorized use of the host computer. The password levels allow users various degrees of control over the host machine.

Password Retry Limitation   This feature prevents hackers from making repeated guesses at passwords in an attempt to break into the system. If more than three unsuccessful entries are made in succession, for example, the host will not accept further incoming calls for a certain preprogrammed length of time, say one hour.

Online Help   Most sophisticated software packages have a provision for calling a toll-free number and getting assistance online. If you are having trouble from a remote computer, you can call the online help number and communicate with technicians in real time.

In Robotics

Mechanical devices such as robots can be operated from a distance by human beings or by computers. A simple example of a remote-control system is the control box for a TV set. Another example is the transmitter used to fly a model airplane. Still another example is an electronic garage-door opener. Most TV controls employ IR beams to carry the data; the model airplane and the garage door receive their commands via radio signals. In this sense, the TV set, the model airplane, and the garage door are all robots.

Robots have been used to explore the bottoms of oceans and large lakes. These machines have traditionally been operated via electric or fiberoptic cable links. A person sits at a terminal in the comfort of a boat or submarine bubble, and operates a robot, watching a screen that shows what the robot sees. The range is limited because it is impractical to have a cable longer than a few miles. Radio waves propagate poorly underwater unless the frequency is so low that the data transfer rate is severely limited. Wireless laser links are possible in clear waters over short distances. Visible red and IR lasers tend to propagate best underwater; visible blue and green propagate less well because the water scatters the rays at those wavelengths. The range is generally shorter than that obtained using electric or fiberoptic cables.

The most dramatic examples of wireless robotic remote control have been provided by "robot astronauts." In 1997, with the rapid advent of the World Wide Web, people throughout the world were able to watch on their home computers as humankind's first true "robot astronaut" probed the surface of Mars in search for evidence of present or past life on the Red Planet. Lay people as well as scientists were able to join in the fascination as the robot got stuck on a rock, and it took the people at mission control days to get it unstuck. The result was a good education for the world concerning the limitations as well as the power of long-distance wireless robot control.

Propagation Delays

When the remote and host machines are separated by a vast distance, signals take a long time to get from one end of the circuit to the other, even if the signals propagate at the speed of light. A remotely controlled robot on the moon is approximately 1.3 light-seconds away. It is roughly 2.6 seconds from the time a command is sent to a computer or robot on the moon until the earth-based operator sees the results of the command.

The propagation delay is greater in interplanetary space. The fringes of the solar system are billions of miles distant. As the Voyager probe passed Neptune and a command was sent to the probe, the results were not observed for hours.

It might seem unlikely today that personal computing will ever involve the remote control of space probes. But in 100 years, consumers might be routinely communicating with computers and machines in deep space. The possibilities are especially intriguing in education. Grade-school children might get a chance to manipulate probes roaming celestial bodies such as the moon, Mars, or the asteroids.

Ultimate Range Limit

There is an absolute limit to the practical distance that can exist between any remotely controlled device and its operator. There is no known way to transmit data faster than the speed of radio waves or visible light in free space (186,282 miles per second). The nearest stars are several light-years away; remote control over these distances would require people and societies to acquire a form of patience that our species does not now possess. Over distances greater than about 50 light-years, it would be impossible for any individual person to remotely control anything, because the results would not be observable within his or her lifetime. But computers, with lifespans on the order of millennia, might control robotic star fleets.

Autonomous Robots

Many researchers believe that computer-controlled robots will someday be household appliances. An *autonomous robot* is a self-contained machine with its own built-in computer system.

Figure 12-4 shows one scheme in which a fleet of autonomous robots might work together. Robots are shown as squares and computers as solid black dots. The computers are connected by radio links (straight lines) into a LAN. This is a peer-to-peer LAN because all the computers have equal authority and intelligence.

Figure 12-4
Autonomous robots in a peer-to-peer wireless LAN.

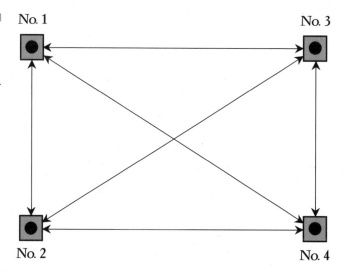

No. 1   No. 3

No. 2   No. 4

The computer in an autonomous robot has about the same amount of memory and operates at about the same processing speed as a typical personal computer. The robot can move around by rolling on wheels or on a track drive, something like a bulldozer or tank. Each *robot controller* (computer) in the fleet can have its own software. Robots can share data with each other, just as computers share information in a LAN.

An autonomous robot is sometimes called a *smart robot*. While several such robots can work as a team if the communications and software are well designed, they can also function independently. These two operating modes allow autonomous robots to do many different things.

Suppose that you assign four autonomous robots to the task of painting your house. Communication is necessary so the robots can coordinate their actions and not interfere with each other. This is an example of autonomous robots working together as a team. Imagine that you use the same four *personal robots* in general home maintenance. One might mow the lawn, the second could trim the bushes, the third could vacuum your floors, and the fourth might wash your windows. In this case they would do their jobs independently, and they would not have to communicate with each other.

Insect Robots

Some robot scientists like the idea of a fleet of "stupid robots" all under the absolute control of a central computer. These machines are called *insect robots* because the whole system functions as a unit, much like an anthill or beehive. In particular, the term is used in reference to systems designed by engineer Rodney Brooks; many of his robots actually look something like insects. He began developing his ideas at Massachusetts Institute of Technology during the early 1990s. The concept of "many robots, one controller," however, is as old as the notion of a robot.

Brooks's insect robots have six legs, and look like mechanical bugs. They can range in size from more than a foot long to less than a millimeter across. Most significant is the fact that they work collectively, rather than as individuals. Discrete units, each with high artificial intelligence (AI), don't necessarily work well in a team. People provide a good example. Professional sports teams have been assembled by buying the best players in the business, but such a team is unlikely to win championships unless the players get along well together. They can be brilliant in the individual, yet be dull in the generality.

Insects, in contrast to humans or individually programmed computers, are stupid in the individual. Ants and bees are like idiot robots. But an anthill or beehive is an efficient system, run by the collective mind of all its members. Such systems are brilliant in the generality. Rodney Brooks saw this fundamental difference between autonomous and collective intelligence. He also saw that most of his colleagues were trying to build autonomous robots, perhaps because of the natural pride of humans, resulting in visions of robots as humanoids. To Brooks, a major avenue of technology was being neglected. He therefore began designing robot colonies consisting of many units under the control of a central computer or AI system.

Brooks envisions microscopic insect robots that might live in your house, coming out at night to clean your floors and counter tops. "Antibody robots" of even tinier proportions could be injected into a person infected with some heretofore-incurable disease. Controlled by a central microprocessor, they could seek out the disease bacteria or viruses and destroy them. The communication between the central controller and the robots would be via wireless, probably at microwave RFs.

Smart Homes

Imagine having all your mundane household chores done without your having to think about them! Your home computer could control a fleet of personal robots via wireless. These robots would take care of cooking, dish washing, the laundry, the yard maintenance, snow removal, and other things. This is the ultimate form of computerized home: having your home's "central nervous system" run by a computer so you can do other things. In the building trade, a computerized home is called a *smart home.*

Figure 12-5A is a block diagram of a central PC and a system of controlled home-convenience devices. The smart home of the future will probably work along these lines. But there is a limit to how many devices people will link to the central computer. Robot cats and dogs, for example, will probably never be common no matter how advanced the technology gets, although they might be fun for children.

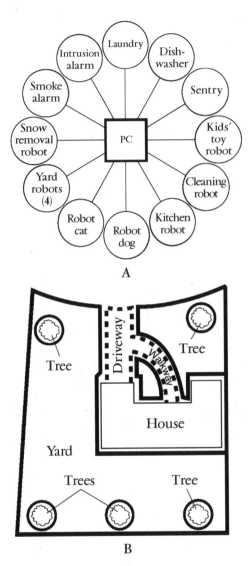

Technology and Ethics

The key to a truly smart home lies in the technologies of robotics and
AI. As these become more available to the average consumer, we can
expect to see, for example, a robotic laundry. We might see robots that
make our beds, do our dishes, vacuum our carpets, shovel snow from
our driveways, clean our windows, and maybe even drive to the store and
shop for our groceries.

Some people question whether computerized, robotized homes are really worth developing. Won't people prefer to spend their hard-earned money in other ways, such as buying vacations or new properties? There are also ethical concerns. Should some people strive for total home automation, when a large segment of society cannot afford to own a home at all?

Let us suppose, for the moment, that we solve the ethical problem, and that everyone has a home with some money to spare. Further imagine that the cost of technology keeps going down, while it keeps on getting more and more sophisticated. The following sections discuss some examples of what the future might hold.

Fire Protection

When people and property must be protected from fire, *smoke detection* is a simple and effective measure. Smoke detectors are inexpensive and can operate from flashlight cells. You probably have one or more of these devices in your home now. (If you don't, get some.)

In a computerized home of the future, a battery-powered smoke alarm might alert a battery-powered, mobile robot via a wireless link. Thus the system could operate even if there were a failure of utility power. Robots are ideal for firefighting, because they can do things too dangerous for humans. The challenge lies in programming the robots to have judgment comparable to that of human firefighters. The robot would locate all the people in the house and alert them to the situation, escort people from the house if necessary, and then put out the fire and/or call the fire department. The robot might even be able to perform some first aid.

Security

Computers and robots can be of immense help around the house when it comes to prevention of burglaries. Security robots have been around for decades. A simple version is the electronic garage-door opener. More advanced systems include intrusion alarm systems and electronic door/gate openers. These devices make it difficult for unauthorized people to enter a property. They can detect the presence of an intruder, usually by means of ultrasound, microwaves, or lasers, and notify police through a telephone or radio link.

The ultimate security system includes a *security robot*. If an intruder enters the property, the security robot might drive the offender away or detain the offender until police arrive. Plenty of people would be scared away by the appearance of a seven-foot-tall humanoid robot with a stun gun and a threatening demeanor. Those who were not frightened off could be captured on video by the robot's high-resolution camera. The robot might employ such advanced devices as *iris-print detection* or *facial feature detection* to identify an intruder even from several meters away. The robot would be linked to the home computer via a broad-band-width, high-speed wireless system.

A *sentry robot* can alert a homeowner to abnormal conditions. It might detect fire, burglars, or water in places it should not be. A sentry might detect abnormal temperature, barometric pressure, wind speed, humidity, or air pollution. A wireless link could alert you via a device similar to a common beeper.

Food Service

Robots have been used to prepare and serve food. So far, the major applications have been in repetitive chores, such as placing measured portions on plates, cafeteria style, to serve a large number of people. But robots can be adapted to food service in common households.

Personal robots, when they are programmed to prepare or serve food, will require more autonomy than robots in large-volume food service. You might insert a disk into a home robot that tells it to prepare a meal of meat, vegetables, and beverages, and perhaps also dessert and coffee. The robot would ask you questions, such as:

1. How many people will there be for supper tonight?
2. Which type of meat would you like, if any?
3. Which type of vegetable(s) would you prefer?
4. What, if anything, would you like to spread on your bread?
5. What beverages would you like?

When all the answers are received, the robot will go through a complex process to prepare the meal. The robot might serve you as you wait at the table, and then clean up the table when you're done eating. It might wash and dry the dishes too. The primary role of wireless in such a system would be twofold: first, the EM link between the robot and the home computer, and second, the audio link between your ears/mouth and the robot's ears/mouth. The robot would be equipped with *speech recognition* and *speech synthesis*, both of which are described later in this chapter.

Yard Work

Riding mowers and riding snow blowers are easy for humanoid robots to use. The robot need only sit on the chair, ride the machine around, and operate the handlebar/pedal controls. Alternatively, lawn mowers or snow blowers could be robots themselves, designed with the applicable task in mind.

The main challenge, once a lawn-mowing or snow-blowing robot has begun its work, is for it to carry out its business everywhere it should, but nowhere else. You don't want the lawn mower in your garden, and there's no point in blowing snow from your front lawn. Wires might be buried around the perimeter of your yard, and along the edges of the driveway and walkways, establishing the boundaries within which the robot must work. An example of this is shown in Fig. 12-5B. Solid, heavy lines are boundary wires for the lawn-mowing robot; dotted lines are boundary wires for the snow-blowing robot. The wires carry alternating current (AC), producing EM fields. The robots have wireless EM sensors that guide them within the boundaries defined by the wires.

Inside the work area, *edge detection* can be used to follow the line between mown and unmown grass, or between cleared and uncleared pavement. This line is easily seen because of differences in brightness and/or color. Alternatively, a *computer map* might be used, and the robot can sweep along controlled and programmed strips with mathematical precision.

Idleness and Faith

If robots can do all the housework, what will be left for people to do? Will homeowners get bored cycling, going to the beach, partying, and otherwise spending time that used to be devoted to maintaining their property? Some people enjoy doing household chores. Robots and computers are capable of doing mundane work for people, but it is not necessary that they be used. Plenty of people will reject them as a needless extravagance.

Another hangup with home computerization is the degree to which people have faith in machines. Many people have trouble entrusting computers with simple things such as storing data. These people would probably never be comfortable going on a vacation and leaving a computer in charge of the house. For them, hiring a human house sitter will be more appealing than relying on a computer to take care of everything. Even if a computer/robot system is better from a statistical standpoint, this type of homeowner will have a happier vacation if a good human friend is watching over the house.

Digipeaters

In recent years, personal computing has acquired an important place in the hobby called *amateur radio*. Many radio amateurs, also called radio "hams," use computers for communications purposes. Radio hams commonly send and receive *electronic mail* (e-mail) from remote sites via radio without incurring the tolls associated with cellular communications services. Some amateur radio operators have sent and received e-mail from places hundreds of miles away from any telephone line.

A pivotal element, equivalent to a communications *node*, in amateur-radio e-mail is known as a *digipeater*. This term is a contraction of the words "digital" and "repeater." Any amateur radio station capable of sending and receiving packet signals can work as a digipeater.

The basic elements of a digipeater are diagrammed in Fig. 12-6. The station *receiver* intercepts the incoming signal (A). This signal might have come from the *source* (originating station), or it might have been relayed from another digipeater. The received signal (B) goes to a *terminal node controller* (TNC), which is similar to a modem. The TNC demodulates the signal, so a computer can understand it. From the TNC, the data (C) goes to the computer, where it might be stored for a short while. Then the computer sends the data (D) back through the TNC, where the signal is changed into a form suitable for radio transmission (E). The *transmitter* modulates a radio signal with this information. The radio signal goes to the transmitting antenna (F). Thus the message is on its way to another digipeater or to the *destination* station.

A digipeater retransmission frequency can be the same as the frequency on which the signal came in, but it does not have to be the same. For example, a digipeater can receive a message on 144 megahertz (MHz), an amateur frequency band near the FM broadcast band, from a source just a few miles away, and resend it on a shortwave frequency, such as 14 MHz, to some other digipeater halfway around the world.

Figure 12-6
Block diagram of a
digipeater.

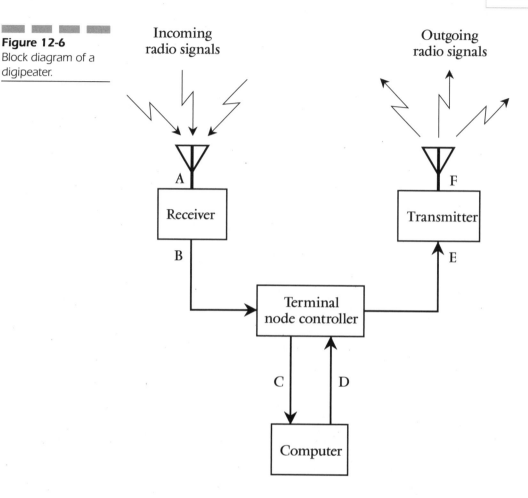

A digipeater will accept only signals that are intended for it. Special data, called a *routing frame,* determines which digipeater(s) the signal is to go through, and in what order. As the signal is routed along, *secondary station identifier* (SSID) codes tell each succeeding digipeater that the signal is intended specifically for it. An amateur-radio e-mail message might pass through several digipeaters on its way from the source to the destination.

Speech Recognition

A microphone can act as a wireless input device for a computer equipped with *speech recognition.* The machine converts acoustic waves, produced by a human mouth (or occasionally by a tape recorder or another computer), into data it can store and manipulate. Until recently, speech recognition software was extremely expensive and would work only on the most sophisticated computers. However, when the Intel Pentium and similar microprocessors became commonplace, software developers became able to produce speech-recognition programs that the average computer user could afford.

Components of Speech

Your voice consists of audio-frequency energy, with components ranging from about 100 hertz (Hz) to several kilohertz. This was known even before Alexander Graham Bell sent the first voice signals over electric wires. Perhaps you have spoken into a microphone that was connected to an oscilloscope, and seen the jumble of waves. How can any computer be programmed to make sense of that? The answer lies in the fact that, whatever you say, it is comprised of only a few dozen sounds, called *phonemes.* A computer can have a *phoneme database* against which it can compare the sounds of an incoming voice, and with which it can generate an outgoing voice.

In communications, engineers have found that a voice can be transmitted quite well if the *passband* is restricted to the range 300 to 3000 Hz. Certain phonemes, like "ssss," contain some energy at frequencies of several kilohertz, but all the information in a voice, including the emotional content, can be conveyed if the audio passband is cut off at 3000 Hz. This is the typical voice frequency response in a two-way radio.

Figure 12-7A is a frequency-versus-time *voice print* of a spoken word with five syllables. Most of the energy is in three frequency ranges, called *formants.* The first formant is at less than 1000 Hz. The second formant ranges from 1600 to 2000 Hz. The third formant is at 2600 to 3000 Hz. These ranges are approximate, and vary somewhat from person to person. Between the formants there are gaps, or ranges of frequencies, at which little or no sound occurs.

The formants, and the gaps between them, stay in the same frequency ranges for any particular person's voice, no matter what is said. The fine details of the voice print determine not only the words but the emotional content of speech. The slightest change in tone of voice will show up in a voice print. Therefore, in theory, it is possible to build a machine that can recognize and analyze speech as well as any human being. The challenge is to put this theory into practice.

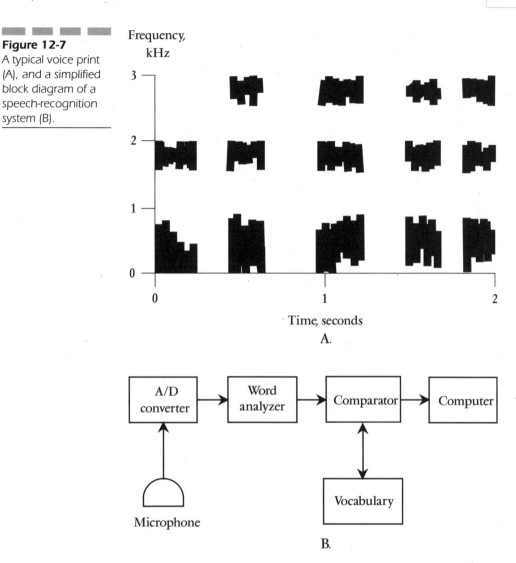

Figure 12-7
A typical voice print (A), and a simplified block diagram of a speech-recognition system (B).

A/D Conversion

The passband can be reduced greatly if you are willing to give up some of the emotional content of the voice in favor of efficient information transfer. In recent years, a technology has been developed and refined that does this very well. It is called *analog-to-digital (A/D) conversion*. An *A/D converter* circuit changes the continuously variable, or analog, voice signal into a series of digital pulses. This is the audio equivalent of *optical scanning*, in which a drawing or photograph is converted into digital signals for storage in a computer. There are several different characteristics of a digital pulse signal that can be varied. These include the amplitude, the duration, and the frequency (number of pulses per second).

A digital signal can transmit a human voice within a passband narrower than 200 Hz—less than 10 percent of the bandwidth required by an analog voice signal. The narrower the bandwidth, in general, the more of the emotional content is sacrificed. Emotional content is conveyed by *inflection,* or variation in voice tone. When tone is lost, the signal resembles written data. But it can still carry some of the subtle meanings and feelings, as you know from reading good books or magazine articles.

Word Analysis

For a computer to decipher the digital voice signal, the computer must have a vocabulary of words or syllables, and some means of comparing this knowledge base with the incoming audio signals. This system has two parts: a *memory,* in which various speech patterns are stored, and a *comparator* that checks these stored patterns against the data coming in. For each syllable or word, the circuit searches its vocabulary until a match is found. The greater the processing speed of the computer, the shorter the delay, in general, until the proper pattern is found.

The size of the computer's vocabulary is directly related to its memory capacity. An advanced speech-recognition system requires a large amount of *random-access memory.* In this sense, speech recognition is the audio equivalent of visual *object recognition.*

Context and Syntax

The output of the comparator must be processed in some way, so that the machine knows the difference between words or syllables that sound alike. Examples are "two/too," "way/weigh," and "not/knot." For this to be possible, the *context* and *syntax* must be examined. This analysis places extreme demands on the computer.

Consider the following three sentences:

I have one, too.

I have won two.

I have won, too.

Each of these can make sense in a certain situation or context. To some extent, a computer might figure out which of the above three sentences is being spoken in a given situation, by analyzing surrounding words and sentences. But the amount of data to be analyzed, and the speed with which it must be done, puts many scenarios beyond the interpretation abilities of computers in existence today.

There must also be some way for the computer to tell whether a group of syllables constitutes one word or two words or perhaps three or four words. The more complicated the words coming in, the greater the chance for confusion. The word "radioisotope" might be confused in several different ways. The word "antidisestablishmentarianism" would befuddle almost any computer.

Any speech-recognition system will sometimes make mistakes, just as people sometimes misinterpret what you say. Such errors will become less frequent as computer memory capacity and operating speed increase. But there might never be a computer that can transcribe speech with 100 percent accuracy in every possible situation. (Even people cannot do that.) A simplified block diagram of a speech-recognition system is shown in Fig. 12-7B.

Emotional Content

The A/D converter in a speech-recognition system removes some of the inflections from a voice. In the extreme, all of the tonal changes are lost, and the voice is reduced to the equivalent of written language. For computers, this is adequate. If a system were 100 percent reliable in simply transcribing the words it received, speech-recognition engineers would be very happy.

When accuracy does approach 100 percent (as it will in time), there will be increasing interest in getting some of the subtler meanings across, too. Take the sentence "You will go to the store after midnight," and say it with the emphasis on each word in turn (eight different ways). The meaning changes dramatically depending on the particular word or words you emphasize.

Tone is also very important for another reason: A sentence might be either a statement or a question. Thus "You will go to the store after *midnight?*" represents something completely different from "You *will* go to the store after midnight!"

Even if all the tones are the same, the meaning can vary depending on how fast something is said. The timing of breaths can make a difference as well.

Speech Synthesis

Speech synthesis is the computerized generation of sounds that mimic the human voice. This technology is somewhat ahead of speech recognition. It is easier for a machine to talk than to listen (just as is the case for most people). Speech synthesis allows a human computer operator to make use of the sense of hearing, as well as the sense of sight, to receive data from the machine. The computer speakers are a form of wireless machine output device; they convert data into acoustic waves.

What Is a Voice?

All sounds, including music, barking dogs, roaring jet engines, ticking clocks, and human speech, are combinations of AC waves, having frequencies between about 20 Hz and 20 kHz. Sounds take the form of vibrations in air molecules. The patterns of vibration can be duplicated as electric currents.

Figure 12-8A shows a simple *compression wave*, of the kind that travels through the air when there is a sound. Also shown is the electric current that produces this wave when the current passes through the coils of a loudspeaker or headset. The maximum positive current, in this example, corresponds to the greatest compression of air molecules. The maximum negative current corresponds to the lowest compression of air molecules. The speaker or headset is an *acoustic transducer* because it changes electric current into audible sound.

In speech, a frequency range (band) of 300 to 3000 Hz is wide enough to convey all the information, and also all of the emotional content, in any human voice. Therefore, a *speech synthesizer* needs only to make sounds within the range 300 to 3000 Hz. The challenge is to produce waves at exactly the right frequencies, at the correct times, and in the correct phase combinations. The inflection, or "tone of voice," and the voice rhythm must be just right. In the human voice, the amplitude and frequency rise and fall in subtle and precise ways. The slightest change in inflection can make a big difference in the meaning of what is said. You can tell, even over the telephone, whether a person is anxious, angry, or relaxed. A request sounds nothing like a command, and a question differs from a statement, even when the words themselves are identical.

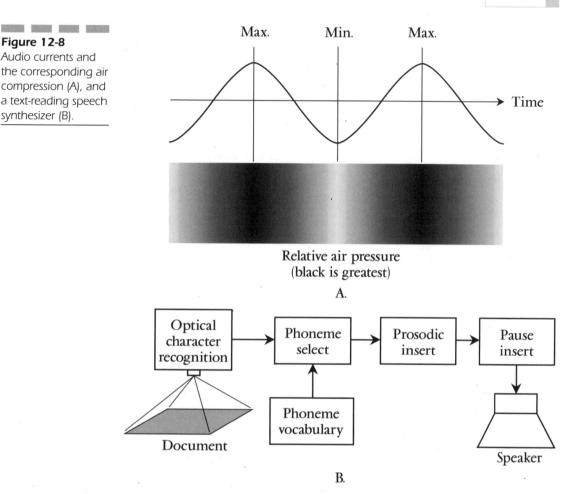

Tone of Voice

In the English language there are 40 phonemes, or elementary sounds. In some languages there are more phonemes than in English; some languages have fewer phonemes. The exact sound of a phoneme can vary, depending on what comes before and after it. These variations are called *allophones*. There are 128 allophones in English. These can be strung together in millions of different ways.

The tone of a voice can depend on whether the person speaking is angry, sad, scared, happy, or indifferent. These subtle inflections are the *prosodic features* of the voice. These depend not only on the actual feelings of the speaker, but on age, gender, upbringing, and other factors. Besides this, a voice can have an *accent*. You can tell when a person is angry or happy, regardless of whether that person is from Texas, Indiana, Idaho, or Maine. Some accents sound more authoritative than others; some sound funny if you have not been exposed to them before. Along with accent, the actual words can be a little different in different regions. This is *dialect.*

You do not want a talking machine to sound angry all the time, or constantly unhappy, or endlessly blissful. You do want the machine to convey the correct mood for a situation, or at least not to convey the wrong mood. Imagine your computerized personal robot rolling up to you and drawling, "You-all gonna *dah* if you don't git outta this house raht *nah-*yow." You'd probably laugh until you realized that the robot was telling you your house was on fire. For engineers, producing a speech synthesizer with credible "tone of voice" is a great challenge. To know how to inflect, a computer must know something about the scenario involved in what is being said. To complicate matters even more, the inflections for some emotions vary from language to language.

Schemes

The simplest type of speech generator is a set of tape recordings of individual words. You have probably heard these in automatic telephone answering machines and services. Some cities have telephone numbers you can call to get local time; some of these services use word recordings. They all have a characteristic choppy sound. There are several drawbacks to these systems. Perhaps the biggest problem is the fact that each word requires a separate recording, on a separate length of tape. These tapes must be mechanically accessed, and this takes time. It is impossible to have a large speech vocabulary using this method. Perhaps the greatest objection is that this is not true speech synthesis but is merely a reproduction of a human's utterances.

A written text can be "read" by a machine and converted into a digital code called ASCII (pronounced "*ask-ee*"). The method used for this is known as *optical character recognition* (OCR). The ASCII can be translated by *integrated circuits* (ICs) into voice sounds. In this way, a machine can read text out loud. Although they are rather expensive at the time of this writing, these machines are being used to help blind people read written text.

Because there are only 128 allophones in the English language, a machine can be designed to read almost any text. But a machine lacks a sense of which inflections are best for various situations. With technical or scientific text, this is not normally important. But when reading a story to a child, mental imagery is crucial. A good story is an imaginary movie, helped along by the emotions of the reader. If these emotions are wrong, or if they do not exist, the listener will not experience the thoughts and feelings that the author intends.

No machine can paint pictures, or elicit moods, in a listener's mind as well as a human being can. These things are apparent from context. The appropriate inflections in a sentence might depend on what happened in the previous sentence, paragraph, or chapter. Technology is a long way from giving a machine the ability to understand and appreciate a good story. But nothing short of that level of computer intelligence will work if a machine's voice is to create a "story daydream" in a listener's mind. Computers will have to develop a genuine appreciation for art!

The Process

There are several different ways in which a machine can be programmed to produce speech. A simplified block diagram of one process is shown in Fig. 12-8B. Whatever method is used for speech synthesis, certain steps are necessary. These are as follows:

1. The machine must access the data and arrange it in the proper order.

2. The allophones must be assigned in the proper sequence.

3. The proper inflections (prosodic features) must be put in.

4. Pauses must be inserted in the proper places.

In addition to these, for enhanced performance, additional features might be included, as follows:

5. The intended mood can be conveyed (joy, sadness, urgency, etc.) at various moments.

6. Overall knowledge of the content can be programmed in. For example, the machine can know the significance of a story, and the importance of each part within the story.

7. The machine can have an interrupt feature to allow conversation with a human being. If the human says something, the machine will stop and begin listening with a speech-recognition system.

This last feature could prove fascinating if two highly advanced machines got into an argument with each other. Some engineers can hardly wait for the day they get to try this. One machine might be programmed to think like a Republican, while the other is programmed to think like a Democrat. The engineer could bring up the subject of taxes, and then gleefully witness the ensuing debate. The result might be an endless loop, as has been the case between human Republicans and Democrats for decades. Or perhaps the machines would arrive at a compromise position, never before imagined by humans, with which Republicans and Democrats could both get along.

Optical Scanning

Optical scanning is an electronic process that converts hard-copy text and graphics into digital form suitable for processing and storage in a computer. Text scanners work in conjunction with OCR software. The optical scanner and OCR software together form the equivalent of audio speech recognition and sound interpretation. Usually, scanners are employed to convert printed matter to digital data. But scanning technology can also be used at a distance, for example, by a robot for the purpose of reading road signs. Reading and looking at graphics are, along with listening to spoken words or other sounds, among the oldest forms of wireless communication. Optical scanning will become increasingly important as computers and smart robots are called upon to do more and more of the tasks previously done by people.

The Charge-Coupled Device

A *charge-coupled device* (CCD) is a video camera that changes images into digital signals. Astronomers use CCDs to enhance views of outer space. Biologists employ CCDs to process the images seen by microscopes. Some robots use them in *machine vision* systems. Meteorologists use them to process satellite photos and radar images. In personal computing, CCDs are used in OCR.

The image you see, focused on the retina of your eye, is an *analog image*. It can have infinitely many configurations and infinitely many variations in hue, brilliance, and saturation. The image on a camera film is the same way. So is the signal that comes out of a TV camera tube. These are all fine images, clear and precise. When a digital computer receives an analog signal, it is like the average American looking at a newspaper in Chinese. A computer needs a *digital image* to make sense of what it sees. Binary digital signals have only two possible states: on and off. These are also called high and low, or 1 and 0. It is possible to render an analog image as a string of high and low signals. Digital impulses can likewise be converted to analog images.

A CCD, in conjunction with software for OCR, works something like a computer graphics program in reverse. It changes an analog image into a digital signal, of the sort that your computer can understand. Patterns in the image are recognized and converted into alphabetic-numeric digital codes. Using OCR software, the images of characters are changed into combinations of ones and zeros. Then, when printed text is "seen" by the CCD, the microprocessor gets the impulses, exactly as if you had typed the text into the computer from a keyboard.

The image on your eye's retina is converted into nerve impulses that your brain can understand; the CCD, with OCR software, is like an eyeball that has been adapted to a computer.

Typically, a CCD has trouble with handwriting. It also has difficulty with some exotic printed fonts. But a CCD with OCR software can resolve standard fonts such as Arial, Courier, or Times Roman. A CCD can read text of any reasonable type size. Newer OCR software can even "learn" to read the handwriting of a particular person, and various nonstandard fonts. A simplified block diagram of a CCD vision system is shown in Fig. 12-9.

Function

Optical scanners can save untold hours of tedious manual retyping. Perhaps you wrote a book several years ago, before you owned a computer. The hard copy of that book sits in your basement, awaiting the massive editing that can turn it into a great novel. It needs the power of your computer's word processor. But you cannot motivate yourself to retype the 1000-page manuscript, and you don't want to pay someone else to do it. An optical scanner can do away with most of the hard labor involved in getting a hard-copy manuscript onto disk.

A good scanner, of the type you would probably want for your novel, might cost $1000 to $2000. For a 1000-page manuscript, that comes to a dollar or two per page, which is a lower price than you are likely to find for any human typist. After you're done scanning your novel, you'll still have the scanner, which you can use for your autobiography, poetry collection, short story anthology, or whatever else.

Figure 12-9
Block diagram of a
charge-coupled
device (CCD) and
computer vision
system.

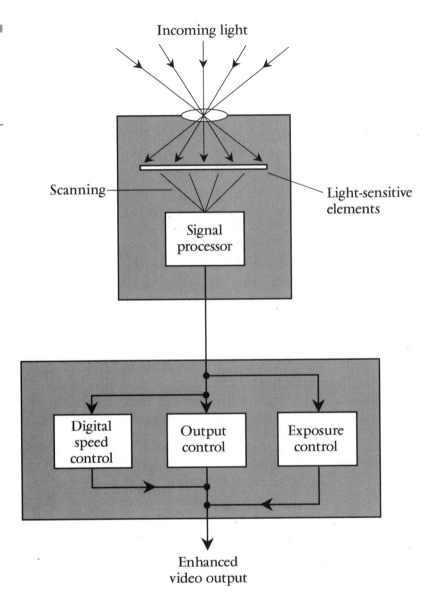

Incoming light

Scanning

Light-sensitive
elements

Signal
processor

Digital
speed
control

Output
control

Exposure
control

Enhanced
video output

The value of optical scanning in small business is hard to measure. For many entrepreneurs, it can make the difference between staying afloat and going bankrupt. It can do the work of one or two full-time typists for a tiny fraction of the long-term cost. Big companies can save money, too. Lawyers and doctors find scanners invaluable for backing up files of all kinds.

Specifications

The simplest optical scanners can convert text, but not pictures, into binary data. These scanners only recognize light and dark regions on a page. Each character is analyzed and converted into binary code. More complex scanners can detect shades of gray and are suitable for converting black-and-white drawings and photos into their digital equivalents for storage on disk. These scanners cost somewhat more than text-only types. The most sophisticated optical scanners can render color images, text, photographs, and everything else needed to make a complete, accurate digital record of any document. Color scanners use three different light beams (red, blue, and green) to get three different images, which are processed and combined in much the same way as a color TV camera works.

The *image resolution* of an optical scanner for printed matter is specified in *dots per inch* (DPI). The higher the DPI, the more detail the scanner can "see." For reliable scanning of text, an image resolution of at least 300 DPI is recommended, and 600 DPI is preferable. Virtually all scanners meet the 300 DPI requirement, and most meet the 600 DPI standard. Some scanners have even higher resolution, such as 900 or even 1400 DPI. When an optical scanner is used for looking at real scenes, such as a robot would employ in navigating along a highway or an iris-scanning automatic teller machine (ATM) would use to identify the person at the controls, the image resolution is generally specified in terms of the number of *pixels* (picture elements) in the image. This figure is a relative, not an absolute, indicator of resolution, because it depends on the size of the image. For any digitized image, greater detail translates into more memory consumed. Color increases the amount of memory or storage that an image takes up, if the image resolution remains constant.

Limitations of the Technology

Scanning technology has historically lagged behind printing technology. Printing is like drifting downstream to the mouth of a river, but scanning is like paddling upstream to the source of the river. "Unprinting" is more difficult than printing for a computer to do. The situation is similar to that for speech recognition versus speech synthesis. Making sense of incoming data is, apparently, more difficult than spewing data out, for machines as well as for human beings.

Even the best scanners make some mistakes. This is especially true if text contains nonstandard symbols. Highly technical material presents the worst problems. Some mathematical symbols are so esoteric that the average person (let alone a machine) is befuddled by them. Ink spots, stray markings, and smudges on a page can cause scanning errors, in much the same way as background noise confuses a speech-recognition system.

A human reader can often tell what a printed character should be, even when it is severely mutilated. But computers lack intuition. To some extent this can be corrected by a built-in spell checker. Some OCR programs have spell checking, but this introduces its own set of problems because it, too, is imperfect.

If a scanner fails to recognize a character, it will usually output a blank space, underline, or default symbol such as the underline (_). These symbols are called *tags*. Scanned text must always be carefully proofread after the data has been saved. Computer programs have been developed that can fill in many of the tags resulting from scanner problems. However, a human being should be somewhere in the quality-control chain, so unforeseen, possibly damaging errors can be caught before they get into media where they might cause embarrassment (or worse).

ELF Fields

An electrical or electronic device always produces an *EM field* because of the AC passing through the wiring and circuitry. The fields have frequencies much lower than those of most radio or TV signals. Sometimes the frequencies are even lower than the *very low frequency* (VLF) radio spectrum and are technically called *extremely low frequency* (ELF). The term ELF has recently been expanded to include all incidental EM radiation from electrical and electronic appliances, even when the frequencies extend into the VLF range.

Electromagnetism

A steady direct current (DC) produces a static *electric field* and a stable *magnetic field*. The higher the voltage, the stronger the electric field. The more current that flows, the stronger the magnetic field. Sometimes you will hear of electric or magnetic fields described in terms of *flux lines*.

The electric and magnetic flux lines from a given source of EM radiation are always perpendicular to each other at every point in space. When current alternates back and forth, as does standard utility current, the intensity of the electric and magnetic fields constantly changes. This generates an EM wave that can travel great distances through space. The phenomenon was explored around the end of the nineteenth century by Nikola Tesla and other scientists. It led ultimately to all forms of EM communications as we know them today. For a detailed discussion of EM fields, see Chap. 1.

An ELF field, like a radio wave, is an EM field. But while radio waves usually have frequencies measured in thousands, millions, or billions of hertz, ELF can range down to only a few hertz. The term "extremely-low-frequency radiation," and the attention it has received in the media, might lead you to think there is something sinister about it. But ELF is nothing like X rays or gamma rays, which can cause radiation sickness. Nor is it like ultraviolet (UV), which can cause skin cancer over long periods. An ELF field will not make you radioactive. There has been no absolute proof, as of this writing, that the ELF field from a computer or any other common appliance is dangerous. But ELF has not been proved completely safe either.

ELF from a CRT Monitor

Most of the ELF you have heard about in relation to computers is emitted by the *cathode-ray tube* (CRT) in the monitor of a desktop workstation. Other parts of the computer are not responsible for much ELF energy. Laptop (notebook) computers produce essentially none.

In the CRT, the characters and images are created as electron beams strike a phosphor coating on the inside of the glass. The electrons constantly change direction as they sweep from left to right, and from top to bottom, on your screen. The sweeping is caused by deflecting coils that steer the beam across the screen. The coils generate magnetic fields that interact with the negatively charged electrons, forcing them to change direction. The result is that the electrons accelerate, and this produces a low-frequency EM field.

Because of the positions of the coils, and the shapes of the fields surrounding them, there is more ELF energy "radiated" from the sides of a computer monitor than from the front. If there is any health hazard with ELF, therefore, it is greater for someone sitting off to the side of a monitor, and less for someone watching the screen from directly in front at the same distance.

Protection from ELF

The best "shielding" from ELF is physical distance. This is especially true for people sitting next to (rather than in front of) a monitor unit. The ELF dies off rapidly as you move away from the monitor. Workstations should be at least 5 feet apart. Stay at least 18 inches away from the front of your own monitor.

Special monitors, designed to minimize ELF fields, are available. They are not cheap, but they offer peace of mind if you're concerned about possible long-term effects of ELF fields. Sweden has the strictest criterion for ELF emissions.

Any monitor can be shut off when it is not in use. This will eliminate ELF fields during downtime. It will also extend the life of the monitor.

With increasing concern about ELF, you will see devices marketed with claims to eliminate or greatly reduce ELF fields. Some of these schemes are effective; others are not. Electrostatic screens that you can place in front of the monitor glass to keep it from attracting dust will not stop ELF fields. However, a magnetism-reducing metal ring or band might bring a monitor into compliance with the *Swedish standard,* even if it does not meet this standard without the device attached.

It's wise to avoid blowing the ELF issue out of proportion, and not to succumb to unsubstantiated hype. Some companies use scare tactics to sell devices of dubious value. If in doubt, consult someone whose word you can trust, such as a computer hardware engineer or an electronics engineer.

CHAPTER 13

Measurement, Monitoring, and Control

Measurement of parameters such as current, voltage, power, and frequency is often necessary in electronic systems, and wireless equipment is no exception. In this chapter, various measurement devices are examined. Wireless apparatus has been used for monitoring and control almost since the invention of radio. This chapter looks at some such systems.

Metering and Test Equipment

Metering refers to the use of electronic or electromechanical instruments to measure and monitor physical quantities. Numerous types of measurement instruments are used in wireless devices and systems.

Concept

Early experimenters with electricity and magnetism noticed that an electric current produces a magnetic field. When a magnetic compass is placed near a wire carrying a direct electric current, the compass needle is displaced. This led to the invention of the *galvanometer* to measure relative current. Refinements of the galvanometer were developed for the purpose of measuring absolute current, voltage, and power.

Electrostatic fields produce forces, just as do magnetic fields. The most common device for demonstrating electrostatic forces is the *electroscope*. It consists of two foil leaves attached to a conducting rod and placed in a sealed container. When a charged object is brought near or touched to the contact at the top of the rod, the leaves repel each other. Refinements of the electroscope were developed for the purpose of measuring absolute direct and alternating voltages.

Another phenomenon, sometimes useful in the measurement of electric currents, is the fact that whenever current flows through a conductor having any resistance, that conductor is heated. The extent of heating is proportional to the square of the current carried by the wire, and also to the wire resistance. By choosing the right metal or alloy, making the wire a certain length and diameter, employing a sensitive thermometer, and putting the entire assembly inside a thermally insulating package, a *hot-wire meter* can be made. The hot-wire meter can measure alternating current (AC) as well as direct current (DC), because the current-heating phenomenon does not depend on the direction of current flow.

A variation of the hot-wire principle can be used by placing two different metals into contact with each other. If the proper metals are chosen, the junction will heat up when a current flows through it. The resulting device is a *thermocouple*. When it gets warm, a thermocouple generates DC. This current can be measured by a conventional DC ammeter. The hot-wire and thermocouple effects are occasionally used to measure current at radio frequencies (RF) from a few kilohertz to many gigahertz.

The Ammeter

An *ammeter* is a device for measuring electric current. There are various types of ammeters. Some measure large currents up to many amperes. Others measure tiny currents down to a fraction of a microampere. Some ammeters measure DC; others measure AC. Some ammeters measure RF current. There are ammeters that show average current, and ammeters that show peak current. Most voltmeters, ohmmeters, and wattmeters are based on the ammeter principle.

Analog Ammeters A magnetic compass, such as the type used in early galvanometers, is inconvenient for use as an ammeter. The compass must lie flat; the coil must be aligned with the compass needle (that is, with the earth's geomagnetic field) when there is zero current. But the magnetic field can be provided by a permanent magnet inside the meter. Such a magnet supplies a stronger field than does the earth, and this makes it possible to detect weaker currents. It also allows the meter to be oriented in any direction. The coil can be attached to the meter pointer, and suspended by means of a spring in the field of the magnet. It is a basic analog device known as a *d'Arsonval meter*. In a variation of the d'Arsonval scheme, the meter needle is attached to a permanent magnet, and the coil is wound on a fixed form surrounding the magnet. The needle/magnet assembly swings on a spring bearing. Current in the coil produces a magnetic field, and this generates a force that deflects the needle. This method works, but not as well as the d'Arsonval method. The mass of the magnet slows down the needle response. The needle also tends to bounce (overshoot) when the current is switched on and off, making it difficult to obtain good readings. An *electromagnet* can be used in place of the permanent magnet in the meter assembly. This electromagnet can be operated by the same current that flows in the coil attached to the meter needle. This eliminates the massive, permanent magnet inside the meter. It also eliminates changes in meter sensitivity that occur as the permanent magnet weakens with age. The sensitivity of a d'Arsonval meter depends on the strength of the permanent magnet, if the meter uses a permanent magnet. The sensitivity also depends on the number of turns in the coil. The stronger the magnet, and the larger the number of turns in the coil, the less current is needed to produce a given magnetic force. If the meter is of the electromagnet type, the combined number of coil turns affects the sensitivity. For a given electric current, the force increases in direct proportion to the number of coil turns.

Digital Ammeters   There are two types of digital ammeter. One type indicates current as a direct numeric readout in amperes, milliamperes, or microamperes. In recent years, numeric digital meters have become common, especially in laboratory test instruments. The other type of digital ammeter consists of a row of light-emitting diodes or a liquid-crystal display. This is known as a *bar-graph meter.* An advantage of digital metering is that it eliminates the need for personnel to interpolate the reading on the scale. There is little chance for error on the part of the technician or engineer, because the readout is straightforward. Another advantage of digital metering is the fact that there are no moving parts to wear out or be damaged by physical shock. Digital meters are difficult to use in situations where frequent adjustments must be made that affect the meter indications, or when the current fluctuates constantly. In a high-fidelity (hi-fi) amplifier, for example, the volume-unit (VU) meters must be analog types, so the operator can readily sense the average and peak sound levels. When controls must be adjusted for a peak (relative maximum) or a dip (relative minimum) in current, an analog ammeter will work better than a digital ammeter.

Wide-Range Ammeters   Sometimes it is desirable to have an ammeter that will allow for a wide range of current measurements. The full-scale deflection of a meter assembly cannot easily be changed, because this would mean changing the number of coil turns and/or the strength of the magnet. But all ammeters have a certain amount of internal resistance. If a resistor of the correct value is connected in parallel with the meter, the full-scale deflection of an ammeter can be increased by a factor of 10, 100, 1000, or even 10,000. The resistor must be capable of carrying the current without burning out. This is called an *ammeter shunt.* Shunts are used when it is necessary to measure very large currents, such as hundreds of amperes. Shunts also allow a microammeter or milliammeter to be used as a multimeter with many current ranges.

Thermal Ammeters   Another phenomenon, useful in the measurement of current, is the fact that when current flows through a resistive wire, the wire heats up. All conductors have some resistance. The extent of heating is proportional to the square of the current being carried by the wire. Heating is directly proportional to the wire resistance if the current remains constant. By choosing the right metal or alloy, making the wire a certain length and diameter, employing a sensitive thermometer, and putting the whole assembly inside a thermally insulating package, a thermal, or "hot-wire," ammeter can be made. Such an ammeter can measure AC as well as DC, because the extent of heating does not depend on the direction of current flow. A variation of the thermal ammeter can be made by placing two different metals in physical contact with each other. If the correct metals are chosen, the junction will heat up when a current flows through it. This is a *thermocouple ammeter.* As with the hot-wire meter, a thermometer can be used to measure the extent of the heating. There is yet another effect. When a thermocouple gets warm, it generates DC. The current can be measured by a conventional analog or digital ammeter. Thus a DC ammeter can be indirectly used to measure AC. This scheme is useful when it is necessary to have a fast meter response time, because the DC meter reacts much more rapidly to thermocouple temperature changes than does a thermometer. Thermal effects are used occasionally to measure RF current from a few hundred kilohertz to several tens of gigahertz.

The Voltmeter

A *voltmeter* is a device that gives an indication of the magnitude of an *electromotive force* or *potential difference.* In an electric circuit whose resistance is constant, the current is directly proportional to the applied voltage. An ammeter can indicate voltage if it is used in series with a fixed, known resistance.

If an ammeter is connected across a moderate source of voltage such as a battery, the meter needle will deflect. But this is unwise and can even be dangerous. Ammeters have low internal resistance; they are meant to be connected in series with other components, not in parallel, and especially not directly across a power supply. A milliammeter or microammeter will probably be damaged if it is connected directly to a source of voltage without any intervening series resistance. The ammeter in effect short-circuits the power supply. In a battery, the electrolyte in the cells can boil out, sometimes violently, under these conditions.

If a large resistor is connected in series with a microammeter, the meter will give a reading directly proportional to the voltage across the meter/resistor combination. The smaller the full-scale reading of the microammeter, the larger the resistance that will be necessary to obtain meaningful indications on the meter. Using a microammeter and a very large value of resistance in series, a voltmeter can be devised that will draw only a tiny current from a power source. This provides an accurate and safe method for measuring moderate voltages. A voltmeter can be made to have various full-scale ranges by switching different resistors in series with a microammeter. The internal resistance of the voltmeter is large because the values of the resistors are large. The greater the supply voltage, the larger the internal resistance of the voltmeter, because the necessary series resistance increases as the voltage increases.

A voltmeter should have the highest possible internal resistance. The power-supply current should go toward working a circuit, not merely into obtaining a voltage reading. This can be facilitated by using a microammeter with the smallest possible full-scale reading, thus requiring the largest possible series resistance. For example, a 0-to-50 microammeter will make a better voltmeter in most situations than a 0-to-500 microammeter. A common specification for voltmeters, especially the type used in laboratories, is the *ohms-per-volt* rating. In most applications, the higher this number, the more accurate the voltmeter will be, and the less it will interfere with the operation of circuits to which it is connected.

An alternative type of voltmeter uses the effect of electrostatic deflection, rather than electromagnetic deflection as is the case with common microammeters. Electric fields produce forces, just as do magnetic fields. Therefore, a pair of metal plates will attract or repel each other if they are electrically charged. An *electrostatic voltmeter* makes use of this phenomenon, taking advantage of the attractive force between two plates having opposite electric charge or having a large potential difference. An electrostatic voltmeter draws almost no current from the power supply, and thus its ohms-per-volt rating is practically infinite. The only medium between the charged plates is the atmosphere. Air is a nearly perfect electrical insulator. One major advantage of an electrostatic voltmeter is the fact that it can be used to measure AC as well as DC voltages. This is not generally true of a microammeter/resistor series combination. But electrostatic voltmeters are fragile and can be easily damaged. Also, mechanical vibration influences the readings.

The Ohmmeter

An *ohmmeter* is a device used for measuring DC resistances. A source of known voltage, usually a battery, is connected in series with a voltmeter, a *zero-adjust potentiometer,* and a pair of test leads for connection to the unknown resistance. A switchable set of internal fixed resistors varies the measurement range. These internal resistors must have very close tolerances for optimum metering accuracy. A schematic diagram of a simple ohmmeter is shown in Fig. 13-1.

Most ohmmeters have several ranges, labeled according to the magnitude of the resistances in terms of the scale indication. The scale is calibrated from 0 to "infinity." Usually zero ohms (representing maximum current) is at the extreme right-hand end of the scale, and "infinity" ohms (representing zero current) is at the extreme left-hand end. The range switch facilitates the measurement of fairly large or fairly small resistances. Most ohmmeters can accurately measure any resistance between about one ohm and several hundred megohms. Specialized ohmmeters are needed for measuring extremely small or large resistances.

Figure 13-1
A schematic diagram
of a simple ohmmeter.

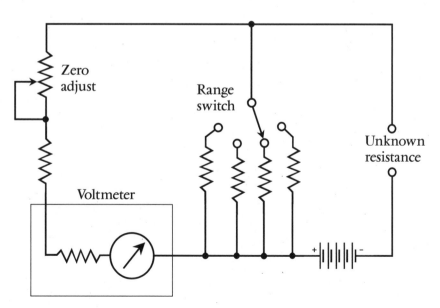

When using an ohmmeter, it is important that the circuit under test not be carrying current. This means the equipment must be switched off, and the power-supply filter capacitors must be allowed to completely discharge. If you try to measure the resistance of a component or circuit when current is flowing through it, the current will cause inaccurate or impossible readings (in some cases "less than zero" or "greater than infinity"). High external voltages can present a shock hazard to you and can also damage the ohmmeter.

The Volt-Ohm-Milliammeter

In electronics labs, a common piece of test equipment is the *volt-ohm-milliammeter* (VOM). As its name implies, the VOM can measure voltage, resistance, or current. The parameter to be measured is selected by a switch.

It is not difficult to adapt a simple microammeter to the measurement of moderate values of voltage, resistance, and current. A typical VOM, available at most electronics stores, costs only a few dollars. Some VOMs have analog readouts; others display values as numerals. They are generally intended for DC applications and will not work in AC or RF circuits without special adapters.

Commercially available VOMs have limits in the values they can measure. The maximum voltage is around 1000 volts (V); larger voltages require special leads and heavily insulated wires, as well as other safety precautions. The maximum current that a common VOM can measure is a few amperes. The highest measurable resistance is several tens of megohms. There are also lower limits on the values a typical VOM can accurately measure. The minimum voltage is a few hundredths of a volt; the smallest current is about one microampere; the lowest resistance is a fraction of an ohm. Measurement of extremely high or low voltage, current, or resistance requires instruments far more sophisticated and expensive than a VOM.

Some VOMs incorporate active amplifiers to enhance the performance of the voltmeter function. A good voltmeter disturbs circuits under test as little as possible, and this requires that the meter have a high internal resistance. One scheme for getting an extremely high effective internal resistance is to sample a minuscule current, far too small for any meter to directly indicate, and then amplify this current so that a meter will show it. When a tiny current is drawn from a circuit, the equivalent resistance is extremely high. The most effective way to accomplish the amplification, while making sure that the current drawn is as small as possible, is to use either a *vacuum tube* or a *field-effect transistor* (FET). A voltmeter that uses a vacuum-tube amplifier to minimize the current drain is known as a *vacuum-tube volt-ohm-milliammeter*. If an FET is used, the meter is called a *field-effect-transistor volt-ohm-milliammeter*. Either of these devices provides extremely high input resistance along with excellent sensitivity. They also allow measurement of lower voltages than conventional VOMs.

The Wattmeter

A *wattmeter* is a device employed to measure electrical power. This generally requires that voltage and current both be known. In some cases, phase difference between current and voltage is involved in the measurement process.

When reactance is not a factor, such as in a DC or utility AC circuit, power is simply the product of the voltage and current. That is, watts equals volts times amperes ($P=EI$). In fact, watts are sometimes called *volt-amperes* in nonreactive circuits. In these applications you can connect a voltmeter in parallel with a circuit to get a reading of the voltage across it, and connect an ammeter in series to get a reading of the current through it, and then multiply volts times amperes to obtain the wattage consumed by the circuit.

Specialized wattmeters are necessary for the measurement of RF power, peak audio power in a hi-fi amplifier, or in certain other specialized applications in wireless devices. Most wattmeters, whatever the associated circuitry, use simple ammeters as their indicating devices.

Specialized Metering Devices

The following are examples of special-purpose meters often encountered in radio and wireless devices, and in the equipment used to test and service wireless systems.

Directional Wattmeter   This device can measure radio transmitter output power and can also give an indication of how well an antenna is matched to a transmission line. Directional wattmeters fall into two categories. One type has a single scale, calibrated in watts, and sometimes also in milliwatts or kilowatts (switch selectable). The meter reads either *forward power* or *reflected power,* depending on the position of a switch or rotatable internal element. Another type of directional wattmeter has two needles in a single enclosure, with a different calibrated scale for each needle. Both of these scales are graduated in watts, and sometimes also in milliwatts or kilowatts. One needle/scale indicates forward power, and the other needle/scale indicates reflected power. There is a third scale, calibrated for the point where the two needles cross. This scale indicates the *standing-wave ratio* (SWR). Such a meter is known as a *crossed-needle meter.*

VU and Decibel Meters In hi-fi equipment, especially sophisticated amplifiers, loudness meters are sometimes used. These are calibrated in *decibels* (dB). One decibel is an increase or decrease in sound or signal level that you can barely detect if you are expecting the change. Audio loudness is expressed in *volume units* (VU), and the meter that indicates it is called a *VU meter.* Usually, such meters have a zero marker with a red line to the right and a black line to the left; they are calibrated in decibels above and below this zero marker. Such a meter might also be calibrated in *root-mean-square (rms) watts,* an expression for audio power. As music is played through the system, or as a voice comes over it, the VU meter needle kicks up. The amplifier volume should be kept down so that the meter does not go past the zero mark into the red range. If the meter goes into the red scale, it means distortion is probably taking place in the amplifier circuit.

S Meter This is a meter in a radio receiver that indicates the relative or absolute amplitude of an incoming signal. Many radio receivers are equipped with S meters. There are several different types. The signal for driving the meter can be obtained from the *intermediate-frequency* (IF) stage of the receiver. The most common method of obtaining this signal is by monitoring the *automatic-gain-control* (AGC) voltage. The stronger the signal, the greater the AGC voltage, and the higher the meter reading. The standard unit of received signal strength is the *S unit.* Generally, one S unit represents a change in signal strength amounting to 6 dB of voltage across the receiver antenna terminals, assuming a constant impedance. This is a signal-strength ratio of 2:1. Thus an increase in signal strength of one S unit means that the signal has become twice as strong. An rms signal level of 50 microvolts (μV) is typically considered to represent nine S units. The S-unit scale in a typical S meter ranges from 1 to 9, or occasionally from 0 to 9. Calibrated S meters are usually marked off in decibels above the S-9 level.

Applause Meter   Sound level in general can be measured by means of this simple device. The instrument gets its name from the fact that it is often used to measure relative levels of applause in auditoriums and theaters. The meter can be uncalibrated, in which case it will give relative but not absolute indications; or it can be scaled in decibels and connected to the output of a precision amplifier with a microphone of known, standard sensitivity. You have probably read or heard that a vacuum cleaner will produce something like 80 dB of sound, and a large truck going by might subject your ears to 90 dB. These figures can be determined by a calibrated applause meter.

Illumination Meter   This instrument, also called a *light meter,* can be made by connecting a milliammeter to a photovoltaic (solar) cell. More sophisticated devices use DC amplifiers to enhance sensitivity and to allow for several ranges of readings. Solar cells are not sensitive to light at exactly the same wavelengths as human eyes. This effect can be neutralized by placing a color filter in front of a solar cell, so the cell becomes sensitive to the same wavelengths, in the same proportions, as human eyes. The meter can be calibrated in units such as *lumens* or *candela.* Sometimes, illumination meters are used to measure infrared (IR) or ultraviolet (UV) intensity. Different types of photovoltaic cells have peak sensitivity at different wavelengths. Optical filters can be used to block out wavelengths that you do not want the meter to detect.

Pen Recorder   A meter movement can be equipped with a marker to produce a graphic hard copy of the level of some quantity with respect to time. A sheet of paper, with a calibrated scale, is taped to a rotating drum. The drum, driven by a clock motor, turns at a slow rate, such as one revolution per hour or one revolution in 24 hours. A device of this kind, along with a wattmeter, might be employed to plot a graph of the power consumed by your household at various times during the day. In this way you might tell when you use the most power, and at what particular times you might be using too much. A computer can be adapted to function as an electronic pen recorder using specialized transducers and software.

Oscilloscope   Another graphic meter is the oscilloscope. This measures and records quantities that vary rapidly, at rates of hundreds, thousands, or millions of times per second. It creates a "graph" by throwing a beam of electrons at a phosphor screen. A *cathode-ray tube* (CRT), similar to a television (TV) picture tube or computer monitor, is employed.

Oscilloscopes are useful for looking at the shapes of signal waveforms, and also for measuring peak and instantaneous signal levels (rather than just the effective levels). An oscilloscope can also be used to approximately measure the frequency of a waveform. The horizontal scale of an oscilloscope shows time, and the vertical scale shows instantaneous voltage. An oscilloscope can indirectly measure power or current, by using a known value of resistance across the input terminals.

Spectrum Analysis

A *spectrum analyzer* is a device that graphically displays signal amplitude as a function of frequency. Engineers use spectrum analyzers in the design, alignment, and troubleshooting of radio transmitters and receivers.

Basic Concept   A spectrum analyzer consists of an oscilloscope and a circuit that provides the spectral display for the CRT. Figure 13-2 shows an example of a typical spectrum-analyzer display. The frequency is rendered horizontally from left to right. Signal amplitude is displayed vertically. In the example shown, there are four signals on the screen. The horizontal line across the display, near the bottom of the screen, is the *noise floor*. In this example, each vertical division represents 10 dB of

Figure 13-2
A spectrum analyzer generates a graph of signal strength (vertical axis) as a function of frequency (horizontal axis).

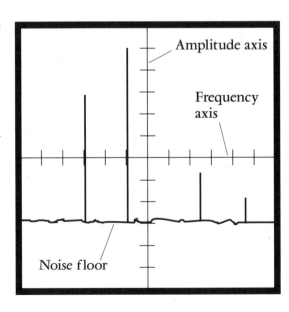

Amplitude axis

Frequency axis

Noise floor

signal level change, and each horizontal division represents 10 kilohertz (kHz) of frequency change. These parameters can be changed at will by the operator. The amplitude display in a spectrum analyzer can be set for a linear scale (say, 0.1 V per division) if desired, rather than the more common logarithmic (decibel) scale. Some spectrum analyzers are designed for the audio-frequency (AF) range. These include narrowband analyzers of fixed bandwidth, such as 5 hertz (Hz), and devices having a bandwidth that is a constant percentage of the frequency of the signal being evaluated. Examples of the latter are octave, half-octave, and third-octave analyzers. Real-time analyzers commonly have visual displays of response in third-octave bands covering the AF range.

Operation   To use a spectrum analyzer, a technician first chooses, by means of a selector switch, the band of frequencies to be displayed. This band might cover the spectrum from DC to 1 gigahertz (GHz) or more, or it might cover a small frequency range, say 10 kHz wide. The *gain* (vertical-scale sensitivity) of the circuit is adjusted as desired. The *resolution* and *sweep rate* must also be adjusted for the application intended. A spectrum analyzer can be useful for determining the levels of *harmonics* and *spurious emissions* from a radio transmitter. In the United States, radio equipment must meet certain government-imposed standards regarding the purity of their emissions. A spectrum analyzer gives an immediate indication of whether or not a transmitter is functioning according to these standards. The analyzer can also be used to observe the *bandwidth* of a modulated signal. Such improper operating conditions as *overmodulation* (in amplitude modulation) or *overdeviation* (in frequency modulation) can be easily seen in the form of excessive signal bandwidth. A spectrum analyzer can be used in conjunction with a device called a *sweep generator* for evaluating the characteristics of a selective filter. You might, for example, need to adjust a bandpass filter to obtain a certain bandpass response at the front end of a radio receiver. You might want to determine whether a low-pass filter is providing the right cutoff frequency and attenuation characteristics. The sweep generator produces an RF signal that varies in frequency, exactly in synchronization with the display of the spectrum analyzer, over a range that the operator can select.

Panoramic Receiver   This is a spectrum analyzer adapted to allow continuous visual monitoring of received signals over a specific band of frequencies. A radio communications receiver can usually be adapted for panoramic reception by connecting a spectrum analyzer into the IF chain. In the panoramic receiver, a display (usually a CRT) shows signals as vertical pips along a horizontal axis. The signal amplitude is indicated by the height of the pip. The position of the pip along the horizontal axis indicates its frequency. The frequency to which the receiver is tuned appears at the center of the horizontal scale. The frequency increment per horizontal division can be set for spectral analysis of a single signal (for example, 0.5 kHz per division), or it might be set for observation of a specific range of frequencies (for example, 10 kHz per division). The maximum possible frequency display range is limited by the characteristics of the receiver IF stages. The amplitude change per vertical division can be set for a certain number of decibels (usually 3 or 10 dB), or it can be set for a linear indication.

Simple Monitoring and Control Systems

The simplest wireless monitoring and control systems consist of a radio or IR transmitter and receiver, and associated mechanical hardware that is operated via signals reaching the receiver. In the case of a monitoring system, the transmitter is located at the site to be "bugged" and is often mobile or portable; the receiver is located at the place where the eavesdropping is done and is usually fixed. In control systems, the transmitter is in a mobile or portable control box or at a fixed control station, and the receiver is in the device or system to be controlled.

Garage-Door Opener

Most people are familiar with *remote-control garage-door openers*. A battery-powered radio transmitter is contained in a box about the size of a pack of cigarettes. It sends a low-power signal, usually in the very-high-frequency (VHF) or ultra-high-frequency (UHF) portion of the radio spectrum. The antenna is generally contained within the box. The receiver is located inside the garage and is connected to an electric motor that opens or closes the garage door upon receipt of a signal from the transmitter. Different units operate on different frequencies, to minimize the possibility of neighbors' units interfering with one another. The transmitter will open and close the door at distances up to about 200 feet (ft).

When radio-controlled garage-door openers first became popular, there were problems with interference. Garage doors would open without any apparent reason. People would come home to find their garage doors open and no one in the house. This was a risk to property. With attached garages in very cold or very hot weather, it greatly compromised the energy efficiency of the house. Although less often a problem, a garage door deliberately left open might close without any command from the owner. Interference problems can usually be eliminated by modulating the transmitter signal with an audio tone having a specific frequency, and by designing the receiver so it will actuate the motor only when a signal is received that has the proper modulating tone.

Some homeowners desire additional security and use systems that require a specific identification code to be transmitted. This code might consist of several digits generated by a telephone tone dialer. The receiver will not actuate the door unless and until a signal comes in that is modulated with the correct digit tone pairs in the correct order.

Maximum remote-control garage-door-opener security is obtained by simply disabling the receiver. During extended trips when a home will be left vacant for some time, the receiving unit can be unplugged, or its battery can be disconnected.

Baby Monitor

A short-range, amplitude-modulated (AM) or frequency-modulated (FM) radio transmitter and receiver can be used to listen at a distance to the sounds in an infant's room. The transmitter contains a sensitive microphone, a whip antenna, and an AC power supply. The unit can be placed on a table or desk or even on the floor near the baby's crib. The receiver is similar to a handheld "walkie-talkie." It is battery-powered and can be carried around. It has an inductively loaded, short "rubber duckie" antenna similar to the antennas on cordless telephone sets. The receiver can pick up signals from the transmitter at distances of up to about 200 ft. The RF signals pass easily through walls, ceilings, and floors.

The so-called baby monitor is subject to interference from other units that might be operating nearby on the same channel. Sometimes *radio-frequency interference* (RFI) is experienced from high-power radio transmitters in the vicinity. But such cases are relatively rare. There are several channels to choose from. If interference occurs, you can switch to another channel or exchange the unit at the place of purchase for one that works on a different channel. Communications privacy and security are not a major concern with baby monitors; the important thing is that the person carrying the receiver knows if the baby starts crying or makes other sounds indicating distress.

Home Entertainment Remote Box

In recent years, control boxes have become standard accessories for TV sets, videocassette recorders (VCRs), and hi-fi sound systems. Most of these devices operate using IR transducers. As a result, the signals do not pass through walls, ceilings, or floors. An IR control box usually works only within a single room. In some cases control is possible from an adjacent room if the door is open, even when the control box and the controlled equipment do not lie on a line of sight. Sufficient IR energy can be reflected from walls, ceilings, floors, and furniture to allow some operating flexibility. Interference is practically unknown with IR control systems.

The control box itself, which looks something like an electronic calculator, is battery-powered and contains one or more IR-emitting diodes (IREDs). These are usually protected by a red plastic window. The diodes transmit IR signals, which are modulated with information concerning the channel or frequency, volume level, and other particulars of the system to be controlled. Some control boxes are sophisticated, and it can take awhile to become proficient at using them.

Intermediate Monitoring and Control Systems

Somewhat more advanced monitoring and control systems employ sensors, computers, and simple robotic devices. Often, radio or IR links are used as well. In many intermediate systems, monitoring and control functions are combined. Such systems can not only inform people of what is taking place at a remote location, but they can allow, or cause, corrective measures to be taken if the situation is not right. Some examples follow. (There are, of course, many other applications besides those mentioned here.)

Smoke Detector

When people and property must be protected from fire, smoke detection is a simple and effective measure. *Smoke detectors* are inexpensive and can operate from flashlight cells.

Smoke changes the characteristics of the air. It is often accompanied by changes in the relative amounts of gases. Fire burns away oxygen and produces other gases, such as carbon dioxide, carbon monoxide, and/or sulfur dioxide. The smoke itself consists of solid particles. Air, and in fact any substance that is a poor conductor of electricity, has a property called the *dielectric constant.* This is a measure of how well the atmosphere can hold an electric charge. Air also has an *ionization potential:* the energy needed to strip electrons from the atoms. Many things can affect these two properties of air. Common factors are humidity, pressure, smoke, and changes in the relative concentrations of gases.

A smoke detector can work by sensing a change in the dielectric constant and/or the ionization potential of the air. Two electrically charged plates are spaced a fixed distance apart (Fig. 13-3). These two plates form an air-dielectric capacitor. A battery is connected to the plates. Normally the plates retain a constant charge and the current in the circuit is zero. But if the properties of the air change, the capacitance will change. This will cause a small electric current to flow in the circuit for a short time.

Figure 13-3
Smoke causes the plates to charge or discharge, actuating the detection circuit and triggering the alarm and robotic fire-protection devices.

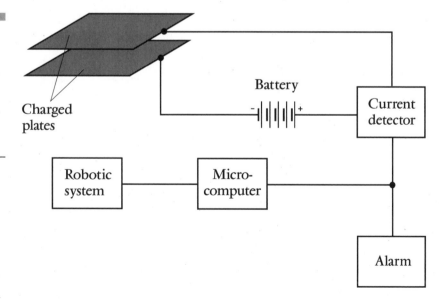

This current can be detected, and the resulting signals can actuate an alarm. The signals might also actuate a simple robotic system, such as a water sprinkler to extinguish fires.

Quality Control

One application of wireless technology, the *laser,* can be found in *quality control* (QC) checking of bottles for height as they move along an assembly line. A laser/robot combination can find and remove bottles that are not of the correct height. The laser device is a pair of *electric eyes* positioned in a precise way. The principle is shown in Fig. 13-4. If a bottle is too short, both laser beams will reach the photodetectors. If a bottle is too tall, neither laser beam will reach the photodetectors. In either of these situations, a robot arm/gripper picks the faulty bottle off the line and discards it. Only when a bottle is within a narrow range of heights (the acceptable range) will the top laser reach its photodetector while the

Figure 13-4
Parallel laser beams
check bottle height.

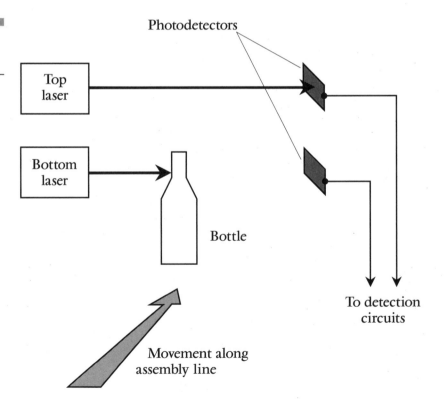

bottom laser is blocked. Then the robot does nothing, and the bottle is allowed to pass.

Keeping Track of Convicts

In recent years, the concept of house arrest has gained popularity because wireless systems make it easy to monitor compliance. This frees jail space for serious and violent offenders, while allowing less dangerous people to continue serving society in a useful way. The behavior of parolees and probationees can also be monitored.

Beeper   Suppose a convict is sentenced to house arrest except during working hours, say from 6:00 P.M. to 8:00 A.M. Monday through Friday, and all day Saturday and Sunday. One method of checking for compliance is to have the convict carry a conventional beeper. An officer can page the convict at random times; the convict must then call the officer within a certain length of time. The call can be traced, and the location of the telephone (home or work) verified. To some extent the process can be automated. Obviously, this scheme has limitations, but it can be useful as a low-security provision. Periodic random physical visits by law enforcement can minimize the possibility that the convict will simply give the beeper to a friend and "play hooky." Another measure, which is controversial, allows for the beeper to be surgically implanted in the convict's body, making it impossible for the convict to give the unit to someone else. The controversy arises from the argument that surgical invasion of the body for nonmedical purposes is Orwellian and sets a dangerous precedent.

Short-Range Transmitter   A somewhat more secure method of ascertaining that a convict is at a certain place at a certain time is by means of a short-range radio transmitter and receiver. The convict wears the transmitting unit. Receiving units are placed at the convict's home, in the car, and in the place of work. The transmitter range is on the order of 100 to 200 ft. Receiver signals are sent to a central monitoring point. The signals are encoded so the monitoring personnel (or computers) know whether the convict is at home, in the car, or at work. Any deviation from normal patterns can be detected. Of course, to avoid "false positives," the transmitter battery must be replaced or recharged at regular intervals. As with the beeper system, periodic random visits by officers can minimize "truancy." Surgical implantation is also possible.

Electromagnetic Location Detection   Another way to keep track of convicts is to use *radiolocation,* or even the *global positioning system* (GPS). A transponder can be carried or worn by the person to be tracked; continuous signals can be sent to the unit asking for a position fix, and the unit can respond via a wireless network such as the cellular telephone system. Special cell phones have been distributed to safety-conscious people in some high-crime areas; these units automatically dial 911 and inform the operator of the location of the phone set. The same technology can be employed to keep track of the whereabouts of convicts.

Amateur Radio

It is not necessary to be physically present at the location of an *amateur radio,* or "ham," station in order to operate that station. A *repeater* is an example of a ham station that does not always have an operator at the controls. A "ham shack" can be located hundreds or even thousands of miles away from its operator, provided a suitable remote control system is employed.

Assets   Remote control has advantages for the city dweller. A ham station can be located at a rural place, where large antennas can be erected and where RFI will not be as likely. Then the operator can work the station from almost anywhere: a downtown condominium, a car, or even a handheld unit on a bus or train. No matter where the remote location might be, the fixed station is always in the ideal location.

Limitations   The main disadvantage of remote control is that, if something goes wrong at the remote station, it cannot be conveniently repaired. Another problem is that, if the remote control is via long-distance telephone, the phone bill can be enormous. And finally, no matter how sophisticated remote control linkups become, there will always be some sacrifices that result from the fact that the operator is not physically present at the remote station.

Local-Site Equipment   At the operator's location, also known as the *local site,* a computer allows use of peripheral apparatus with the telephone or radio linkup unit. A microphone, keyer, speaker, and modem (if telephone control is used) or terminal (if radio control is used) are standard. In addition, a telephone set or transceiver is needed. An antenna is needed for the link if radio control is used.

Remote-Site Equipment   At the remote site, where the large fixed station is located, there is a telephone or transceiver to handle the commands from, and send telemetry to, the local site. There is a modem or terminal. With a telephone link, an automatic answerer is needed. With a radio link, a link antenna is necessary. There is a computer, similar to or identical with the one at the local site. There is a station (main) transceiver, connected to the main antenna. There might also be rotators, antenna switches, and transmatches.

Bugged!

Everyone has heard of the eavesdropping device called a "bug." Such a system consists of a tiny radio transmitter that can be hidden in a room, placed in a shirt pocket, or planted in a car. The antenna is a length of wire. A receiver can be located nearby. The transmitter range is necessarily short; the device must operate at a low RF power level (on the order of a few milliwatts) to conserve battery energy. At the receiving end, a person can listen to the sounds in the vicinity of the transmitter. More often, a sound-actuated tape recorder is connected to the receiver. The tape can be played back later, and the sound-actuation circuit eliminates the long silences that make up most of what the "bug" actually "hears."

The proliferation of cellular repeaters, and similar wireless communication devices, greatly extends the operating range of the traditional bug. It is not necessary for the receiver to be within the radius of the original transmitted signal. If the transmitter is near a wireless repeater that connects into a larger system, eavesdropping can be done anywhere within the coverage of the wireless system. With the advent of low-earth-orbit (LEO) satellite systems, it is theoretically possible to literally bug a room on the other side of the world. Of course, however advanced the technology might become, it will always be necessary for someone to physically get onsite to install the transmitter.

Sophisticated bugs can be hard-wired to utility power lines (if the unit can be hidden in a ventilator shaft or other permanent, secluded place). This facilitates the use of a transmitter that can run at a higher RF power level than a battery-operated unit. Useful operating life is not of immediate concern. Utility power also allows, within certain limits, the use of video as well as audio surveillance. Digital cameras, which can capture nonmoving images every few seconds, can be smaller than a golf ball and provide surprising *image resolution* (detail).

Bugging is not a job for amateurs. Eavesdropping-related laws vary from one country to another. In the United States, it is generally necessary to get a court order to legally carry out a bugging operation. The law is similar to that for a telephone wiretap. People in general need not fear bugs in their homes, offices, or cars. It is nevertheless a good idea to avoid slandering, making threats, or otherwise talking about people in a grossly negative way in any place where a bug could be hidden. (It's good etiquette, too.) Depending on the future of communications laws and the general mood of society, the probability of common people being bugged might increase during the twenty-first century. But one limiting factor overrides all others regardless of laws and technology: in order to effectively bug someone, a human being must be employed, and presumably paid, to listen to or watch hours of signals or tapes.

Electromagnetic bugs can be detected rather easily, although reliable detection equipment is expensive. The presence of a suspicious RF field gives away the existence of the transmitter. A person concerned about being bugged can install a *spectrum analyzer* and connect it to a computer to constantly monitor the RF spectrum in the vicinity. Previously nonexistent signals will show up in the spectral plot, whereupon the computer will inform the operator. A broadband receiver can then be tuned to the frequency at which the new signal appears, and the modulation on the signal can be checked.

Eye-in-Hand System

To assist a robot gripper, or "hand," in finding its way, a camera can be placed in the gripper mechanism. The camera must be equipped for work at close range, from about 3 ft down to a fraction of an inch. The positioning error must be as small as possible. To be sure that the camera gets a good image, a lamp is included in the gripper along with the camera (Fig. 13-5). This so-called *eye-in-hand system* can be used to precisely measure how close the gripper is to whatever object it is seeking. It can also make positive identification of the object, so that the gripper does not go after the wrong thing.

The eye-in-hand system uses a *servo*. The robot is equipped with or has access to a controller (computer) that processes the data from the camera and sends instructions back to the gripper. While most eye-in-hand systems use visible light for guidance and manipulation, it is also

Figure 13-5

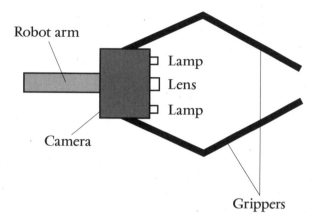

Figure 13-5
A robotic eye-in-hand system. The lamps and camera can function at visible or IR wavelengths.

possible to use IR. This can be useful when it is necessary for the robot hand to sense differences in temperature.

Flying Eyeball

In environments hostile to humans, robots find many uses, from manufacturing to exploration. One such device has been called a *flying eyeball.* This robot can see very well underwater and can also move around. But it cannot manipulate anything; it has no arms or grippers.

A cable containing the robot in a special launcher housing is dropped from a boat. When the launcher gets to the desired depth, it lets out the robot, which is connected to the launcher by a tether. The tether and the drop cable convey data back to the boat. In some cases, the tether can be eliminated and a wireless link can be used to convey data from the robot to the launcher. This allows the robot to have more freedom of movement, without the concern that the tether might get tangled up in something. This link is usually in the red portion of the visible spectrum, or in the IR region.

The robot contains a TV camera and one or more lamps to illuminate the undersea environment. It also has a set of thrusters, or propellers, that let it move around according to control commands sent via the link between the boat and the robot. Human operators on board the boat can watch the images from the TV camera, and guide the robot around as it examines objects on the sea floor.

Advanced Monitoring and Control Systems

Advanced wireless monitoring and control systems almost invariably involve the use of computers. Text, telemetry, image data, animated video, sounds, and programs can be transmitted between or among central and remote locations.

Teleoperation and Telepresence

Teleoperation is the technical term for the remote control of autonomous robots. A remotely controlled robot is sometimes called a *telechir*. Teleoperation is used in robots that can look after their own affairs most of the time but occasionally need the help of a human operator. In a teleoperation system, the human operator can control the speed, direction, and other movements of a robot from some distance away. Signals are sent to the robot to control it; other signals come back, telling the operator that the robot has indeed followed instructions. These signals are called *telemetry*.

Some teleoperated robots have a limited range of functions. A good example is a space probe, such as Voyager, hurtling past some remote planet. Earthbound scientists sent telemetry to Voyager, aiming its cameras and sometimes even fixing minor problems. Voyager was, in this sense, a teleoperated robot. The link was at RF.

Telepresence is a refined, advanced form of teleoperation. The robot operator gets a sense of being on location, even if the telechir and the operator are miles apart. Control and feedback are done via telemetry sent over wires, optical fibers, radio waves, or, in some cases, IR or visible line-of-sight links.

In a telepresence system, the telechir is an autonomous robot and often resembles a human being. The more humanoid the robot, the more realistic the telepresence. The control station consists of a suit that you wear or a chair in which you sit with various manipulators and displays. Sensors give you feelings of pressure, vision, and sound. You wear a helmet with a viewing screen that shows whatever the robot camera sees. When your head turns, the robot head, with its vision system, follows. Thus you see a scene that changes as you turn your head, just as if you were in a space suit or diving suit at the location of the robot. *Binocular machine vision* gives you a sense of depth. *Binaural machine*

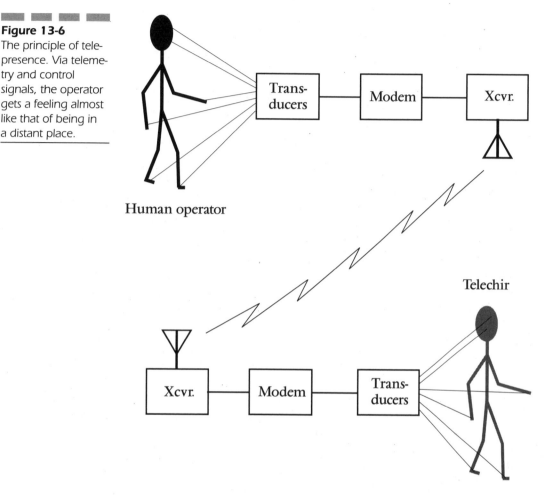

Figure 13-6
The principle of tele-
presence. Via teleme-
try and control
signals, the operator
gets a feeling almost
like that of being in
a distant place.

Human operator

Telechir

hearing lets you perceive sounds just as your ears would hear them if you were on site.

A block diagram of a telepresence system is shown in Fig. 13-6. The data flows in both directions. At the telechir end of the circuit, the transducer converts sounds, images, and mechanical resistance into electrical impulses to be sent to the operator, and also converts electrical impulses from the operator into mechanical motion. At the operator end, the transducer converts mechanical motion into electrical impulses to be sent to the telechir and converts electrical impulses from the telechir into sounds, images, and mechanical resistance. Modems modulate and demodulate signals for and from the transceivers.

The robot propulsion system can consist of a track drive, a wheel drive, or "robot legs." If the propulsion uses legs, you propel the robot by walking around a room. Otherwise you can sit in a chair and "drive" the robot as if it were a cart. The robot has at least one arm, each with grippers resembling human hands. When you want to pick something up, you go through the motions. Back-pressure sensors and position sensors let you feel what's going on. If an object weighs 10 pounds (lb), it will feel as if it weighs 10 lb, but it is as if you are wearing thick gloves. You cannot feel texture.

You might throw a switch, and something that weighs 10 lb would feel as if it weighs only 1 lb. This might be called STRENGTH X 10 mode. If you switch to STRENGTH X 100 mode, a 100-lb object would seem to weigh 1 lb. This would require a robot with great mechanical and structural strength. The link between the robot and the operator might be via cable if the separation is relatively small and there is no great need for mobility. However, the preferred method would be wireless. Because of the complexity of the telemetry signals, large bandwidth would be required.

There are many uses for telepresence systems. Plausible applications include:

1. Working in extreme heat or cold

2. Working under high pressure (on the ocean floor, for example)

3. Working in a vacuum (in outer space, for example)

4. Working where there is dangerous radiation

5. Disarming bombs

6. Handling toxic substances

7. Serving as a police robot

8. Serving as a robot soldier

9. Assisting in neurosurgery

Of course, the robot must be able to survive conditions at its location. Also, it must have some way to recover if it falls or gets knocked over.

The technology for telepresence currently exists. But there are some problems that will be difficult, if not impossible, to overcome. The most serious limitation is the fact that telemetry cannot and never will travel faster than the speed of light in free space. The delay between the transmission of a command and the arrival of the return signal must be less than 0.1 second (s) if telepresence is to be realistic. This means that the robot cannot be more than about 9300 miles (mi), or 15,000 kilometers (km), away from the control operator. Another problem is the resolution of the robot's vision. A human being with good eyesight can see things with several times the detail of the best fast-scan TV sets. To send that much detail, at realistic speed, would take up a huge signal bandwidth. There are engineering and cost problems that go along with this. Still another limitation is best put as a question: How will a robot be able to "feel" something, such as roughness or smoothness, and transmit these impulses to the human brain?

Applications of Wireless Telepresence

Wireless-controlled robots are well suited to handling dangerous materials. This is because, if there is an accident, no human lives will be lost. In the case of radioactive substances, robots can be used and operated at a distance, so that people will not be exposed to the radiation. The remote control is accomplished by means of teleoperation and/or telepresence.

Robots have been used for some time in the maintenance of nuclear power plants. One such machine, called ROSA, was designed and built by Westinghouse Corporation. It has been used to repair and replace heat-exchanger tubes in the boilers. The level of radiation is extremely high in this environment. It is hard for human beings to do these tasks without endangering their health. If people spend more than a few minutes per month at such work, the accumulated radiation dose will exceed safety limits.

Robots can be used to disarm nuclear warheads. If an errant missile came down without exploding the warhead, it would be better to use machines to eliminate the danger, rather than subjecting people to the risk and stress of the job.

In space missions, it is often necessary to perform repairs and general maintenance in and around the spacecraft. It is not always economical to have astronauts do this work. For this reason, various designs are under consideration for a *flight telerobotic servicer* (FTS).

The FTS is a wireless-controlled robot. The extent to which it is controlled depends on the design. The simplest FTS machines use software from a computer in a spacecraft or on earth. More complex FTS devices make use of telepresence. Because of the risk involved in sending humans into space, scientists have considered the idea of launching FTS-piloted space shuttles to deploy or repair satellites.

Human underwater divers cannot go deeper than approximately 1000 ft. Rarely do they descend below 300 ft. Even at this depth, after a long dive, a tedious period of decompression is necessary to prevent illness or death from the bends. Not surprisingly, there is great interest in developing robots that can dive down more than 1000 ft, while doing all, or most, of the things that human divers can do.

The ideal *submarine robot* would use wireless telepresence. The link would probably be IR or visible light in the red part of the spectrum. Ultrasound might also be used.

Imagine a treasure-hunting expedition, in which you salvage diamonds, emeralds, and gold from a sunken pirate ship 3000 ft down, while sitting warm and dry in a *telepresence-control suit* with a glass of soda beside you! Or you might test shark repellents without fear. You could disarm a nuclear warhead at the bottom of a deep bay or repair a deep-sea observation station.

The *Titanic,* the "unsinkable" ocean liner that sank, was found and photographed by an undersea robot called a *remotely operated vehicle* (ROV). This machine did not employ wireless telepresence but was maneuverable and did provide many high-quality pictures of the wrecked ship.

Robotic Space Missions

The U.S. space program climaxed when Apollo 11 landed on the moon and, for the first time, a creature from the earth walked on another world. Some people think the visitor could have been and should have been a robot remotely controlled from the earth via a wireless link.

Spacecraft have been remotely controlled for decades. Communications satellites use radio commands to adjust their circuits, and sometimes even to change their orbits. Space probes, such as the Voyager that photographed Uranus and Neptune in the late 1980s, are controlled by radio. Satellites and space probes are crude robots. Space probes work something like other hostile-environment machines.

Some people say that robots should be used to explore outer space, while people stay safely back on earth and work the robots via remote control. A human being can wear a telepresence control suit and have a robot mimic all movements. The robot might be some distance away. Ultimately, with technology called *virtual reality* (VR), it might be possible to duplicate the feeling of being in a place to such an extent that the person can imagine he or she is really there. Stereoscopic vision, binaural hearing, and a crude sense of touch could be duplicated. Imagine stepping into a gossamer-thin control suit, walking into a chamber, and existing, in effect, on another planet, free of danger from extreme temperatures or deadly radiation!

If robots are used in space travel, with the intention of having the machines replace astronauts, then the distance between the robot and its operator cannot be very great. The reason is that the control signals cannot propagate faster than 186,282 miles per second (299,792 kilometers per second), the speed of electromagnetic (EM) fields in free space. The moon is 1.3 light-seconds from earth. (A light second is 186,282 mi or 299,792 km.) If a robot, rather than Neil Armstrong, had stepped onto the moon on that summer day in 1969, the robot's control operator would have had to deal with a delay of 2.6 s between command and response. It would take each command 1.3 s to get to the moon, and each response 1.3 s to get back to the earth. True telepresence is impossible with a delay that long. Experts say that the maximum delay for true telepresence is 0.1 s. The distance between the robot and its controller cannot be more than 0.05, or 1/20, light-second. That is about 9300 mi or 15,000 km, slightly more than the diameter of the earth.

Property Protection

Wireless technology is extensively used in intrusion-prevention devices, circuits, and networks. Often, such systems are computer-controlled. Systems range from simple machines, such as card readers or pushbutton-code devices, to sophisticated electromechanical networks. They can vary greatly in complexity. Comparatively simple (low-end) systems are adequate for homeowners and small businesses. Large corporations, government agencies, and businesses with high-value inventory are more likely to need advanced (high-end) security systems.

Knowledge-Based

The *knowledge-based identification system* is comparatively simple. Authorized people are issued numerical codes. The entrances to the property are equipped with locks that disengage when the proper sequence of numbers is punched into a keypad. This keypad can be hard-wired into the system, or it can be housed in a box about the size of a typical garage-door opener or home-entertainment remote control. It works like a bank automatic-teller machine (ATM) personal identification code.

There might be only one access code, given to all the personnel in the department, company, or organization, or there might be numerous codes, a different one issued to each authorized person. The term "knowledge-based" arises from the fact that, in order to gain entry, a person must know a specific piece of information (in this case the access code).

One of the main advantages of this type of system is that the codes cannot easily be guessed. The authorized people should memorize their access numbers. The numbers should never be written down in any form that will give away their meaning or purpose. Another asset of knowledge-based security systems is their relatively low cost.

A disadvantage of this scheme is that access codes occasionally leak out. People tend to give secrets away when situations arise that make it expedient to do so (or if it becomes inexpedient not to). Also, codes are sometimes forcibly stolen. Once a code is stolen, anyone who has it can get into the property until that code is invalidated.

Possession-Based

A more sophisticated scheme is the *possession-based security system*. This gets its name from the fact that authorized people must possess some object that unlocks the entry to a property.

Magnetic cards are a popular form of possession-based security device. You insert the card into a slot, and a microcomputer reads data encoded on a magnetic strip. This data can be as simple as an access code, of the sort you punch on a keypad. Or it might contain many details about the card bearer. A so-called *smart card* can be used for security purposes. So can bank ATM cards, credit cards, and PCMCIA standard adapter cards.

The *passive transponder* provides a wireless form of possession-based security system. These are magnetic tags that can be carried by, worn by, or in some cases planted in the bodies of, authorized personnel. They're the same little things that department stores employ to deter petty thieves. The transponder need not be inserted in a slot; it can be read from several feet away. If you've ever been leaving a store or library and been detained because an electronic detector beeped at you, you've experienced a passive transponder at work.

A *bar-code* tag is another form of passive transponder. You have seen bar-code labels or tags in stores, where they are used for pricing merchandise and keeping track of inventory. Bar coding allows instant *opto-*

electronic identification of objects. A bar-code tag has parallel bands of various widths. A laser rapidly scans across the bands. The bands absorb the laser light, while the white regions between the bands reflect the light back to a sensor. The sensor thus receives a binary data signal that is unique to the pattern on the tag. This signal can contain considerable information. A bar-code reader does not have to be brought right up to the tag. Nor does the item have to be oriented in any special way.

The main advantage of possession-based security systems is convenience. You need not worry about forgetting a code number. But this is more or less nullified by the fact that little plastic cards can easily get misplaced, and they are easy to steal. Although *subcutaneous transponders* (devices inserted under the skin) can circumvent these problems, most people recoil from the notion of having electronic apparatus installed in the human body for any nonmedical purpose.

Biometric

An advanced intrusion-prevention scheme, the *biometric security system,* gets its name from the fact that it ascertains, or measures, biological characteristics of the people who are authorized to enter a property. Such a machine can employ vision systems, object recognition, and/or pattern recognition to check a person's face. The machine might use speech recognition to identify people by the waveforms of their voices. It might record a hand print, fingerprint, or iris print. It could employ a combination of all these things. A powerful computer analyzes the data obtained by the sensors and determines whether or not the person is authorized to enter the premises.

Biometric systems are essentially foolproof. However, they must be backed up by threat of force if they are to be of real use. Wherever security requirements are so strict that a biometric security system is needed, there will probably be a few brilliant and reckless people intent upon defeating it and getting into the premises. An American government installation in a hostile country is a prime example of such a property. While the system can be almost impossible to fool, it must also be set up so that it cannot be overcome by a brute-force surprise attack. History has shown that it is almost impossible to do this and at the same time ensure that no authorized personnel are falsely arrested or injured as a result of a system error.

For homeowners and small businesses, biometric systems are generally too expensive. But there are some exceptions. Top-secret archives, priceless works of art, and some scientific experiments (and research personnel) justify the most sophisticated security systems available, no matter what the cost.

Intrusion and Fire Detectors

Visible, IR, and ultrasonic wireless devices are extensively used in home and business security systems. Most devices are designed to detect motion; a few can detect body heat.

Electric Eye   The simplest device for detecting an unwanted visitor is an electric eye. Narrow beams of IR or visible light are shone across all reasonable points of entry, such as doorways and window openings. A photodetector receives energy from the beam. If, for any reason, the photodetector stops receiving the beam, an alarm is actuated. A person breaking into a property cannot avoid breaking a beam, especially if large openings have two or three electric eyes spaced at suitable intervals. The main problem with this system is that a power failure can trigger the alarm unless the energy source has a backup battery or the system has a microprocessor that allows it to differentiate between an intrusion and a power failure.

IR Motion Detector   Many popular intrusion alarm devices make use of IR motion-detection transducers. Two or three wide-angle IR pulses are transmitted at regular intervals; these pulses cover most of the room in which the device is installed. A receiving transducer picks up the returned IR energy, normally reflected from the walls, the floor, the ceiling, and the furniture. The intensity of the received pulses is noted by a microprocessor. If anything in the room changes position, there will be a change in the intensity of the received energy. The microprocessor will notice this change and will trigger the alarm (Fig. 13-7). These devices consume very little power in regular operation, so batteries can serve as the power source. A typical alarm system of this kind uses six or eight AA or AAA alkaline electrochemical cells, which will operate the device continuously for several months.

Radiant-Heat Detector   Infrared devices can detect changes in the indoor environment in another way: the direct sensing of IR (often called *radiant heat*) emanating from objects. Humans, and all warm-blooded animals, emit some IR energy. So, of course, does fire. A simple

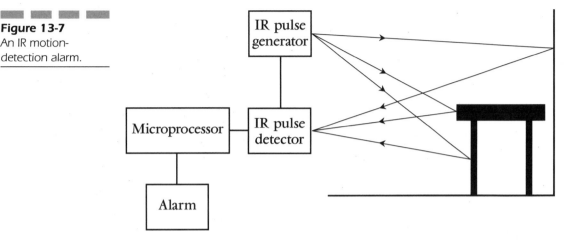

Figure 13-7
An IR motion-
detection alarm.

IR sensor, in conjunction with a microprocessor, can detect rapid or sudden increases in the amount of "heat" present in a room. The time threshold can be set so that gradual changes, such as might be caused by the sun warming a room, do not trigger the alarm, while rapid changes, such as a person entering the room, will trigger it. The temperature-change (increment) threshold can be set so that a small pet will not actuate the alarm, while a full-grown person will. This type of device, like the IR motion detector, can operate from batteries, and in fact consumes even less power than the motion detector because no IR pulse generator is required. The main problem with radiant-heat detectors is that they can be fooled. False alarms are a risk; the sun might suddenly shine directly on the sensor and trigger the alarm. It is also possible that a person clad in a winter parka, boots, hood, and face mask, just entering from a subzero outdoor environment, might fail to set off the alarm. For this reason, radiant-heat sensors are used more often as fire-alarm actuators than as intrusion detectors.

Ultrasonic Motion Detector   Motion in a room can be detected by sensing the changes in the relative phase of acoustic waves. An ultrasonic motion detector employs a set of transducers that emit acoustic waves at frequencies above the range of human hearing (more than 20 kHz). Another set of transducers picks up the reflected acoustic waves, whose wavelength is a fraction of an inch. If anything in the room changes position, the relative phase of the waves, as received by the various acoustic pickups, will change. This data is sent to a microprocessor, which can trigger an alarm and/or notify the police.

World-Scale Security

Wireless technology will play a major role in the future of world security. Of special relevance is communications among computers (and, more or less directly, computer operators). It might never be possible to link all the computers in the world together in such a way as to combine their processing power; propagation delays ensure that the processing speed of such a machine would be limited. However, the specter of computers "running amok" as they collectively share data via satellite links is not too farfetched. There will always be a need for human supervision of large computer systems and networks.

National Defense

Computers have many different possible applications in warfare, both conventional (nonnuclear) and nuclear. Computers can be used to plan invasion strategies as well as to decide where to place soldiers and equipment for defense. Computers are already in use by the Army, the Navy, the Air Force, and the Marine Corps of the United States. Computers are also employed in large-scale nuclear defense.

During the years of the Cold War between the United States and Russia, the United States developed a strategic defense system in which a large computer played a major role. That system is still in use. Information from radar stations is sent to the computer. The computer interprets this information and figures out the probability that a nuclear attack is in progress. If it appears that an attack is taking place, the human supervisors are consulted, and a careful double-checking process is followed. If the human supervisors conclude that no attack is taking place, the system is reset. But if the human supervisors become convinced that a nuclear attack is in progress, they tell the computer to go ahead and activate one or more missile silos.

Obviously, the stakes in this process are extremely high. During the Cold War, the leaders and people in both the United States and Russia were fearful that a single computer (or human) error could result in the launching of an offensive nuclear missile for no reason. The scenario was painted in science-fiction novels and movies. The "action scenes" are not terribly difficult to imagine.

Some researchers have suggested that we ought to give computers complete control of all the weapons in the world. Proponents of this idea say that there could be no war if the master computer or network were sufficiently "smart," that is, if *artificial intelligence* (AI) were advanced enough. Computers would see the illogical nature of war and would supposedly never engage in it. They could also prevent humans from making wars with each other, but their methods might not be especially well liked by humans who revere their right to free choice.

Other researchers suggest that powerful computers could annihilate humanity if they malfunctioned. The slightest glitch, either in hardware or in software, can cause a computer to go haywire. Nearly everybody has experienced this. Have you ever been unable to mail a package at the post office, or make an airline reservation, or use your credit card? Have you ever been composing a story, magazine article, or book chapter on a computerized word processor and had it "lock up" so you lost work? These events are not unusual. In theory, a single transistor in an integrated circuit (IC) could short out, and a computerized strategic defense system might think it was under attack. Or a computer might actually become paranoid, like "Hal" in *2001: A Space Odyssey.* A computer virus could perhaps make a computer "go insane."

A more chilling prospect is this: What if a smart, logical, all-powerful computer, with complete control of all weaponry, decides there are too many people in the world, and that we humans would be healthier and happier if our numbers were cut by a factor of, say, 10? It would therefore be logical to exterminate 90 percent of the population. Maybe the computer would take it a step further and decide that human beings are a danger to the earth and must be eliminated altogether, allowing evolution to get a fresh start on its quest for the ultimate living being! Whether or not the methods employed were "humane" would be of no concern to a machine. There are people who have presented rational arguments for these kinds of actions; we should not be surprised if a computer comes to similar conclusions.

Colossus

Suppose that a single intelligent computer were given control of all the nuclear weapons systems in the world. If that could be done, then a nuclear war could never occur because of human error. It would have to be executed by the computer.

To some extent, computers already control the nuclear defenses in the United States. A *fail-safe* system makes it impossible for any one person to press a button and cause a nuclear holocaust. But some scientists think the current systems don't go far enough. They argue that a computer, seeing how illogical nuclear war is, would not let it happen. A computer would, they say, therefore be the ideal controller for nuclear arsenals.

Such is the scenario for a 1960s novel and movie called *Colossus: The Forbin Project*. The U.S.-designed computer, named *Colossus*, discovers that there is a similar system in Russia. The two computers interconnect and begin to share data and learn from each other. Their AI grows more and more rapidly until, before very long, they are smarter than any human being. They are also supremely powerful, because they hold control of all the nuclear missiles in both the United States and Russia.

This intelligence and power having been achieved, the computers decide that they will not only outlaw war but they will control people's lives in every detail. *Colossus* tells people when, and how much, they may eat, work, sleep, and even exercise. If anyone disobeys *Colossus*, the computer might retaliate by launching a missile and killing several million humans. To show that it means business, the computer does exactly this on one occasion.

Many researchers in AI believe that the *Colossus* scenario is realistic. Nowadays, with the Cold War between the United States and Russia apparently over, there seems to be less risk of a massive nuclear holocaust than there was in the 1960s. But there are still great numbers of nuclear missiles in the world. If we ever have the opportunity to give computers absolute control over all nuclear weapons, we had better be sure there's a way to "pull the plug." If the Internet, supported by hundreds of wireless satellites circling the globe, has been made sufficiently resistant to sabotage, "pulling the plug" might be extremely difficult if not impossible. The Internet was designed, after all, to be resistant to a shutdown.

The End Point

Wireless links between and among people, computers, and robots are extensively used in military defense or aggression. The extent of electronic automation can vary from simple communications systems to "star wars" fought among machines in the upper atmosphere, in outer space, or in the deep sea.

Military minds have long dreamed of ways to alleviate the human suffering involved in war. This was the major motivation behind the so-called Strategic Defense Initiative (SDI) that received attention during the presidency of Ronald Reagan. The idea was to employ robot spacecraft to intercept and destroy enemy missiles in space. Some military experts think it is possible to build wireless-controlled robot soldiers. These non-human warriors could not only fight on the ground, but they could pilot aircraft, helicopters, and boats. Presumably, they would not feel the pain, boredom, loneliness, sorrow, and other agonies that go along with fighting in a war. Artificial intelligence could help generals plan their strategies and could control whole armies of *insect robots*.

Maybe someday the leaders of the world will agree to do away with military hardware and fight wars with software instead, so they can spend more resources for food, shelter, and birth control. Conflicts between nations could be resolved by having generals play computer games on the wireless Internet. This might sound ridiculous, but it makes more sense than the current situation, in which world leaders authorize the expenditure of vast amounts of money and human energy for the sole end purpose of devising high-tech ways to murder ever larger numbers of innocent people.

14

Security and Privacy in Communications

Security and privacy are increasingly critical issues in communications, especially now that portions of most circuits are wireless. Even a conventional telephone connection is likely to employ a wireless link over part of the route. But security and privacy issues are not limited to wireless communications, or even to communications per se. Everyone ought to be concerned about the ever-increasing amounts of information floating around in cyberspace as well as in real space, and about the ever-expanding opportunities for such information to be confused, misused, or falsified for some people's gain and at other people's expense.

Wired versus Wireless Eavesdropping

Wireless eavesdropping differs from *wiretapping* in two fundamental ways. First, eavesdropping operations are easier to carry out in wireless systems than in hard-wired systems. Second, eavesdropping of a wireless link is generally impossible to detect (until vital information has been leaked), whereas the existence of a tap can usually be detected in a hard-wired system by means of specific tests.

Wireless links are much easier to "tap" than hard-wired circuits because hard-wired signals are electric currents confined to solid conductors, but wireless signals propagate in the air, through walls, around obstructions, and sometimes even into outer space. It takes fairly sophisticated apparatus to carry out a wiretap on a phone line. There must be a physical modification to the system. This requires that the eavesdropping equipment be located in a particular fixed place.

Eavesdropping on the cellular bands requires nothing more than a radio receiver capable of picking up the signals. The receiver can be located anywhere that the signals penetrate. Cellular eavesdropping equipment can be installed in a car or van, which can be driven around until it is within direct-link distance of the cellular telephone unit to be monitored.

An especially troublesome feature of radio-frequency (RF) links is that, if someone is eavesdropping on a conversation or transaction, there is no way for the operators of the transmitting station (*source*) or receiving station (*destination*) to know about it. In a hard-wired system, a tap will cause a change in the impedance of the circuit, and this change can often be detected by equipment either at the source or at the destination. But a

Figure 14-1
Eavesdropping on RF
links in a telephone
system. Heavy,
straight lines repre-
sent wires or cables;
zigzags represent
RF signals.

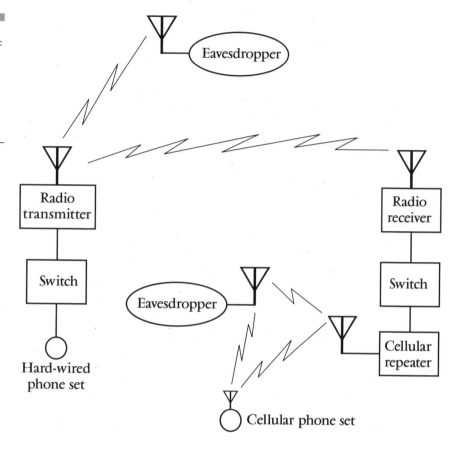

Figure 14-1
Eavesdropping on RF links in a telephone system. Heavy, straight lines represent wires or cables; zigzags represent RF signals.

radio receiver has no effect on the impedances of the source or destination antenna systems. If any portion of a communications link is done via wireless, then an eavesdropping receiver can be positioned within range of the RF transmitting antenna (Fig. 14-1) and the signals intercepted. The existence of such a *wireless tap* will cause no change in the electronic characteristics of any equipment in the system.

Some RF links are more difficult to tap than others, because of the nature of the medium. Microwave systems, which transmit signals in narrow beams, require that an eavesdropping receiver be located within the beam, or else right up next to the transmitting tower. Infrared (IR) or visible-light laser links have beams so narrow that it is very hard to place a receiver in the path of the signal. In addition, the existence of a receiver in the path of a laser beam would cause a reduction in signal strength at the destination, and this reduction might be detectable, giving away the existence of the tap. Unfortunately, laser communications has not become widespread, largely because the equipment is expensive and difficult to align and maintain. Moreover, laser systems are influenced by the weather far more than any other type of communications medium. Rain, fog, or snow can render a laser communications system useless.

Causes for Concern

Most people do not worry that their telephone conversations might be monitored. This is probably because only a small minority engage in activities that make government agencies suspicious and raise the specter of court-ordered eavesdropping. There is, in the back of almost everyone's mind, a vague fear that "Big Brother is watching and listening," but few people really care. Such nonchalance might have been valid in the middle of the twentieth century, but it is naive as we embark on the twenty-first. It neglects to take into account the real and growing threat of criminal activity by nongovernment entities. Communications systems are becoming more complicated, and people are relying more heavily on them, day by day. This trend shows no sign of slowing down. Each new device, and each incremental increase in public dependence on technology, creates new opportunities for criminals to defraud the rest of the population, individually and collectively.

The Criminal Element

Although eavesdropping makes big news at times, a government agency (such as the local police) must get a court order to legally monitor a telephone line. The prevailing public attitude is something like, "I'm not committing any crimes. Why should I have anything to worry about?" Sometimes this even extends as far as, "Anyone who listens in on my phone conversations will die of boredom." But anyone can, in fact, have his or her telephone line monitored. There are occasional instances in which government agencies eavesdrop on phone conversations illegally. The resolution of those cases rests with the courts, and involves debate over minutiae of the Constitution, the rights of the police, and the rights of the common citizen. The greater concern should be that illegal eavesdropping is done every day by criminals who have no regard for the courts, the Constitution, the police, or the common citizen.

It is possible to inflate the fear of eavesdroppers to the point of paranoia. After all, there are millions of telephone subscribers in the United States alone, and probably over 1,000,000,000 in the world. It is physically impossible for more than a tiny fraction of these subscribers to be monitored on a continuing basis, simply because there are not enough "telephone spies" to go around. The ratio of interesting-to-boring data is exceedingly small. With the growth of the Internet and its abundant frivolity, the ratio of useful to useless data, from the point of view of a potential criminal, is getting smaller every day. Nevertheless, there are handsome (if illegal) profits to be made, and some people get a thrill from the technical challenge of defeating or outsmarting a complex electronic system. The following are some of the things that can happen to ordinary citizens as a result of the interception of data from wireless communications links.

Impersonation

You probably have, at one time or another, placed a credit-card order for merchandise via an 800 number. How do you know that nobody was listening to your conversation and was carefully noting, or perhaps even tape recording, the credit card number and expiration date, along with the exact way you pronounced your name? Chances are that no one was listening on any particular occasion. But there are myriad credit-card orders placed every day by thousands of people. It is certain that some of these calls are intercepted, and likely that the information is later used for fraudulent purchase attempts. More and more telephone connections involve wireless links; this increases the risk unless wireless communications are encrypted to an extent that foils even expert wireless network crackers.

There are other, perhaps more dangerous, ways in which information about you can be used by impostors to impersonate you. If you use a cell phone, the data that identifies your set can be copied into other cell-phone sets. This is called "cloning" and has caused havoc in some regions of the country. Billing errors can occur, but it is also possible that you might take the blame for a crime committed using a cell phone with copies of your set's identifying data. Fortunately, cellular manufacturers and providers have taken this threat seriously in recent years, if for no other reason than the fact that a few massive lawsuits could bankrupt them. Nevertheless, for every security measure devised, someone eventually finds a way around it. The result is a never-ending game of "cat and mouse." This game drains resources (but it is not altogether bad; the reasons are discussed later in this chapter).

The Internet

Breaches in or lack of security can take place in hard-wired systems as well as in wireless systems. This is especially true of the Internet.

How much information about you is generally available? You might be surprised. You can get on the Internet and use one of the powerful search engines to find out. At the time of this writing, HotBot and AltaVista are excellent search engines for the World Wide Web (usually called "the Web"). They both allow you to search for Web sites containing the names of millions of individuals, and they can help you conduct a massive search. Countless street addresses are available to Internet users. Nationwide and worldwide telephone directories take data from listings in local directories and post them on the Web. If you don't want people to be able to easily discover where you live, you might ask the telephone company not to list your address in the local directory. In the extreme, you can get an unpublished number.

In most cases, if you find information about yourself on the Web that you do not want there, you can electronically request that it be deleted, and your wishes will be honored. The problem is that you cannot discover every single site at which information about you appears on the Net. The contents of the Net changes from hour to hour. A full-time private investigator probably couldn't snare every bit of data about you that hangs in cyberspace. Even if such a thing were successfully done today, something new about you might appear during the night and be on the Web tomorrow. It would, in a figurative sense, be like trying to rid the world of cockroaches.

As wireless systems such as low-earth-orbit (LEO) satellite networks begin to proliferate on the Net, security problems will likely increase. It will be easier for unauthorized people to "clone" your Internet or online-service username and password, especially if this information is programmed into wireless transceivers intended specifically for use with personal computers. Additional security measures, besides simple password protection, will be needed. There will have to be some way for the system to know that the person using your screen name and password is in fact you.

Misuse of Information

What will people do with information about you, assuming they find it? That depends on the reason they were looking you up in the first place. In most cases, simple curiosity is the motive. For example, an old friend might wonder where you are, and go looking for your e-mail address or phone number on the Web. Another way people might come across data about you is by chance. If you have certain keywords or phrases in a Web site you have set up, say "science fiction writing" or "abstract painting," people might come across your site as a result of searches on phrases. Anyone who has a Web site probably receives e-mail occasionally from companies wanting to sell things; the marketing can be tailored by electronically seeking sites that have certain words in them. This sort of thing is comparatively harmless, although if it gets out of control, it can fill mailboxes with unwanted messages and make it difficult for Internet users to efficiently use their e-mail.

Consider the following scenario. Imagine you are self-employed and have an individual health insurance policy. Suppose the health insurance company learns that you live alone, and that you buy three dozen eggs and 15 packs of cigarettes a week? How might it discover this? Computer networks contain more information about people than individuals generally suspect, and this information can be transferred in ways most people don't even think is possible, let alone likely. Many people use automatic-teller-machine (ATM) cards for common purchases. The receipts for such purchases often contain lists of the items that were bought. This and other information can be made available to anyone who wants it. It is not inconceivable that the health insurance company might raise someone's premium or cancel coverage altogether if it suspects that person might be eating five eggs and smoking two packs of cigarettes every day! The insurance company might even do it if the goods were not actually all consumed by the person in question. It might even happen if there was a computer error and some identification numbers got mixed up.

Recently there has been talk of the advent of "smart cards" that will contain all kinds of data about their owners on little magnetic strips that can be run through little transducer boxes in stores all over the whole world. Cash, some people believe, will eventually become obsolete. You'll pay for your food, toiletries, hotel stays, meals out, gasoline, and everything else using the same little card with the same little magnetic strip on it. Records of what you bought, when you bought it, and where you bought it will be entered into a massive network of computers and will theoretically be available to anyone who wants it. If your health insurance company believes it would be in its best financial interest to drop you because you eat cholesterol and smoke cigarettes, it might do exactly that, even if the law has to be stretched to the limit (or beyond).

There will doubtless be people who read these paragraphs and insist that this sort of thing cannot happen in a country like the United States, until they see a newspaper article or television special about it. Or, perhaps, until it actually happens to them. But the odds are that if something of this nature does happen to you, it will be so subtle and will seem so ordinary that you won't suspect anything is wrong until it is too late. Maybe you'll never know about it; your record will be somehow damaged, or you will be defrauded, and you will never notice the fact that you have in fact been libeled or robbed. Many "electronic criminals" rely on escaping detection of any kind. Their intent is to make money illegally, and if they can do it without causing anybody pain, so much the better!

Key Escrow

The most sensational publicity surrounding privacy and security involves espionage, sabotage, and plots that might be carried out by corporations, governments, terrorists, or militia groups. This has split people into two camps: those who favor strict government control of electronic security schemes and systems, and those who favor little or no government control. Official policy varies from country to country. The controversy is likely to continue indefinitely.

Corporations usually favor minimal government restriction of international communications security. More and more corporations are becoming multinational, and feel the need to carry out private transactions across national borders. They fear that if governments have the ability to eavesdrop on corporate communications, then criminals or adversaries will be able to eavesdrop as well, because the criminals or adversaries will simply steal the decryption programs, or *keys*, from the government. Many private citizens, in contrast, tend to favor some degree of government control over security systems, to ensure that the Internet and other communications media cannot easily be used by rogues and terrorists. The basic idea is to provide the government with keys to all encryption programs, according to certain laws that would allow the government to eavesdrop on communications or transactions after getting a court order for a particular case. This is the so-called *key escrow* controversy.

There are two relevant questions to consider on this issue. First: Should the keys to all ciphers be held in escrow by the government, for use if necessary via court order? Second: Is it actually possible for decryption programs to be held in escrow by any government with zero risk of theft by people or entities who have malicious intent? The answer to the first question depends on who you ask; many knowledgeable people will say "Yes" and many other knowledgeable people will say "No." The answer to the second question is, from the standpoint of anyone in touch with technology and even remotely familiar with the criminal mentality, a definite "No." A government can enact a million laws in an attempt to protect its people from crime; but such laws by themselves have little or no positive effect, because criminals, by definition, disregard the law. In fact, such laws might increase the danger to the public, because criminals could steal keys from the government and gain access to data exchanged by law-abiding people and corporations who have faith in the system and who do not suspect or believe the ugly truth about the system's vulnerability.

The ideal encryption scheme would be operable in practice by everybody for whom it is meant to be used, and usable in practice by nobody for whom it is not meant to be used. In other words, if a criminal tried to use a perfectly designed encryption scheme, the scheme would fail to work; it would electronically "know" that the application was inappropriate or illegal. Well-trained and well-compensated software engineers and mathematicians might at least approach this ideal; lawmakers and police departments can never hope to come close.

How great is the danger that superterrorists might, for example, use encrypted communications to facilitate the construction and deployment of a nuclear bomb, and thus coordinate a plot to hold a city hostage? Would legalizing the international exchange of *strong encryption*—that is, ciphers that are essentially unbreakable—materially increase the risk of such a scenario? Many people believe so; many others think not. Conversely, if no one is allowed to use encryption that cannot be broken by any government, how great is the risk of "Big Brother" eavesdropping on private citizens for any number of hostile reasons? Some people think the risk is great; others think it is small or nonexistent. This debate will probably never end, although a single superterrorism event, in which strong encryption is shown to have been used, will doubtless swing public opinion in favor of laws attempting to strengthen government control over encryption schemes. The primary danger would then be that too many people might believe that such laws, all by themselves, would actually prevent such an event from taking place again.

Computer Security

The term *computer security* refers to the protection of computer data against unauthorized snooping and tampering. Security programs are available for personal computers. Some utility packages contain security software.

Passwords

A *password* is a form of security that is used to restrict access to a computer, system, or network. A password need not (and should not) be a word; it can consist of any string of characters up to a certain maximum length.

The best way to choose a password is to make something up at random. Or you can look through a dictionary and find a strange word. Once this has been done, one or more of the letters can be changed to numbers; for example, "LOGICAL" might be changed to "L0G1CAL" (change letter O to numeral zero, and letter I to numeral one). It's a bad idea to use things like your birth date (for example, 072462 representing July 24, 1962). You should avoid codes like "PASSWORD," "COMPUTER," your name, or your telephone number. Pure letter or pure numeral passwords are not good. Never use the same password on more than one computer; if you do, and a hacker gets the password, it leaves multiple machines open to unauthorized use.

Once you have decided on a password, write it down someplace if you must. But if possible, commit it to memory and don't write it down anyplace. If you have to record a password, put the record somewhere obscure but permanent. Don't enter it anywhere in your computer system. Don't indicate, next to the password, what it represents. One good scheme is to make the password part of a bogus piece of information. For example, you might write in your address book:

John P.
17274 N.W. 344 Street
Rochester, MN 55901

Jane W.
22020 Lee Boulevard
Vero Beach, FL 47633

Both of these are nonexistent addresses. Possible passwords might be the character sequences "17274NW" or "22020LEE" from the street addresses. But you, and only you, would know this. Other people might see your data and think, "I didn't know Rochester was so big," or "Don't ZIP codes in Florida start with 3?" You can even make up other dummy addresses that do not contain the password, further confusing would-be hackers (and perhaps yourself eventually).

Some systems allow limited access to data, based on partial entry of a password. Suppose you have personal data, some of which you are willing to let other people see. A few of your friends might have the privilege of entering new data into your files, for example, to write down a reminder to call them on the phone. But only you have the ability to delete or change information that is already there. This is called a *hierarchical password* scheme.

You might choose a long password that consists of several independent parts. If you want someone to be able to look at your data but not modify it in any way, you give them part, but not all, of the password. To let someone look at your data and also enter new information, you might give them a larger partial password. It should be impossible to guess, from either of the partial passwords, anything further within the password. Partial passwords like "WASH" "WASHING" are too easy to extrapolate into a complete password such as "WASHINGTON."

Many systems have a *retry limitation* security feature. This allows the user a certain number of attempts, such as three, to get the password right. After that, the system will lock the user out even if he or she inputs the correct password. This prevents people from breaking in by repeated guessing.

Note: Do *not* use passwords suggested here. Invent your own!

Other Features

Further safeguards against tampering and snooping can be added to a security scheme. Sometimes, even authorized personnel must be kept from installing, copying, or sabotaging data and software.

Encryption is the conversion of data into a nonstandard form. It adds an additional security step, after use of a password, to keep data from getting into the wrong eyes. If a password is broken or stolen, a *decryption* program must be used to make sense of encrypted information. Each user must have, in addition to a password or set of passwords, a decryption package. Encryption is discussed in more detail later in this chapter.

Hard disk lockout keeps people from storing (writing) anything on a computer hard disk. The main reason for having this feature is to keep hackers from installing a *Trojan horse* or *virus* in the computer. Such tampering can disable the operating system. In the worst case, the result can be a catastrophic loss or leakage of secured information. With lockout, it is possible to read data on the hard disk, provided the user has gotten through the password and decryption barriers.

Even people with the passwords and decryption codes—that is, supposedly authorized personnel—cannot always be trusted. *Copy protection* keeps them from putting secured data on diskettes, and thereby helps to ensure that the information does not leak. Copy protection can also keep people from transmitting the data out of the system via modem.

Every time anyone tries to gain access to secured data, the date and time can be recorded by the computer. *Access attempt logging* can be useful in case secured information does leak out. Irregularities in the use pattern often show up in an access attempt log.

High-Level Security

For the protection of especially sensitive archives and classified information, a complete, high-level, computerized security system can be installed. It works in the same way as a property-protection security system, except that data, not material, is the object of the security effort.

With a high-level security system, you can not only restrict people's ability to access data from a computer, but also keep unauthorized people from physically getting near the machine. In addition, every person who logs onto the system can be positively identified, along with the date, time, and filenames accessed.

Hackers and Crackers

A person who enjoys tinkering with computer hardware and software is sometimes called a *hacker.* These people spend large amounts of time at computers. Many of them work in the computer industry, and then come home and spend the evening with a computer.

So-called good hackers have been responsible for advancement in the computer industry. Technically inclined amateur radio operators, if they are interested in computers, can become hackers who pioneer communications schemes. For example, radio amateurs have set up wireless *packet communications* networks that use the radio instead of the telephone lines.

Software hackers are constantly trying to create better computer programs. Some sophisticated hackers are working with robotics and artificial intelligence (AI), with the intent of building so-called personal robots. Others enjoy working with computer music; more than a few of them have started up bands to test their new sounds on the public. Good hackers are spurred on by the fact that they can make money, and advance their careers, from their schemes. Although these people are sometimes called "nerds" or "geeks" in high school and college, the mockery often turns to envy later, when they have become successful engineers or have started their own computer companies.

The question has been asked, "Why is it that if you give two people baseball bats, one will use it to hit home runs while the other will use it to smash windows?" Some computer wizards use their skill to harm people, corporations, and governments. A "bad hacker" is known as a *cracker.* (This expression derives from an analogy to "safe cracking" by burglars, and has nothing to do with people from the southern United States.)

Many computer networks are protected by security schemes, such as passwords or format requirements. Crackers find ways to break, or crack, into these networks and alter the data. Some of these people are software vandals. Some of them want to steal money. Thus you might get a credit-card bill for $50,000 that you never racked up, or find that your credit report is inaccurate. You might get a telephone bill two inches thick for one month, with a balance of $85,000. You might end up with a false arrest record. When things like this happen, the public pays for it.

An especially destructive and pointless sort of cracking takes the form of a *Trojan horse* or *virus.* One of these occasionally gets into software, especially the kind that you can download, install, and employ without paying a fee. It can disrupt a computer system; in some cases the data on the hard disk is so badly mutilated that it is no longer of any use.

Cracking is against the law. The penalties vary. Often, because of the interstate nature of computer networks, the crime falls under the jurisdiction of the federal government. Several states have introduced legislation mandating prison sentences and/or fines for people convicted of distributing programs with the intention of sabotaging computers.

The Virus

A virus is a malicious computer program, or a fragment of mischievous programming code, that causes a computer to do strange and unexpected things. A virus can alter or erase the data on a hard disk. This makes the computer less efficient or inoperative.

A virus gets into a computer when "infected" software is downloaded from some other system. A virus can reproduce (make copies of itself). It can attach to all kinds of different programs within a computer system. Then, if any of these programs is downloaded into some other computer system, that system will also become "infected." It is like a disease epidemic. A computer virus might remain latent for some time before it springs into action. This is similar to the way a disease virus behaves in the human body. Some command, or the execution of some program or part of a program, will activate the virus. But even during the latency period, the virus can make copies of itself, some or all of which end up on diskettes that have been used with the infected computer.

Virus creators are almost always computer experts. They give various explanations for why they do what they do. Here are four categories into which these people might be pigeonholed.

Graffiti Writers Some people create viruses because they aren't supposed to. They are naughty children who never grew up. These people's viruses do not necessarily harm a computer. They might cause the colors on your screen to change suddenly, or make the machine play a musical tune, or display a message on the screen for a second.

Cyberpaths These people are bitter and cynical. They suffer from serious mental or emotional disorders. They often appear in *cyberspace* (online) with screen names taken from science fiction or horror stories. Many come from eastern Europe and underprivileged countries. Their viruses are intended to wreak the greatest possible havoc on the largest possible number of computers. As worldwide computer networks grow, the dangers posed by cyberpaths will probably increase.

Revenge Takers This type of person often holds a good job, but might produce a virus and keep it stored on a diskette at home. In subtle ways, such a person might let the boss know that he or she is not the sort of person it would be wise to lay off or fire. In the event he or she is actually laid off or fired, the virus can be uploaded or otherwise installed in the company computer, resulting in costly loss of data.

Whiz Kids These are generally children or teenagers who write viruses just to see if it can be done, and to witness the results. This is a more or less innocent motive, although the consequences, at least within the child's own computer, can be demoralizing. Because most new computers now have preinstalled modems and software for online services, there is some danger that viruses written by curious children might get into cyberspace and cause widespread problems.

The best protection against viruses is to avoid "infections" in the first place. Some steps you can take are as follows.

1. Be wary of mail-order software; the seller must guarantee that the software is free of viruses.

2. Don't download software unless you have acquaintances who have used it recently and are having no symptoms of viral "infection" in their computers.

3. Never use software that has been, or that you suspect might have been, pirated (illegally obtained or copied).

4. Never download software directly onto a hard disk. Put it on diskettes instead, and keep it off the hard disk until it has been tested with a *vaccine* (antivirus program).

5. Buy a vaccine, and use it to check all new software before first use.

In AI, a virus would have the effect of doing "progressive brain damage." The machine would lose its intelligence until it was a complete idiot compared to its former self. Or it might become, in effect, a lunatic machine. The possibilities are as bizarre, hilarious, and horrible as you care to imagine.

As computers become more important in society, the potential danger posed by viruses gets greater. Imagine a virus that could slowly infect an AI system in charge of Wall Street! There might be a strange "bull market," during which everyone would be happy (except the doomsayers, who are always miserable). Then, one morning, we would awaken to the "big crash." It might be months before the cause was determined. Maybe nobody would ever know what set the catastrophe in motion. Conceivably, it could be brought about by some 13-year-old sitting in a basement with a computer: a lonely, frustrated genius just looking for some way to have fun.

The Trojan Horse

A *Trojan horse* is a program that is written into software with the intent of disrupting a computer's operating system. It is something like a virus. But while a virus can replicate itself, spreading through networks and affecting vast numbers of machines, a Trojan horse usually does its damage within one computer.

The name "Trojan horse" comes from a legend about the ancient city of Troy. In those times, cities were surrounded by high walls to resist invasion. Troy was known for its seemingly impenetrable wall. Anyone who climbed the walls was thrown back down by defenders. Greek invaders tried to find ways to get through the barrier. Because brute force did not work, the warriors resorted to trickery. The Greeks built a huge, hollow wooden horse on a wheeled platform. A number of soldiers crawled inside. The idol was left at the gate of the city, supposedly as a peace offering. The Greeks who had not entered the idol returned to their ships and retreated temporarily out of sight. The people of Troy, noting the absence of Greek ships and soldiers, opened the gate to their city, wheeled in the wooden horse, and put it in the center of town. That night, the Greek ships returned, the soldiers crawled out of the horse and opened the city gate from inside, and Troy was defeated.

The foregoing analogy is an excellent depiction of the way a Trojan horse gets into a computer and incites chaos. A diskette, advertised as having some unique and fabulous contents, is sold or given away. The data might also be available online, so computer users can download it onto their hard disks. Within the software, there is a program that can mutilate or erase data on a computer's hard disk, sabotaging the operating system.

The best way to protect your computer against Trojan horses is to avoid downloading online software directly onto your hard disk. You should be skeptical of "too-good-to-be-true" software sold cheaply, or given away, online or on diskettes. Always download software onto diskettes or some medium other than your hard disk. A vaccine program can then be used to test the software, finding and erasing any Trojan horses or viruses before they can reach the hard disk.

Vaccines

A vaccine is a program or utility that searches for, and usually eliminates, a Trojan horse and/or virus from a diskette or hard disk. Sometimes you'll hear it called an *antivirus program* or *antivirus utility.* You can buy vaccines all by themselves, but advanced operating systems have them already included.

When downloading files or software from online, always put them on a diskette. Don't download anything directly onto your hard disk. After you have downloaded the data and/or programs onto a diskette, run the diskette through a vaccine. The vaccine will notify you if it has found anything suspicious. If the vaccine finds something out of the ordinary, it will let you know. Some vaccines give you the option of leaving the diskette as is (and taking your chances with it) or cleaning it up. Other vaccines will automatically erase anything suspicious.

Some vaccines can be installed so they run automatically every time you power up. Some run constantly in the background, so if anything suspicious appears, you are notified immediately.

Once a vaccine has cleaned up the diskette and has indicated to you that there is nothing apparently wrong with the data, then you can transfer the data to your hard disk if you want. There is still some risk involved with this, however, because vaccines are not perfect. New viruses and Trojan horses are constantly being deployed by those people who feel compelled to cause havoc in other people's computers.

Figure 14-2 shows a typical data vaccination process. Circles represent steps taken by you, the computer operator. Rectangles are steps taken by the vaccine program or utility. Diamonds are branch points, at which the procedure can take either of two different paths. The steps proceed in the following fashion.

Download.   You take a file or program from online, and transfer it to a diskette. This data can be anything: a text file, a musical tune, a photograph, a drawing, a game, a program, etc. If the data won't fit onto a single diskette, then it must be broken up into pieces, each of which will fit onto a diskette. Then each diskette must be vaccinated individually. Alternatively, a high-capacity medium, such as a magneto-optical cartridge, can be used.

Start vaccine.   You give a command or make a menu selection telling the computer to begin running the vaccine program or utility.

V/TH scan.   The vaccine searches the diskette for a virus (V) or Trojan horse (TH).

Anything odd? (Y/N).   If something unusual or suspicious is detected on the diskette, the vaccine will give you a warning to that effect. If nothing strange is found, the vaccine tells you that, too.

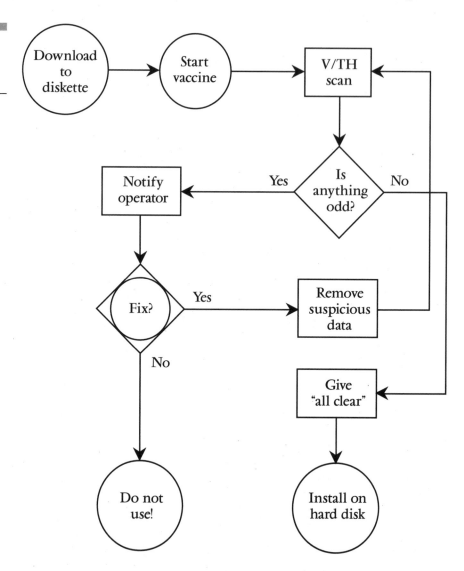

Figure 14-2
Typical vaccination
process for a
computer diskette.

Notify operator. You are told that something is out of the ordinary with
the data on the diskette.

Fix? (Y/N). You can choose whether you want to erase the strange data,
or leave it intact. If you decide to leave the suspect data intact, you
use it at your own risk. (It is recommended that you don't use it.)

Remove suspicious data. The vaccine erases the suspect item from the
diskette, and then goes back to the V/TH scan to look for more suspicious data.

Give all clear.   You are told that nothing appears out of the ordinary with the data on the diskette.

Install on hard disk.   You can, if you wish, transfer some or all of the diskette data to your computer's hard disk.

The creators of viruses and Trojan horses are clever. They are always developing new "bugs" that escape detection and/or eradication by vaccines. It is quite possible that a vaccine will fail to find a virus or Trojan horse even if one exists. It is also possible that the program will find one and then fail to get rid of it, even though the vaccine tells you the "bug" is gone. Viruses can sometimes be extremely difficult to eliminate. This author was told of a case in which a computer user went so far as to reformat an "infected" hard disk, and discovered the virus still there afterward. This is like burning down a house to get rid of a rat, and then finding the rat in the ashes, alive and as healthy as ever. Formatting a hard disk is an unwise thing for a computer novice to do. If you encounter a virus that a vaccine cannot eliminate, you should get expert assistance.

Vaccines are updated periodically in an attempt to keep pace with the evolution of viruses and Trojan horses. If you plan to use your computer online, and especially if you plan to download files and/or programs, you should consider buying a vaccine that is updated regularly.

Levels of Security

In a communications system, the extent of security can be expressed in terms of degrees or levels. In purely electronic terms, there are four levels of security and privacy, ranging from zero (no security at all) to the maximum amount that existing technology allows. In practice, certain codes, such as Morse, offer limited privacy even in nonsecure systems, because casual listeners usually cannot "understand the language."

No Security (Level 0)

In a zero-security or *nonsecure communications system,* anyone can listen in on a conversation at any time, provided they are willing to spend the money and/or time to obtain the necessary equipment.

The most obvious examples of zero-security links are amateur ("ham") radio and citizens band (CB) voice communications. A simple shortwave receiver, also called a *general-coverage receiver* or *communications receiver,* can be used at frequencies below 30 megahertz (MHz). Above 30 MHz, a *very-high-frequency (VHF) receiver* can be used. Below 30 MHz, amateur radio operators almost always use *single-sideband* (SSB) for their voice communications; CB'ers use both SSB and conventional *amplitude modulation* (AM). On the amateur bands above 30 MHz, narrowband *frequency modulation* (FM) is the mode of choice for voice communications, although SSB, AM, and some specialized modes such as *spread spectrum* are occasionally encountered.

Ham radio and CB communications are legally public; there are no laws barring people from eavesdropping. Amateur radio operators, in particular, are aware of this, and voluntarily regulate their content accordingly. In practice, specialized digital codes such as Morse, Baudot, or ASCII provide a certain measure of privacy even in communications that are legally public. This is because a typical communications receiver does not include the ability to decode digital transmissions in these forms. You can go to a local electronics store and purchase a small short-wave receiver for $100 or so, but if you do not know the Morse code and do not care to spend the additional money for a Morse decoding circuit or computer program, you will not be able to effectively eavesdrop on radio hams who communicate via this old-fashioned but nevertheless popular mode.

Early cellular and cordless telephone systems were notorious for their lack of security and privacy. Pressure from consumers has gradually been bringing about change. The earliest cordless phone sets operated on shortwave and VHF frequencies. There were only a few frequency combinations assigned to the sets; people could (and did) drive their cars around with cordless handsets activated, searching for dial tones. When a dial tone was heard, calls could be made using someone else's telephone line. The hapless victim would likely remain ignorant of the situation until the arrival of the next long-distance phone bill.

Police communications in most cities and towns are "scrambled" to prevent people from listening in with VHF receivers. However, in a few locales, police and sheriff's departments still use unscrambled FM. Scanner radios and illegally converted business or amateur radio transceivers can be used to eavesdrop on local law-enforcement action in such places. Criminals—burglars, for example—are obviously not deterred by laws that make such listening illegal, and the ease of eavesdropping has helped some crooks evade law-enforcement personnel and make getaways.

The lack of privacy in conventional radio communications is compounded by the fact that if someone is eavesdropping, none of the communicating parties can detect the intrusion by any known electronic means. Radio receivers can "snatch" a signal from the airwaves without affecting the behavior of the transmitter or the nature of the signal at the intended receiver(s).

Wire-Equivalent Security (Level 1)

An old-fashioned telephone connection is hard-wired from end to end. That is, wireless links are not used anywhere in the circuit. Some such telephone connections still take place, but they are becoming rare. Yet this scheme has undeniable assets when it comes to privacy and security. An end-to-end hard-wired connection requires considerable effort to tap, and sensitive detection apparatus can usually reveal the existence of any wiretap. An effective wiretap causes an "impedance bump" in a telephone cable. The result is that some of the transmitted signal is reflected back toward the source. This is the same effect that takes place when an impedance discontinuity occurs in an RF transmission line. Even a tiny amount of reflected signal energy can be detected rather easily, and this will arouse the suspicions of anyone equipped with detection instruments. An exceptionally well-engineered wiretap might evade detection by equipment designed to reveal changes in impedance continuity, but the cost of setting up and maintaining such a wiretap acts as a barrier to all but the most determined and technically savvy snoops.

Today, microwave towers dot the countryside; some of the signals hopping among them contain ordinary telephone conversations. Geostationary satellite links are widely used, especially in long-distance telephone service. True wireline-equivalent security is not technically possible with such systems, because a tap in the wireless portion of the link will not leave any "footprint." However, encryption of the data itself can make it difficult for a would-be eavesdropper to listen to the conversations taking place. This can be carried out in two basic but different ways: *end-to-end encryption* (over the entire circuit, including hard-wired portions) or *wireless-only encryption.*

A wire-equivalent security system must have certain characteristics in order to be effective and practical.

Cost The system must be affordable to the people who will use it. Generally these are ordinary citizens engaging in conversations and transactions that are not necessarily important to would-be eavesdroppers but that could be embarrassing to the people who take part. Examples are discussion of irritable bowel syndrome, astrological profiles, and any of the myriad Internet "chat rooms," newsgroups, or Websites having content that is controversial or offensive to some people.

Private The system must be reasonably safe for conversations such as credit-card mail-order purchases via 800 numbers. Few people would make such transactions if they believed that anyone with a scanner (legal or not) could listen in and record the credit card number, expiration date, mother's maiden name, and the manner in which the purchaser pronounced his or her own first and last names. If everyone suddenly lost faith in the telephone ordering process, companies who sell via mail order would suffer. Thus wire-equivalent security is in the interest not only of the consumer but of many businesses.

Consistent When the system traffic is heavy (many subscribers are using it at the same time), the degree of privacy afforded to each individual should not be any less than is the case when the system traffic is light. That is, the security feature should not be compromised in order to increase the number of conversations that the system can simultaneously handle.

The encryption used for wire-equivalent security should be unbreakable for at least 12 months, and preferably for 24 months or more, in the opinions of engineers. The technology should be updated at least every 12 months, and preferably every 6 months.

Security for Commercial Transactions (Level 2)

In the event a wire-equivalent security system fails and a credit-card transaction is intercepted, most of the resulting loss is "eaten" by the bank issuing the credit card. The consumer might have to pay an amount up to a certain limit, such as fifty dollars; but in general, an individual need not fear a financial catastrophe resulting from a single security breach. However, some forms of financial data are more sensitive and warrant protection beyond wire-equivalent. An example is the so-called *electronic transfer of funds*. Another example is the discussion of trade secrets among companies. Yet another example is the discussion of the details of an important court case between a lawyer and client.

Even today, many companies and individuals refuse to transfer money by electronic means because they fear that a criminal will gain access to an account and "clean it out." While such events are rare and can usually be resolved so the customer does not end up being devastated, it is nevertheless a traumatic experience to discover that one's bank account balance has suddenly and inexplicably dropped to zero. It can appear that one's life savings have evaporated into cyberspace. Cases of this sort can make lawyers rich and journalists famous, but a few well-publicized "horror stories" can cause a loss of consumer confidence all out of proportion to the actual risk. This fear, itself, is destructive to the general economy.

The encryption used in commercial transactions should be of such a nature that engineers believe it would take a hacker at least 10 years, and preferably 20 years or more, to break the code. The technology should be updated at least every 10 years, but preferably every 3 to 5 years, and more often if possible.

Mil-Spec Security (Level 3)

Some governments don't want strong encryption to be legally available to everyone who can afford to buy it. The reasons for this have already been outlined. In the case of military operations, however, such restrictions do not apply. Governments take a "no holds barred" attitude when it comes to the protection of their national interests and their citizens. Thus, security to military specifications (*mil-spec security*) involves the most sophisticated forms of encryption that a given government can muster. The technologically advanced countries, and those with economic power, have an advantage. These countries have the best engineers, or at least the money to hire them.

One means by which a government might try to optimize the security of its communications is to limit the extent of security legally available to nongovernment entities. In the United States, attempts have been made to criminalize the export of strong encryption, on the basis that such codes are, in effect, usable as weapons. Domestic laws are not likely to prove effective against hostile nations and terrorists. These entities will acquire the strongest encryption that their resources will allow. Most funds earmarked for enforcement would, in this writer's opinion, be better spent on developing encryption schemes that mathematically and electronically protect legal or friendly communications, and fail to protect illegal or hostile communications. The importance of this will become vividly apparent as the twenty-first century unfolds. As technology gains ever more power over human activities, aggressor nations and terrorists will probably try to bring down powerful nations by seeking out and striking at the weakest points in communications infrastructures. It has been suggested that certain governments might wage electronic wars against the United States by introducing viruses, false data, and other disruptive elements into cyberspace. America will be in a strong position if it has a solid technological defense. But little, if any, protection will be afforded by legislative "paper tigers."

Extent of Encryption

Security and privacy are obtained by *digital encryption.* There are so many possible digital encryption schemes that their number can, for practical purposes, be considered infinite. The basic theory of digital encryption is discussed later in this chapter. The idea is to make signals unreadable to everyone except those people who have the necessary *decryption algorithm,* or key.

For wire-equivalent (level 1) communications security, encryption is required only for the wireless portion(s) of the circuit. (The hard-wired parts of the circuit provide wire-equivalent security by default.) The cipher used for encryption of wireless links at level 1 should be such that, to break the code, a would-be eavesdropper must expend roughly the same effort as would be required to tap the hard-wired portion

Figure 14-3A

Wireless-only encryption. Heavy, straight lines represent wires or cables; zigzags represent RF signals.

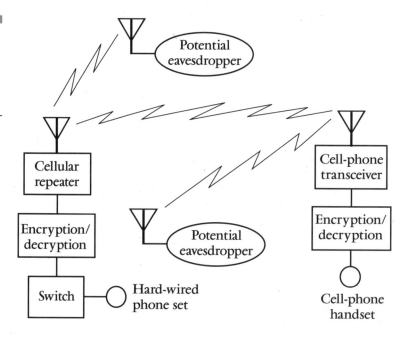

of the circuit. This is not a well-defined criterion, but as a rule, the strongest affordable encryption scheme should be used, and the cipher should be changed at regular intervals to keep it "fresh." The block diagram of Fig. 14-3A shows wireless-only encryption for a hypothetical cellular telephone connection. For this scheme to work, the cellular set must be equipped with an encryption/decryption circuit and software.

For security at levels 2 and 3 as defined above, end-to-end encryption is necessary. This means that encryption/decryption circuits and software must be placed at the source and destination. The signal will therefore be encrypted at all intermediate points, even those for which signals are transmitted by wire or cable. The diagram of Fig. 14-3B shows this scheme in place for the same hypothetical cellular connection as depicted at A. The only difference is in the placement of the encryption/decryption apparatus at the hard-wired end of the circuit. With the scheme shown in Fig. 14-3B, even if someone succeeds in tapping the hard-wired part of the circuit, eavesdropping will not be possible unless that person has the decryption key. In this example, the same cipher is used for the entire connection. It is possible, however, to change ciphers at intermediate points in the circuit, if it is believed that this will offer greater security than the use of a single cipher from end to end. The strongest possible encryption software should be employed, and the cipher should be changed often.

Figure 14-3B
End-to-end encryption. Heavy, straight lines represent wires or cables; zigzags represent RF signals.

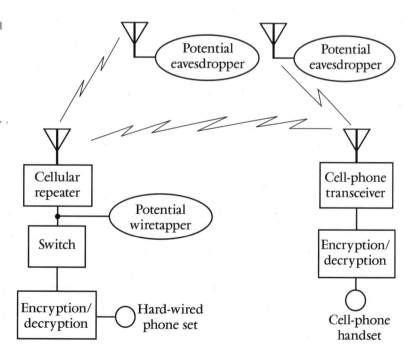

Wireless Telephone Security and Privacy

Security features have been incorporated into cordless telephones in recent years. These measures generally stop short of voice encryption because of the cost factor, although some high-end digital cordless phones employ voice encryption.

Ease of Intrusion

Most cordless phones are designed to make it difficult for unauthorized people to pirate a telephone line for the purpose of placing illegal or long-distance calls. Prevention of eavesdropping is a lower priority, except in the most expensive cordless systems.

Eavesdropping   Listening in on one end of a cordless phone conversation was, at one time, extremely easy, and eavesdropping on both ends was not particularly difficult. The earliest cordless phones operated at relatively low radio frequencies. A general-coverage radio receiver, capable of demodulating AM signals, could be tuned to approximately 1.7 MHz (just above the standard AM broadcast band), and signals could be intercepted up to a mile away. Nowadays, it is harder to eavesdrop on the link between a cordless handset and its base unit, because the frequency is much higher. In addition, many cordless telephones have short operating ranges, minimizing the possibility that someone might randomly come across a given conversation. Nevertheless, if someone knows the frequencies at which a cordless handset and base unit operate, and if that person is determined to eavesdrop on conversations that take place via that system, it is possible to place a wireless tap on the line. Such a system is a sophisticated, miniature *repeater.* It contains two radio receivers, one for the handset signal and one for the base-unit signal. It also contains a demodulator and a radio transmitter that operates at some frequency far removed from the telephone unit frequencies. The conversation can be intercepted and recorded at a remote site (Fig. 14-4). An electronics technician with reasonable skill can design and build a wireless tap, provided he or she knows the frequencies (channels) at which the cordless device operates. Fortunately, most cordless telephones these days employ multiple channels selected at random. It is more difficult (but by no means impossible) to design and construct a reliable wireless tap for a multiple-channel cordless set, as compared with a single-channel set.

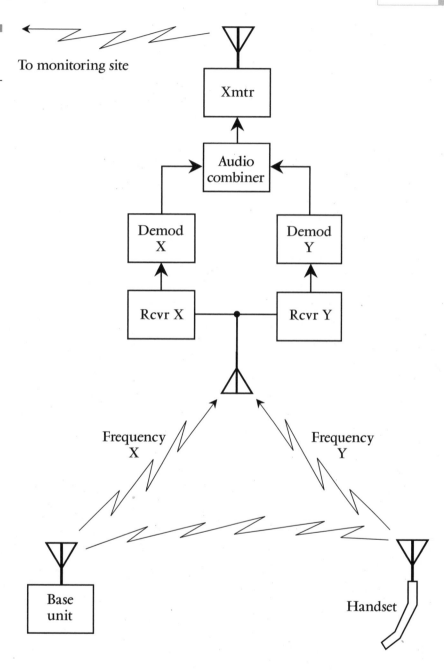

Figure 14-4
Wireless tapping of a
cordless telephone.

To monitoring site

Dialing In   Early cordless telephones were notorious for the ease with which unauthorized people could pirate the line. In the mid-1980s, this writer lived in a suburban home in a Florida town. The neighborhood was typical middle-class. A cordless phone was purchased and installed. The system had three different channels, which had to be selected both on the base unit and on the handset. As an experiment, I disconnected the base unit from the line and checked each of the three handset channels in turn. On one of these channels there was a dial tone, accompanied by receiver hiss indicating that the handset was near the limit of coverage of the vulnerable base unit. A brief stroll in the street revealed, within a couple of minutes, the residence in which the base unit was located. Returning to my house, I dialed my own telephone number, and the hard-wired units in the house rang. I had picked up the receiver of one of these sets, and there I was on the other end of the line! At this point, I decided it was a good idea to avoid using the cordless set at that location. If I had continued to use the cordless set, it would have been as easy for the residents of the other house to pirate my line as it would have been for me to pirate theirs.

Solutions

The designers and manufacturers of cordless telephones have responded to public concern by incorporating features to minimize the possibility of eavesdropping and unauthorized access to subscriber lines. These measures also reduce interference among cordless phones in close proximity.

Manual Channel Switching   If you are using a cordless phone and experience interference from another cordless phone in the vicinity, you might not be able to understand the voices in the other conversation, but you will hear "monkey chatter," squealing noises, and/ or sounds resembling sferics. These noises will usually be severe enough to distract both you and the party with whom you are talking, and in some cases they will obliterate the conversation. When this type of interference occurs, you can punch a "channel" button and switch your cordless phone to a different pair of frequencies. Most cordless phones have at least half a dozen channels; some have 10 or more. In apartment and condominium buildings, interference among cordless phones is a common annoyance, and the channel switching feature usually eliminates problems when they occur. Even in single-family-home residential areas, interference among cordless phones takes place, because of the greater use of long-range systems in these districts.

Random Channel Initialization   Many cordless phones have a microprocessor that assists in security. The microprocessor selects a specific pair of frequencies every time the unit is taken off-hook. This pair of frequencies is chosen at random, and there are thousands of possible channel combinations. (One channel is used to send signals from the base unit to the handset; the other channel is used for sending signals from the handset to the base unit.) Phones that use this scheme effectively foil wireless taps tuned to a fixed pair of frequencies. A sophisticated wireless tap might use a scanner to locate the signals, but the scanner might not always "land" on the correct frequencies. For example, a scanner might come across another cordless phone conversation, or even a signal from a different communications service. To foil such scanners, "decoy" signals can be transmitted at various frequencies near, but not exactly at, the handset and base-unit channels. The decoys contain no meaningful data, but only a test signal or no modulation at all. Random channel initialization makes it practically impossible for a rogue handset to establish contact with the base unit and initiate a call, even if the rogue handset is for the same make and model of cordless phone as the authorized handset. This is because the microprocessor will recognize only the channel code signals from its authorized handset, and no others.

Encryption/Decryption   The most effective, but also the most expensive, way to guarantee privacy and security with cordless phones is digital encryption and decryption of the voice signals, both at the base unit and at the handset. Wireless taps are foiled unless the potential eavesdropper has the decryption key. Unauthorized use of a line is impossible as well, because random channel initialization can be used in addition to encryption/decryption. When deciding whether or not to purchase a cordless phone that has the encryption/decryption feature, one must weigh the cost of the feature against the actual need for the cordless phone. If a particular call is likely to contain information so sensitive that encryption/decryption is deemed necessary, you might want to forgo the use of the cordless phone and make the call with a hard-wired set. You will also want to be sure that the party to be called is aware of the nature of the call and can deploy adequate security measures at his or her end.

Cell Phones

Cellular telephones are, in a sense, super-long-range cordless phones. The long range dramatically increases the likelihood that eavesdropping and unauthorized use will occur. In recent years, cell-phone vendors have begun advertising their systems as "snoop-proof" and "clone-resistant." Some of these claims have more merit than others. The word "proof," in particular, should be regarded with skepticism.

The cellular channels are well known, and anyone with a reasonable amount of technical skill can convert a conventional scanner radio to cover the cellular bands. A competent technician can build a scanner that will intercept digital as well as analog cellular signals. Laws forbidding the sale and distribution of cellular-capable scanners might make it less convenient and more costly to eavesdrop, but determined offenders will not be stopped by laws. The nature of the medium renders such laws unenforceable. Encryption is the only truly effective way to maintain privacy and security of cellular communications.

It is important that the access and privacy codes as well as the actual voice or data communication, be encrypted if a cell-phone system is to be 100 percent safe. This is because, once an unauthorized person knows the codes via which your cell phone accesses the system—in other words, the "name" of your set—countless other sets can be programmed to fool the system into thinking that they belong to you. Many hapless cell-phone users have found out that this type of fraud, known as *cell-phone cloning,* is easy to do when people leave themselves vulnerable to it. Not only can it cause you to receive a grossly inflated cell-phone bill, but it can get you in trouble with the law if cloners use their sets in the commission of crimes. The security of a cellular telephone network is directly proportional to the strength of the digital encryption that it employs.

In addition to digital encryption of the data, some form of user identification (*user ID*) must be employed. The simplest is a *personal identification number* (PIN) similar to that used with bank ATM cards and some credit cards. More sophisticated systems might employ *voice-pattern recognition,* in which the cell phone will function only when the designated user's voice speaks into it. *Hand-print recognition* or *iris-print recognition* might also be employed. Any and all user ID data must be encrypted along with voice signals, if cloners are to be prevented from intercepting the user ID data by eavesdropping. Digital information, once obtained, is easy to clone to perfection. This is one of the so-called *revenge effects* of digital technology. The hardware and software that facilitates convenient data backup and archiving also makes it possible to forge signatures, voice prints, hand prints, iris prints, and any other form of digitized information, no matter how complex. Insofar as a digital system is concerned, if encryption is not used or is not strong enough, someone else can steal your identity and become your "evil twin."

Two-Way Radio Security and Privacy

The users of two-way radios, such as amateur and CB transceivers, realize that their communications are generally not private or secure, and operate accordingly. These services are sometimes criticized for their meaningless "contacts," in which the weather, electronic equipment, and other trivial topics dominate conversations. However, there are sometimes legitimate reasons to restrict access to repeaters, or to seek some measure of privacy in communications via two-way radio.

Tone Squelching

Subaudible tones, or audible *tone bursts,* are sometimes used in repeaters to limit the access to certain persons, such as members of a repeater club. Such measures can also be used to keep a receiver from responding to unnecessary or unwanted signals. These schemes in general are known as *tone squelching.* For the receiver squelch to open in the presence of a signal, that signal must be modulated with a tone having a certain frequency. Tone-squelch systems are most often used with FM. This form of communications is used by radio amateurs in certain parts of the United States and in some other countries. It can also offer a limited degree of security in general wireless communications.

In an FM subaudible tone-squelch system, the carrier is continuously sine-wave modulated at a deviation ranging from plus or minus 500 hertz (Hz) to plus or minus 1.5 kilohertz (kHz). The tone cannot be heard on the signal as it is received, because the frequency is below the response cutoff of the voice-amplifier chain. In European countries, and also in some U.S. communications systems, tone bursts of higher frequency are often used for tone-squelch purposes. The burst might consist of one tone, or it might consist of several tones in sequence, each having a different frequency. Tone-burst systems are used by some police departments and business communications systems in the United States;

in recent years, *tone-burst squelch* has gained popularity among radio amateurs as well.

Spread Spectrum

The theory of *spread spectrum* was discussed in Chap. 7. Signals in this mode require special receivers. A conventional receiver or scanner cannot intercept spread spectrum unless the receiver follows exactly along with the changes in the signal frequency. The frequency-variation sequence can be complicated, and in its most sophisticated form is almost as good as digital encryption for security purposes.

Spread spectrum is regulated by laws that vary from one country to another. In the United States, this mode is legal only at the higher radio frequencies. As its name implies, spread spectrum in effect gives a signal a much wider bandwidth than a "normal" signal requires. At low frequencies, employing this mode is a wasteful misuse of spectrum space. Besides this, spread-spectrum engineering is more difficult, from a purely technical standpoint, at low frequencies than at higher frequencies.

Audio Scrambling

Analog radio voice communications can be altered in a way that does not require any shifting of the transmitted carrier frequency. You have probably heard of *audio scrambling.* It is commonly used by many police and fire departments, and has been in existence for decades.

A voice channel, or *passband,* is typically 2700 Hz wide, containing audio frequencies between 300 and 3000 Hz. (Actually, a human voice contains energy at frequencies below 300 and above 3000 Hz, and high-fidelity transmission requires a passband from a few hertz up to about 20 kHz; but for communications purposes, the 2700-Hz standard is adequate.) As you speak into the microphone, the audio signal contains components that fluctuate within this range. Sibilants, such as the sound you make when you say "sss," produce audio energy at the upper end of the passband range. Vowels and most consonants produce energy in the middle and lower portions of the passband range. By inverting these frequency components, a voice can be rendered unintelligible, yet it will contain all the original information.

An amateur-radio SSB transmitter and a general-coverage or amateur shortwave receiver can be used to demonstrate the principle of *audio-*

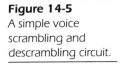

Figure 14-5
A simple voice
scrambling and
descrambling circuit.

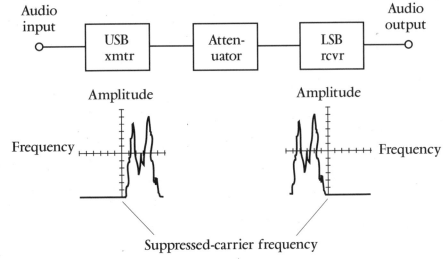

frequency (AF) scrambling. The arrangement is diagramed in Fig. 14-5. The transmitter is set for the lowest possible power output; the final amplifier should be disabled if possible. The transmitter output is connected to a dummy antenna or attenuator, so the receiver input circuitry will not be overloaded or damaged, and so no signal will be transmitted over the airwaves. The transmitter is set for upper sideband (USB) and the frequency is selected and noted. The receiver is set for lower sideband (LSB) and is set for a frequency exactly 3000 Hz higher than that of the transmitter.

Suppose, for purposes of illustration, that the suppressed-carrier frequency of the USB transmitter is set to 3.800 MHz. This will produce a spectral output similar to that shown in the graph on the left in the illustration. Each horizontal division on the frequency scale represents 500 Hz; each vertical division on the amplitude scale represents 5 dB. The LSB receiver is tuned to receive a signal whose suppressed-carrier frequency is 3.803 MHz. The signal energy from the transmitter will fall into the receiver passband, as shown by the graph on the right in the illustration. However, the AF components will be "inverted." A transmitted audio tone with a frequency of f_{tx} Hz will result in a received audio tone having a frequency of $f_{rx} = 3000 - f_{tx}$ Hz. Audio tones of 1500 Hz will not be changed in frequency, but all the other audio components will be "mirrored" within the passband relative to the "center" frequency of 1500 Hz. The low-frequency "silent zone" in the transmitted signal, ranging from zero to 300 Hz in most SSB transmitters, will be moved to the top of the passband at 2700 to 3000 Hz. The resulting audio output is commonly called "monkey chatter" and is familiar to amateur radio operators who use SSB for communications.

The audio output from the circuit of Fig. 14-5 can be tape-recorded, and this recorded signal can be applied to the input. The inversion process will be repeated, and the "upside-down" signal will be "right-side-up" again. The audio output will then be understandable as a human voice. Thus the circuit shown in Fig. 14-5 can serve as an AF voice descrambler as well as a scrambler. This is a primitive example, and is shown here merely to illustrate how analog AF scrambling can work. More sophisticated scramblers split up the audio passband into two or more subpassbands, inverting some or all of these segments and perhaps also rearranging them frequencywise within the main passband.

Digitization

Less common digital codes such as Morse code and Baudot offer a limited amount of privacy and security. Unless a person can read Morse code, either "in the head" or with the help of a Morse-reading machine or program, signals sent in this old-fashioned mode appear alien. The same holds true for Baudot, a largely outdated digital code that has been replaced by ASCII (American Standard Code for Information Interchange) in computer systems. Morse and Baudot codes do not constitute true encryption, however. They are codes, not ciphers.

Voice and image signals can be digitized, and in recent years this has become common practice. In advanced computer communications, such as *Internet telephone* and *videoconferencing*, voices and/or images are converted to strings of binary digits 1 (high) and 0 (low) for transmission over telephone lines, cable systems, and wireless networks. While it is easy to listen in on an analog telephone conversation, digitization presents a barrier to casual eavesdroppers, in the same way as do Morse or Baudot codes. However, the extent of privacy is limited by the wide availability of computers, and the affordability of Internet telephone and videoconferencing software.

All binary digital modes, be they Morse, Baudot, ASCII, or any other scheme that makes use of the high/low dichotomy, can be privatized by means of computer programs called *digital encryption algorithms*. The better the cipher, the greater the privacy and security.

Strong Encryption

The most effective way to keep communications secure and private is to use digital encryption at all subscriber points. This means that the signals in the medium must be digital (rather than analog). For inherently analog data such as voices and pictures, *analog-to-digital (A/D) conversion* must be used with all transmitters, and *digital-to-analog (D/A) conversion* must be used with all receivers. These converters are in addition to the encryption and decryption algorithms.

What Is a Digital Cipher?

A *digital cipher* is a mathematical scheme that changes the arrangement of high (logic 1) and low (logic 0) bits in a digital signal, in such a way that the signal becomes difficult or impossible to receive and understand without a decryption key. The cipher must not affect the signal in such a way that any of the necessary intelligence is lost, but it must change the signal enough so that the casual hacker will not be able to intercept it.

Digital encryption generally falls into two categories, known as *weak encryption* and *strong encryption*. As its name implies, a weak cipher can be broken with some effort, even by a person or agency lacking a decryption algorithm. Of course, the "weakness" of a cipher is a relative thing. Some ciphers can be broken by a smart 12-year-old with a personal computer; others require the concentrated efforts of a team of programmers. A strong cipher cannot be broken by anyone who does not have the necessary decryption algorithm or key, unless extreme effort is expended.

From the foregoing, it might appear that weak encryption is inherently useless because a weak cipher can be broken by potential illegal eavesdroppers. However, the cost of obtaining and implementing a digital cipher is in more or less direct proportion to the strength of the cipher. For some applications, for example, casual telephone conversations, the use of strong encryption would be a waste of money. If you're calling to have a pizza delivered to a party, you are not likely to be concerned that a criminal might intercept the call and then show up at your house to steal the pizza.

There are some people who will settle for nothing less than the strongest cipher that technology, and the laws in their region, will allow, no matter how trivial the contents of the communication. In light of the increasingly complex nature of communications systems, and almost daily reports of strange and unforeseen fraud schemes and terrorist acts, this sentiment is understandable.

Expectations and Requirements

A strong encryption system must have certain characteristics in order to be effective and reliable. Some of the more important criteria are as follows.

Key Escrow In some countries, the government has the right to eavesdrop (with appropriate legal reason or a court order). In the case of strong encryption, this means that the government must have access to decryption keys for any and all encrypted communications taking place under its jurisdiction. In such a country, it is illegal to use strong encryption for which the government does not have the key. This issue has caused controversy in the United States and in some other countries; details were discussed earlier in this chapter. Because of the international nature of many communications, the situation can become exceedingly complicated and the legal nature obscure. If in doubt, consult an attorney.

Export and Travel It is illegal to take certain forms of encryption across national borders. For example, in September 1997, I downloaded a Web browser from a major online service to a notebook computer equipped with Windows 3.11. The software was encrypted to protect users against possible tampering by malicious hackers; the online service had strongly recommended that I use this encrypted version rather than the general Web browser supplied for use with their system. But before the software could be downloaded, I had to agree to certain terms, one of which was that I would not use, or take, the program outside the United States. Basically, I interpreted this to mean that I could not legally carry the notebook computer out of the country without first erasing the Web browser program and overwriting it with something else (so it could not be undeleted). The reason for this restriction is that, at the time of the download, the U.S. government considered strong encryption a form of munition, akin to military secrets or hardware. By the time you read this, the situation might have changed, but you should check the laws before carrying any encrypted files or programs, or any encryption schemes, across any national border. In light of growing concerns over terrorist activity, governments are inclined to take violations very seriously.

User Anonymity   When you initiate a communication, your identity should not be revealed to the party you are calling. Further, that party should not be able to obtain your identity by any means; if the party does not wish to receive anonymous communications, your communication can be shut out. In recent years, the proliferation of *caller ID* in telephone systems has been met with an increasing demand for telephone lines with *anti-caller-ID.* In cellular systems, the cell-phone unit identification code should be encrypted, so it cannot be picked up and fraudulently used by cloners or hackers. With data communications, especially Internet connections, you might desire anonymity for at least three reasons: (1) minimization of junk e-mail resulting from Website detection of subscriber identities, (2) prevention of harassment by people or organizations hostile to your perceived interests or associations, and (3) a guarantee that your monthly bill will represent only the calls you have made, and not fraudulent calls placed by unauthorized people.

User Location   When you use a cell phone, you should have the option of concealing your location. In some situations, such as 911 calls, it is important that authorities be able to pinpoint where you are. In other cases, you might not want anyone to know where you are. The location can generally be concealed by strong encryption of the user ID codes. *Radio direction finding* (RDF) is theoretically possible with any signal, encrypted or not, but if the person using the RDF equipment does not know who is transmitting the signal or what the call is about, direction finding is of no practical use.

Message Content   Voices, images, e-mail, and general computer files and programs should be encrypted so hackers cannot intercept and misuse any of the data. Eventually, all communications (hard-wired and wireless) will probably be digitized, and analog systems will be relics. Increasingly powerful personal computers, and correspondingly more sophisticated programs for breaking ciphers, will evolve. This will necessitate the strongest possible encryption for sensitive information. Thus a conflict is inevitable between the private and public sectors. How important is the right of an individual or corporation to carry on communications that cannot be intercepted by law-enforcement or other government personnel? How important, in contrast, is the right of the public (that is, the government) to have decryption keys to intercept possible criminal or terrorist-related communications? These questions do not lend themselves to easy answers. If trends are any indicator, these issues will become more pressing as technology advances, and as people desire both greater freedom and enhanced security.

Effective Lifetime The perfect strong cipher would be unbreakable by any known or anticipated schemes, for the foreseeable future. But it has become apparent that technological events are hard to predict more than a couple of years in advance. In some realms, such as the World Wide Web, the prediction horizon is only a few months distant. As has been mentioned, the governments of technological powers such as the United States are in the best positions to develop near-ideal ciphers, because they have vast economic resources and have a tendency to go to any lengths to protect their national security. Some multinational corporations are attaining technological power comparable to governments. An optimist would suggest that a strong cipher ought to be unbreakable for 20 years. That means that if all the resources of a powerful government or corporation were bent to the task of breaking the code without the aid of the decryption key, and taking technological advancement into account, it would be 20 years, on the average, before a single encrypted message could be intercepted against the will of the communicating parties.

15

Noise and Interference

Noise and *interference* are the bane of all electronic communications, wired and wireless. These phenomena degrade speed, reduce accuracy, and increase frustration.

Noise, which can never be completely eliminated, is unwanted energy that comes from natural sources or from noncommunications devices and systems. Noise has one universal characteristic: It is less orderly than desired signals. This fundamental difference is at the heart of various methods by which noise is suppressed, allowing signals to get through more clearly.

Interference, which can often be avoided, comes from communications equipment, sometimes of the same type and used for the same purpose as the device(s) being interfered with. Some types of interference are easily dealt with; other types are difficult to reduce. The most troublesome form of interference is deliberate, malicious interference such as *jamming*.

Noise and interference occur at all electromagnetic (EM) wavelengths, from the longest radio waves to infrared (IR), visible light, and ultraviolet (UV).

External Radio Noise

In radio-frequency (RF) systems, noise that comes from outside the receiver is known as *external noise*. There are many possible sources of such noise. In general, the more sensitive the receiving equipment, and the longer the distance over which communications is attempted, the more significant external noise becomes. To some extent the significance of external noise varies with the communications frequency as well.

Cosmic Noise

Noise from outer space is known generally as *cosmic noise*. It occurs at all wavelengths from the very-low-frequency (VLF) radio band to the X-ray and gamma-ray spectra. At the lower frequencies, the ionosphere of our planet prevents the noise from reaching the surface. At some higher frequencies, tropospheric absorption prevents the noise from reaching us. But at many frequencies, cosmic noise arrives at the surface with little or no attenuation.

Cosmic noise limits the sensitivity obtainable with receiving equipment because this noise cannot be eliminated. Radio astronomers deliberately listen to cosmic noise in an effort to gain better understanding of our universe. To them, it is human-made noise and interference, rather than cosmic noise, that adversely affects receiving equipment.

Cosmic noise is easy to mistake for *tropospheric noise*, but cosmic noise can be identified by the fact that it correlates with the plane of the Milky Way galaxy. The strongest *galactic noise* comes from the direction of the constellation Sagittarius, because this part of the sky lies on a line between our solar system and the center of the galaxy. Perhaps the most intriguing form of cosmic noise arrives with equal strength from all directions. In 1965, Arno Penzias and Robert Wilson of Bell Laboratories observed faint cosmic noise that seemed to be coming from the entire universe. All other possible sources were ruled out. Most astronomers now believe that the noise originated with the fiery birth of our universe, in an event called the *Big Bang.*

Galactic Noise

Noise from our galaxy, the Milky Way, was first observed by Karl Jansky, a physicist working for Bell Telephone Laboratories, in the 1930s. It was an accidental discovery. Jansky was investigating the nature of atmospheric noise at a wavelength of about 15 meters (m), or 20 megahertz (MHz). Jansky's antenna consisted of a rotatable array, not much larger than the Yagi antennas commonly used by radio amateurs at the wavelength of 15 m.

Galactic noise contributes, along with noise from the sun, the planet Jupiter, and a few other celestial objects, to most of the cosmic radio noise arriving at the surface of the earth. Other galaxies radiate noise, but since the external galaxies are much farther away from us than the center of our own galaxy, sophisticated equipment is needed to detect the noise from them.

Solar Flux

The amount of radio noise emitted by the sun is called the *solar radio-noise flux,* or simply the *solar flux.* The solar flux varies with frequency. However, at any frequency, the level of solar flux increases abruptly when a solar flare occurs. This makes the solar flux useful for propagation forecasting.

A sudden increase in the solar flux indicates that ionospheric propagation conditions will deteriorate within a few hours.

The solar flux is commonly monitored at a wavelength of 10.7 centimeters (cm), or a frequency of 2800 MHz. At this frequency, the troposphere and ionosphere have no effect on radio waves, so the energy reaches the surface at full strength.

The 2800-MHz solar flux is correlated with the 11-year *sunspot cycle.* On the average, the solar flux is higher near the peak of the sunspot cycle, and lower near a sunspot minimum.

Sferics

Electromagnetic noise is generated in the atmosphere of our planet, mostly by lightning discharges in thundershowers. This noise is called *sferics.* In a radio receiver, sferics produce a faint background hiss or roar, punctuated by bursts of sound we call *static.* An example of sferics, as they might appear on the display of an oscilloscope connected to the intermediate-frequency (IF) section of a radio receiver, is shown in Fig. 15-1.

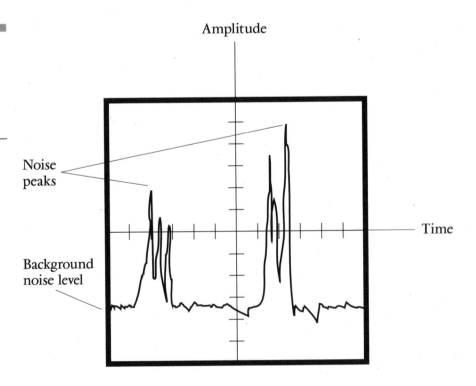

Figure 15-1
Sferics as seen on an oscilloscope display. Time is on the horizontal scale; amplitude is on the vertical scale.

A huge potential difference exists between the surface of the earth and the ionosphere. The earth's surface and the ionosphere behave like concentric spherical surfaces of a massive capacitor, with the troposphere and stratosphere serving as the dielectric. Sometimes this dielectric develops "holes," or pockets of imperfection, where discharge takes place. Such "holes" are usually associated with thundershowers. There are normally about 700 to 800 such areas at any given time, concentrated mostly in the tropics. Sand storms, dust storms, and volcanic eruptions also produce some lightning, contributing to the overall sferics level.

An individual lightning stroke produces a burst of RF energy from VLF through the microwave region. Energy is also produced in the IR, visible, and UV portions of the EM spectrum.

The current that flows across the atmospheric capacitor averages 1500 amperes (A). At very low and low radio frequencies, the sferics propagate around the world, as the EM field is trapped between the "plates" of the capacitor. As the frequency increases, sferics become a more local phenomenon until, at the upper end of the radio spectrum, sferics travel only a few miles. This is why the general level of external background noise decreases as the frequency gets higher, making low-noise circuit design of such importance in the very-high-frequency (VHF), ultra-high-frequency (UHF), and microwave spectra.

Sferics are not confined to the earth. Some noise is generated by storms in the atmosphere of the giant planet Jupiter. Astronomers have heard this noise with radio telescopes. Sferics probably also occur on Saturn, and perhaps on Uranus, Neptune, Venus, and Mars. In the cases of Venus and Mars, dust storms and volcanic eruptions would be the cause of sferics.

You can hear the sferics from a distant thundershower on the standard amplitude-modulation (AM) broadcast band. If you have a general-coverage communications receiver, you can listen at progressively higher frequencies as the storm or storm system approaches. When you hear the static bursts at 30 MHz, the storm is probably less than 100 miles away.

All storm systems produce sferics, although the amount of noise varies. Using directional antenna systems and special receivers, some meteorologists have been able to locate and track large storm systems. A receiver designed especially for listening to atmospheric noise is called a *sferics receiver.* In normal radio communications, however, sferics are nothing but a nuisance.

Precipitation Static

Precipitation static is a form of radio interference caused by electrically charged water droplets or ice crystals as they strike objects. The resulting discharge produces wideband noise that sounds similar to the noise generated by electric motors, fluorescent lights, or other appliances.

Precipitation static is often observed in aircraft flying through clouds containing rain, snow, or sleet. But occasionally, precipitation static occurs in radio communications installations. This is especially likely to happen when it is snowing; then the noise is called *snow static*. Dust storms can also cause precipitation static. Precipitation static can be severe at times, making radio reception difficult, especially at the very low, low, and medium frequencies.

A *noise blanker* or *noise limiter* can be effective in reducing interference caused by precipitation static. A means of facilitating discharge, such as an inductor between the antenna and ground, can be helpful. Improvement can also be obtained by blunting any sharp points in antenna elements.

Corona

When the voltage on an electrical conductor, such as an antenna or high-voltage transmission line, exceeds a certain value, the air around the conductor begins to ionize. The result is a blue or purple glow called *corona*. This glow can sometimes be seen at the ends of an antenna at night when the transmitter has high RF power output. The ends of an antenna carry the highest voltages. Corona is more likely to occur when the humidity is high than when it is low, because moist air has a lower *ionization potential* than dry air. It causes a characteristic strong, broadband hiss that can render a wireless communications receiving system ineffective.

Corona can occur inside a cable just before the dielectric material breaks down. Corona is sometimes observed between the plates of air-variable capacitors handling large voltages. A pointed object, such as the end of a whip antenna, is more likely to produce corona than will a flat or blunt surface. Some inductively loaded whip antennas have *capacitance hats* or large spheres at the ends to minimize corona.

Corona sometimes occurs as a result of high voltages caused by static electricity during thunderstorms. Such a display is occasionally seen at the tip of the mast of a sailing ship. Corona was observed by seafaring explorers hundreds of years ago. They called it *Saint Elmo's fire*.

Impulse Noise

Any sudden, high-amplitude voltage pulse will cause RF energy to be generated. *Impulse noise* is produced by household appliances such as vacuum cleaners, hair dryers, electric blankets, thermostat mechanisms, and fluorescent-light starters. This type of noise is most severe at very low and low radio frequencies. But serious interference can occur in the medium-frequency (MF) and high-frequency (HF) ranges. Impulse noise is seldom a problem above 30 MHz. Impulse noise can be picked up by high-fidelity (hi-fi) audio systems. The greater the number of external peripherals (such as tape players, microphones, and speakers) that exist in a hi-fi system, the greater is the susceptibility to this type of noise. Wireless microphones and headsets are especially vulnerable.

Impulse noise in a radio receiver can be reduced by the use of a good ground system. *Ground loops* should be avoided. A noise blanker or noise limiter can be helpful. A receiver should be set for the narrowest response bandwidth consistent with the mode of reception. Impulse noise in a hi-fi sound system can be more difficult to eliminate. An excellent ground connection, without ground loops, is imperative. It might be necessary to shield all speaker leads and interconnecting wiring.

Ignition Noise

Ignition noise is a wideband form of impulse noise, generated by the electric arc in the spark plugs of an internal-combustion engine. Ignition noise is radiated from many different kinds of devices, such as automobiles and trucks, lawn mowers, and gasoline-engine-driven generators. Figure 15-2 is an example of ignition noise as it might appear on an oscilloscope connected in the IF chain of a radio receiver.

Ignition noise can usually be reduced in a receiver by means of a noise blanker. The pulses of ignition noise are of very short duration, although their peak intensity can be considerable. Ignition noise is a common problem for mobile radio operators, especially if communication in the HF range (below 30 MHz) is contemplated. Ignition noise can be worsened by radiation from the distributor to wiring in a truck or automobile. Sometimes special spark plugs, called *resistance plugs*, can be installed in place of ordinary spark plugs, and the ignition noise will be reduced. An automotive specialist will know whether or not this is feasible in a given case. An excellent vehicle-chassis ground connection

Ignition noise as seen on an oscilloscope display. Time is on the horizontal scale; amplitude is on the vertical scale.

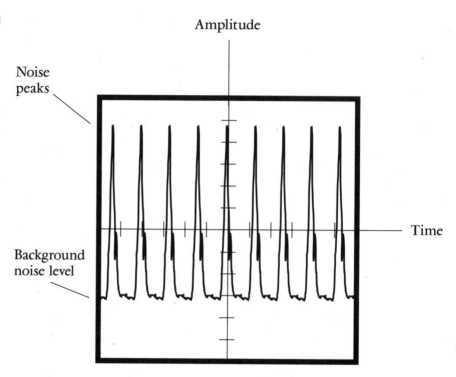

is imperative in any mobile installation. Mobile receivers for HF work should have effective, built-in noise blankers.

Ignition noise is not the only source of trouble for the mobile radio operator. Noise can be generated by the friction of the tires against pavement. High-tension power lines often radiate significant impulse noise at HF. The vehicle's alternator can also cause noise in the form of a whine that changes pitch as the car accelerates and decelerates.

Power-Line Noise

Utility lines, in addition to carrying the 60-hertz (Hz) alternating current (AC) that they are intended to transmit, carry other currents. These currents have a broadband nature. They result in an effect called *power-line noise*. The currents usually occur because of electric *arcing* at some point in the circuit. The arcing might originate in appliances connected to the terminating points; it can take place in faulty or obstructed transformers; it can even occur in high-tension lines as a corona discharge into humid air. The broadband currents cause an EM field to be radiated from the power line, because the currents flow in the same instantaneous direction along all line conductors.

Power-line noise sounds like a buzz or hiss when picked up by a radio receiver. Some types of power-line noise can be greatly attenuated by means of a noise blanker. For other kinds of noise, a limiter can be used to improve the reception. *Phase cancellation,* in which the noise picked up by an auxiliary antenna is used to null out the noise from the main receiving antenna, can reduce the level of received power-line noise in some cases.

Internal Circuit Noise

Some noise is produced by the active components within electronic equipment. This is known as *internal noise.* At very low, low, and medium frequencies, external noise is almost always stronger than internal noise. For this reason, internal noise normally does not play a significant part in limiting the sensitivity at frequencies below a few megahertz. In the high-frequency portion of the radio spectrum, external noise becomes less intense. In the very high, ultra-high, and microwave parts of the spectrum, internal noise is much more important than external noise. Low-internal-noise receiver design is of major concern at frequencies above approximately 30 MHz. It is important that internal noise be kept as low as possible in the early stages of a multistage amplifier chain, because any noise generated in one amplifier will be picked up and amplified, along with the desired signals, in the succeeding stages.

The molecules of all substances, including the metal and other materials in an electronic circuit, are in constant motion. The higher the temperature, the more active the molecules. This generates noise in any amplifier. As the electrons in a circuit hop from atom to atom, or impact against the metal anode of a vacuum tube, noise is generated. The larger the amount of current flowing in a circuit, in general, the more noise there will be. Noise is also produced by the vibration of the circuit components in an amplifier. This can cause problems in mobile equipment, such as radio transceivers used in cars, trucks, and aircraft. Sturdy construction and, if needed, shock-absorbing devices, reduce the problem of *mechanical noise.* While nothing can be done about the *thermal noise* at a given temperature, the equipment can be cooled to extremely low temperatures to minimize the movement of the molecules and thus decrease the level of thermal noise. In theory this scheme will work in any electronic circuit. But in practice it is expensive, and maintenance of the cooling system is complicated. Internal noise can be reduced by the use of specialized amplifying devices, such as the *gallium-arsenide field-effect transistor* (GaAsFET). Elaborate low-noise amplifier designs, such as the *maser* used in radio telescopes, generate far less noise than ordinary amplifiers.

Thermal Noise

In all materials, the electrons are in constant motion. Because the electrons move in curved paths, they are always accelerating, even if they stay in the same orbital shell of a single atom. This acceleration of charged particles produces EM fields over a wide spectrum of wavelengths. To some extent, the random movement of the positively charged atomic nuclei has the same effect. In electronic circuits, this charged-particle acceleration causes thermal noise.

The level of thermal noise in any material is proportional to the *absolute temperature* in kelvins. (A kelvin is the same "size" as a Celsius degree; however, the kelvin scale starts at absolute zero. Any temperature expressed in kelvins is 273 greater than the same temperature expressed in degrees Celsius.) The higher the temperature gets, the more rapidly the charged particles are accelerated. The lower the temperature becomes, the more slowly the particles move. As the temperature approaches absolute zero—the coldest possible condition—the particle speed and acceleration approach zero, and so does the level of thermal noise.

Thermal noise imposes a limit on the sensitivity that can be obtained with electronic receiving equipment. This noise can be minimized by placing a *preamplifier* circuit in a bath of liquid gas, such as helium or nitrogen. Helium has a boiling point of only a few kelvins. Such cold temperatures cause the atoms and electrons to move very slowly, thus greatly reducing the thermal noise compared with the level at room temperature.

Shot-Effect Noise

In any current-carrying medium, the individual charge carriers cause noise impulses as they move from atom to atom. The individual impulses are extremely faint, but the large number of moving carriers results in noise that can be heard in any amplifier or radio receiver. This effect is called *shot effect*. The noise is known as *shot-effect noise* or *shot noise*.

Shot-effect noise limits the ultimate sensitivity that can be obtained in a radio receiver. This is because a certain amount of noise is always produced in the *front end*, or first amplifying stage, and this noise is amplified by succeeding stages along with the desired signals. The amount of shot-effect noise that a device produces is roughly proportional to the current that it carries. In recent years, low-current solid-state devices such as the GaAsFET have been developed for optimizing the sensitivity of a receiver front end by minimizing the *noise figure.*

Hiss

A form of audio noise, in which the amplitude is peaked near the midrange or treble regions, is often called *hiss*. Hiss is always heard in active audio-frequency (AF) circuits. The sound of hiss is familiar to anyone who has worked with hi-fi equipment, public-address systems, or radio receivers.

In audio equipment, the hiss is generated as the result of random electron or hole movement in components. The hiss generated in the early stages is amplified by subsequent stages. In a radio receiver, hiss can be generated in the IF stages and in the front end, mixers, and oscillators. Some hiss even originates outside the receiver, in the form of thermal noise in the antenna conductors and atmosphere.

Broadband hiss, especially in audio circuits, is sometimes called *white noise*. It gets its name from the fact that white light contains energy at all visible wavelengths, and thus is wideband energy; wideband noise is therefore, in a sense, "white." White AF noise consists of a more or less uniform distribution of energy at wavelengths from 20 Hz to 20 kilohertz (kHz). If energy is wideband in nature but concentrated toward one end or the other, the noise is called *pink noise* or *violet noise*. At low levels, white noise can be pleasing to the ear, and is used in specialized audio devices to simulate the sounds of the ocean or of wind through trees, helping people relax or sleep.

Conducted Noise

When noise in a communications system comes through the power supply, it is said to be *conducted noise*. This can be a problem with fixed as well as with mobile wireless communications stations.

In a fixed station, conducted noise can come from the AC utility mains. Some appliances generate high-frequency noise over a broad spectrum. This is radiated by the power lines in the vicinity of the appliance, and is also conducted along the power lines. Although utility transformers choke off most of this noise, there usually are several households on a single power transformer. Therefore, your neighbor's (as well as your own) hair dryer, vacuum cleaner, electric blanket, light dimmer, or other transient-generating device can interfere with your reception via conducted noise.

Conducted noise in a fixed station can be suppressed by inserting RF chokes in series with both power leads from the station to the wall outlet. These chokes must be rugged enough to carry the current needed by the station. You can wind them on toroidal or solenoidal powdered-iron cores using No. 12 or No. 14 soft-drawn, insulated copper wire. Inductances of about 1 mH are usually sufficient.

In a mobile station, conducted noise comes from the alternator and from the spark plugs, through the power leads and into the radio. Conducted noise in a mobile station can be minimized by connecting the radio power leads to the battery, rather than through the cigarette lighter. Filtering the alternator leads using capacitors of about 0.1 microfarad (μF) will help reduce alternator whine. Resistance wiring in the ignition system sometimes helps with spark-plug noise problems.

Noise Reduction

In communications systems, there are various ways to reduce the level of noise compared with the level of the desired signals. This improves the signal-to-noise ratio. When conditions are marginal, a small improvement in this ratio can result in a dramatic decrease in the number of errors, and/or a substantial increase in communications speed.

Noise "Randomness"

Certain forms of EM noise, especially human-made noise, exhibit a definite pattern of amplitude versus time. Other types of noise show no apparent relationship between amplitude and time; this is sometimes called *random noise*. An example of *nonrandom noise* is the impulse type. Examples of random noise are sferics, thermal noise, and shot-effect noise.

Random noise is more difficult to suppress than noise having an identifiable waveform or repetitive pattern. If noise shows discernible amplitude-versus-time patterns, it is possible to design a circuit that "knows" the noise behavior and that can act accordingly to get rid of it. But no circuit exists that "knows" the patterns of random noise, because there are no patterns.

A limiting circuit, which at least allows the desired signal to compete with the noise, is a good defense against strong random noise. The use of narrowband emission, in conjunction with a narrow receiver bandpass, is also helpful in dealing with strong random noise. *Frequency modulation* (FM) can give better results than *amplitude modulation* (AM) if the receiver is equipped with an effective *limiter* or *ratio detector*. Directional or noise-canceling antenna systems can sometimes improve communications in the presence of random noise.

Noise versus Frequency

An often-overlooked way to reduce the effects of EM noise is to choose an operating frequency at which the noise level is low. Although noise usually occurs over a wide band of frequencies, most noise decreases in intensity as the operating frequency increases. Amateur radio operators know this. During a severe thunderstorm watch, for example, sferics make local emergency voice communication difficult at 75 m (approximately 3.9 MHz), but the 2-m voice band (approximately 146 MHz) is almost unaffected. Given a choice, most amateur radio operators will therefore conduct severe-weather voice communication at 2 m rather than at 75 m.

The next time there are heavy thundershowers in your area, try listening to a portable radio that can receive both AM and FM broadcast. You'll notice that the sferics are much less severe in the FM band. This is largely because of the difference in frequency, a ratio of about 100 to 1. The modulation method is also a factor.

Various electrical and electromechanical devices produce EM noise. The most common offenders, as far as radio reception is concerned, are internal-combustion engines, vacuum cleaners, hair dryers, light dimmers, and some electric motors. This noise, like its natural counterpart, usually occurs with greater intensity at lower frequencies than at higher frequencies. In some cases there exist one or more frequencies at which the noise level reaches a peak. When this happens, it is usually because utility wires are acting as resonant transmitting antennas for the noise.

Noise versus Modulation

It is not always possible to change the operating frequency to reduce or eliminate interference caused by noise. In such situations, the proper choice of modulation can make a big difference.

Radio modulation can be broadly classified as either *analog modulation* or *digital modulation.* Examples of analog modulation include conventional AM and FM broadcast, *single sideband* (SSB), conventional *fast-scan television* (FSTV), and most *slow-scan television* (SSTV). Digital modulation modes include *continuous-wave* (CW) *radiotelegraphy, frequency-shift keying* (FSK), *digital television,* and various forms of *pulse modulation.* In general, digital-modulation systems offer better noise immunity than analog-modulation systems.

There are two basic methods of achieving analog voice communication: AM and FM. (Single-sideband emission is a special type of AM.) Virtually all noise is characterized by fluctuations in amplitude, with the frequency ill defined and spread out. Because of this, FM offers better noise immunity, all other factors being equal, than AM, provided the FM receiver is equipped with some kind of limiter circuit to nullify variations in amplitude. This fact, in addition to the difference in frequency, accounts for the huge difference in sferics between the standard AM broadcast band and the standard FM broadcast band during times of heavy thunderstorm activity. The same applies to amateur radio voice communications at 75 m (usually done in SSB) versus 2 m (usually done in FM).

Noise versus Bandwidth

Noise is almost always wideband in nature. Signals, in contrast, usually occupy a narrow band of frequencies. The proper choice of receiver *selectivity* can affect the extent to which the receiver will discriminate between a desired signal and unwanted noise.

Figure 15-3 shows two simplified spectral illustrations of a hypothetical signal as it might be received using two different *passband filters.* The peak signal level is slightly higher than the peak noise level, so the signal is readable in either situation. At A, the receiver passband is several times as wide as the bandwidth of the modulated signal. At B, the receiver passband is only a little wider than the bandwidth of the signal. In both situations, the total energy contained in the noise is represented by the area of the noise rectangle, and the total energy in the signal is represented by

Figure 15-3
At A, the passband is
much wider than the
signal; at B, the
passband is only a
little wider than
the signal.

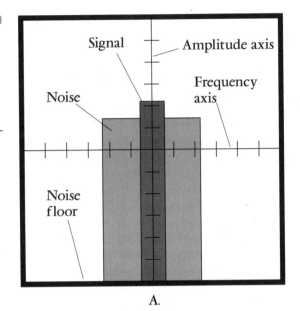

A.

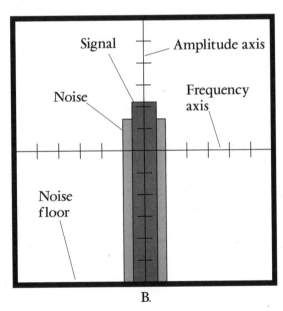

B.

the area of the signal rectangle. The signal energy is the same in both instances, but the noise energy is lower in the case shown by Fig. 15-3B, in which the receiver passband is narrower. The ratio of signal energy to noise energy is higher in case B, and the signal is therefore easier to "copy." It would be possible to narrow the receiver bandwidth a little further than is shown in Fig. 15-3B, affording slightly more noise reduction. But if the receiver passband is made narrower than the signal bandwidth, reception will be degraded because the receiver will discriminate against the signal as well as against the noise.

Some signals occupy a wide band of frequencies if evaluated over time but have narrow bandwidth at any given instant. This is known as *spread-spectrum communications*. Provided that the receiver frequency follows along with the transmitter frequency, narrowband filters will work for noise reduction in spread-spectrum mode in the same manner as they work for conventional narrowband signals.

Noise Blankers

A *noise blanker* is a noise-reducing circuit commonly used in radio receivers at very low, low, medium, and high frequencies. The noise blanker operates by discriminating between short, intense bursts (typical of some types of noise) and the more uniform characteristics of a desired signal. The noise blanker is usually installed in one of the IF amplifiers of a superheterodyne receiver, or just prior to the detector stage in a direct-conversion receiver.

Under low-noise conditions, the noise blanker has no effect on the signals passing through the IF amplifier stage. But when a sudden, high-amplitude, short-duration pulse occurs, the noise blanker cuts off the stage. The receiver output thus contains a brief moment of silence (in which no signal or noise is received) instead of a loud pop or click. The receiving operator might notice this silence, but it will usually not be such as to cause interference to the desired signal.

Noise blankers are most effective against impulse noise, in which the interfering pulses are regular and of short duration. Noise blankers can also work against *key clicks* caused by an improperly operating radiotelegraph transmitter, and against some *woodpecker* signals from over-the-horizon radar systems. Noise blankers do not work very well against the more continuous noise resulting from thundershowers or power-line transformer arcing.

Noise Limiters

A *noise limiter*, also called a *noise clipper*, is a circuit that prevents externally generated noise from exceeding a certain amplitude. The circuit can

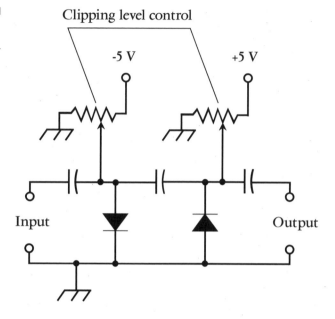

Figure 15-4
A simple noise limiter.

consist of a pair of diodes with variable bias for control of the clipping level (Fig. 15-4). The bias is adjusted until clipping occurs at the signal amplitude. Noise pulses then cannot exceed the signal amplitude. This makes it possible to receive (with some difficulty) a signal that would otherwise be obscured by the noise. The noise limiter is generally installed between two IF stages of a superheterodyne receiver. In a direct-conversion receiver, the best place for the noise limiter is just before the detector stage.

A noise limiter can use a circuit that sets the clipping level automatically, according to the strength of an incoming signal. This is a useful feature, because it relieves the receiver operator of the necessity to continually readjust the clipping level as the signal fades. Such a circuit is called an *automatic noise limiter* (ANL).

Limiters do not work as well as blankers against impulse noise. But a limiter can sometimes provide relief from interference caused by sferics or transformer arcing, when a noise blanker would be ineffective.

Line Filters

A *line filter* is a passive circuit, usually consisting of capacitance and/or inductance, inserted in series with the AC power cord of an electronic device. The line filter allows 60-Hz AC to pass unaffected, but higher-frequency noise is suppressed. A typical line filter is in effect a *lowpass filter*, consisting of a capacitor or capacitors in parallel with the power leads, and an inductor or inductors in series. The component values are chosen for a cutoff frequency somewhat above 60 Hz. Line filters are of use only when noise enters the receiver via the power line.

Special Antennas

Any receiving antenna with a sharp directional null can be used to reduce the level of received human-made noise. Such an antenna can also suppress strong unwanted local signals. A *ferrite loopstick* or *loop antenna* is especially useful for this purpose. Both of these antennas have sharp directional nulls along their axes. When the null is oriented in the direction of a noise source, the noise level drops. When tuned to resonance by means of a variable capacitor, such an antenna exhibits narrow bandwidth, and this serves to further increase the immunity of the antenna system to broadband noise. The feed line must be properly balanced or shielded if the benefits of these types of antennas are to be realized.

The noise rejection of a small loop antenna can be enhanced by the addition of a *Faraday shield*, also called an *electrostatic shield*, around the loop element. The Faraday shield discriminates against the electric component of an EM field, while allowing the magnetic component to penetrate unaffected. Many noise signals are locally propagated by electrostatic coupling. When the antenna is insensitive to these fields, the signal-to-noise ratio is improved at the receiver input.

The overall noise susceptibility of an antenna system can be minimized by careful balancing or shielding of the feed line, maintaining an adequate RF ground, and minimizing the bandwidth. In general, at the medium and high frequencies, a horizontally polarized antenna is less susceptible to human-made noise than is a vertically polarized antenna.

Phase Cancellation

Human-made noise can be greatly reduced at frequencies below about 100 kHz, where power-line noise is an especially serious problem, by using two antennas that receive the noise at equal amplitudes but that receive the desired signals at different amplitudes. The two antennas are

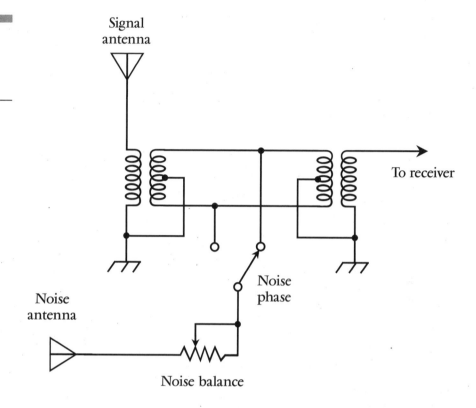

Figure 15-5
A noise-canceling
antenna scheme,
especially effective
at VLF.

Signal
antenna

To receiver

Noise
phase

Noise
antenna

Noise balance

connected together so the noise signals cancel. An example of this technique is illustrated in Fig. 15-5.

The signal antenna and the noise antenna must be reasonably near each other, so they pick up local human-made noise whose characteristics are as nearly identical as possible. But they should not be too close together, or signals will be canceled along with the noise. The signal antenna should be mounted in a location favorable for the reception of radio signals, but the noise antenna should be placed so it will receive as much of the unwanted noise as possible, and relatively little of the desired signal. For example, the signal antenna might be a quarter-wave vertical operating against an earth ground with radials in the back yard; the noise antenna can be a short, random length of wire run on the floor inside the house.

The potentiometer serves to adjust the level of signals and noise from the noise antenna, while having a minimal effect on the overall signal level from the two antennas combined. The phase switch allows for injection of the noise from the noise antenna either "right-side-up" or "upside-down," so that you can be sure to get it out of phase with the noise from the signal antenna. At frequencies below approximately 30 kHz, where wavelengths are measured in kilometers, the noise from the two antennas will be either almost exactly in phase or exactly out of phase. Thus, by choosing the correct switch position (by trial and error), the potentiometer can be adjusted until cancellation is achieved. At VLF, human-made external noise is generally worse than it is anywhere else in the radio spectrum. Thus the fact that cancellation works best at these wavelengths is a welcome blessing.

The author used this scheme for receiving signals at VLF, in conjunction with a VLF converter and a shortwave receiver, with excellent results. The signal antenna was a 16-foot (ft), ground-mounted vertical antenna fed with coaxial cable. The noise antenna was a length of wire 30 feet long, run through the hallway of the house. At the low end of the band, between 9 and 15 kHz, power-line noise could be reduced by 20 to 30 decibels (dB), while signal levels remained unaffected.

As the frequency is increased, this scheme gets less effective, because the noise impulses at the two antennas are no longer either in phase or out of phase unless the antennas are so close together that the circuit cancels the desired signals as well as the noise. This method is most effective against power-line noise and ignition noise; it does not work very well against sferics.

Cryogenics

Cryogenics is the science of the behavior of matter and energy at extremely low temperatures. When the temperature of a conductor is brought to within a few degrees of absolute zero, the conductivity increases dramatically. If the temperature is cold enough, a current can be made to flow continuously in a closed loop of wire. This is called *superconductivity.* Supercooling of a receiving antenna allows the use of amplifiers with greater gain than is possible without such cooling. Thermal noise is thereby reduced, and this improves the signal-to-noise ratio.

Cryogenic devices are not widely used in consumer communications equipment because the cost is high and because maintenance of the system is difficult and complicated. However, the technology offers some promise in *space communications,* where signal levels tend to be extremely low, and internal noise reduction becomes a critical issue. Cryogenic receiving preamplifiers have been used in *radio astronomy* to detect energy from distant celestial objects.

Synchronization

One of the most important problems in communications is the maximization of the number of signals that can be accommodated within a given band of frequencies. This has traditionally been done by attempting to minimize the bandwidth of a signal. There is a limit, however, to how small the bandwidth can be if the information is to be effectively received.

Digital signals, such as CW radiotelegraphy, occupy less bandwidth than analog signals, such as SSB. Perfectly timed Morse code, for example, consists of regularly spaced bits, each bit having the duration of one dot. The length of a dash is three bits; the space between dots and/or dashes in a single character is one bit; the space between characters in a word is three bits; the space between words and sentences is seven bits. This makes it possible to identify every single bit by number, even in a long message. Therefore, the receiver and transmitter can be synchronized, so that the receiver "knows" which bit of the message is being sent at a given moment.

In *synchronized digital communications*, the receiver and transmitter follow each other in lockstep, so the receiver hears and evaluates each bit individually. This makes it possible to use a receiving filter having extremely narrow bandwidth. The synchronization requires the use of an external, common frequency or time standard. The broadcasts of WWV/WWVH can be used for this purpose. Frequency dividers are used to obtain the necessary synchronizing frequencies. A tone is generated in the receiver output for a particular bit if, and only if, the average signal voltage exceeds a certain value over the duration of that bit. False signals, such as might be caused by filter ringing, sferics, or other noise, are usually ignored because they rarely result in sufficient average voltage.

Experiments with synchronized communications have shown that the improvement in signal-to-noise ratio, compared with nonsynchronized systems, can be significant, especially at low to moderate data speeds. The reduced bandwidth allows proportionately more signals to be placed in any given band of frequencies.

Coherent Radiotelegraphy

There exists a little-explored mode of CW radiotelegraphy, in which the receiver and transmitter are synchronized by means of a primary time standard such as WWV. If the receiver "knows" the speed at which the transmitter is sending, the string of bits at the receiver can be synchronized precisely with the string of bits at the transmitter, using the primary time standard and taking propagation delays into account. When this is done, a sensing circuit at the receiver can consider a bit to be high (logic 1, or "on") if there is signal for 50 to 100 percent of the time, and low (logic 0, or "off") if there is signal for 0 to 49 percent of the time. These percentages might be adjusted for further improvement in accuracy; this would require experimentation. This synchronized mode is known as *coherent radiotelegraphy* and is of interest to some radio amateurs and experimenters. Coherent radiotelegraphy makes it possible to greatly reduce the bandwidth needed by a CW signal at a given speed. This, in turn, provides for a substantial improvement in signal-to-noise ratio.

The main problem with coherent radiotelegraphy is that it is difficult to synchronize the receiver with the transmitter. The trouble is compounded when a station calls "CQ" (which means "Calling anyone"), because potential answering stations might be attuned to the wrong speed or be in a location where propagation delays are greater or less than anticipated.

Nowadays, *packet radio* has made coherent radiotelegraphy obsolete from a technical standpoint. Packet offers error-free communications with greater versatility than coherent radiotelegraphy. However, many amateur radio operators prefer to use radiotelegraphy, "copying" signals in their heads rather than having machines print out the data for them. The reason for this preference, and the almost irrational stubbornness with which some radio "hams" cling to radiotelegraphy, can be summed up simply: These people have a lot of fun "operating CW." It's a respite and a change of pace from the high-speed, machine-enhanced digital communications typical of the Internet and other computer networks.

Radio Interference

Radio interference is the presence of unwanted RF signals that increase the difficulty of radio reception. Interference from communications transmitters can be either accidental or intentional. The RF spectrum is heavily used in modern industrialized nations, and a certain amount of accidental interference is to be expected. The narrower the average signal bandwidth, in a given band of frequencies, the lower the probability of accidental interference. Intentional interference is more difficult to predict, and can sometimes be impossible to analyze. One psychologically disturbed person can raise havoc all out of proportion to the mathematical fraction of band space which his or her signal occupies.

Electromagnetic Interference (EMI)

Electronic devices sometimes interfere with one another's operation at short range. For example, an amateur radio transmitter might cause changes in the volume of the audio output from a hi-fi system. A computer can interfere with shortwave-radio reception. Television receivers are notorious for the spurious signals (actually harmonics of the scan signal) they generate at very low, low, and medium frequencies. These phenomena, known collectively as *electromagnetic interference* (EMI), are inevitable in a society saturated with wireless devices and systems.

Good hardware engineering is the most effective defense against EMI. That is to say, "prevention is the best cure." Commonsense measures include the shielding of wires that carry RF signals, RF grounding of all electronic equipment, and limitation of power output to the minimum necessary to obtain the desired end. Chapter 1 contains more information about EMI.

Adjacent-Channel Interference

When a radio receiver is tuned to a particular frequency and interference is received from a signal on a nearby frequency, the effect is referred to as *adjacent-channel interference*. When receiving an extremely weak signal near an extremely strong one, this type of interference is likely, especially if the stronger signal is voice modulated. No transmitter has absolutely clean modulation, and a small amount of off-frequency emission occurs with voice modulation, especially AM and SSB.

Adjacent-channel interference can be reduced by using proper engineering techniques in transmitters and receivers. Transmitter audio amplifiers, modulators, and RF amplifiers should produce as little distortion as the state of the art will permit. Receivers should employ selective filters of the proper bandwidth for the signals to be received, and the *adjacent-channel response* should be as low as possible. A relatively flat response in the passband, and a steep drop-off in sensitivity outside the passband, are characteristics of good receiver design. An example of a good response curve for narrowband FM reception is shown

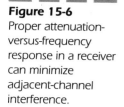

Figure 15-6
Proper attenuation-versus-frequency response in a receiver can minimize adjacent-channel interference.

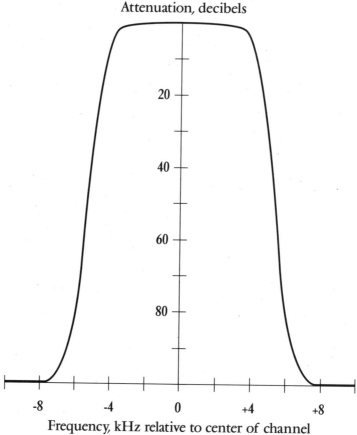

Attenuation, decibels

Frequency, kHz relative to center of channel

in Fig. 15-6. This ensures that desired signals will be received without undue distortion and that unwanted off-frequency signals will be attenuated as much as possible.

Splatter

The colloqualism *splatter* is used to describe the effects of a voice radio signal having too much bandwidth. Splatter can occur with AM, SSB, or FM.

When the amplitude, frequency, or phase of a signal is varied, *sidebands* are invariably produced. The more rapid the instantaneous change in amplitude, frequency, or phase becomes, the farther the sidebands occur from the carrier frequency. If modulation is excessive, *peak clipping* will occur in AM or SSB modes, and *overdeviation* will occur in FM. This causes sidebands to be generated at frequencies too far removed from the carrier.

In a radio receiver, splatter sounds like a crackling noise at frequencies up to several hundred kilohertz from the carrier frequency of the transmitter. Sometimes splatter is mistaken for sferics or human-made impulse noise. Splatter can cause serious interference to communications circuits. In its worst form, a single signal with severe splatter can affect dozens of other conversations on the same frequency band.

In an AM or SSB transmitter, splatter can be eliminated by:

1. Avoiding overmodulation
2. Operating audio amplifier circuits with a minimum of distortion
3. Ensuring proper bias and drive for all transmitter amplifiers
4. Ensuring that external linear amplifiers are operating properly
5. Proper operation of speech-processing circuits and automatic level control

In an FM or transmitter, splatter can be avoided by:

1. Keeping the deviation within rated system limits
2. Ensuring that the audio amplifiers are operating with minimum distortion

Key Clicks

Key clicks are a form of splatter from a CW radiotelegraph transmitter, resulting from excessively rapid rise time or decay time. In a properly adjusted CW radio transmitter, the rise and fall time should be finite, so that the signal amplitude-versus-time function appears similar to Fig. 15-7A. If the decay time or rise time is too short, as shown at B and C, sidebands are generated at frequencies far removed from the signal frequency. These sideband emissions occur at the decay and/or rise instants. Such sidebands sound like clicks or pops to an operator listening away from the signal frequency. These clicks can cause objectionable interference to other communications.

In order to ensure that a CW transmitter will not produce key clicks, it is necessary to provide a *shaping network*. Such a network is usually comprised of series resistors and parallel capacitors in the keying circuit.

■ ■ ■ ■

Figure 15-7
Proper CW keying
(A), excessively short
decay time (B), and
excessively short
rise time (C).

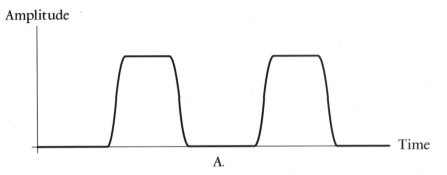

A.

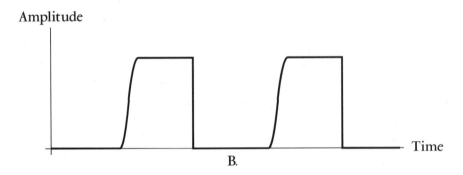

B.

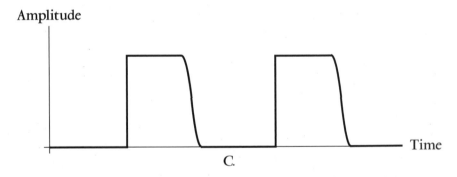

C.

Harmonics

Any signal contains energy at integral (whole-number) multiples of its
frequency, in addition to energy at the desired frequency. The lowest-
frequency component of a signal is called the *fundamental frequency*; all
integral multiples are called *harmonic frequencies*, or simply *harmonics*.

In theory, a pure sine wave contains energy at only one frequency and
has no harmonic energy. In practice, this ideal is never achieved. All signals

contain some energy at harmonic frequencies, in addition to the energy at the fundamental frequency. The signal having a frequency of twice the fundamental is called the *second harmonic*; the signal having a frequency of three times the fundamental is called the *third harmonic*; and so on.

Wave distortion always results in the generation of harmonic energy. While the nearly perfect sine wave has very little harmonic energy, the sawtooth wave, square wave, and other distorted periodic oscillations contain large amounts of energy at the harmonic frequencies. Whenever a sine wave is passed through a nonlinear circuit, harmonic energy is produced. A circuit designed to deliberately create harmonics is called a *harmonic generator* or *frequency multiplier.*

Harmonic output from radio transmitters is usually undesirable, and the designers and operators of such equipment often go to great lengths to minimize this energy as much as possible. Methods of minimizing harmonic radiation from a radio transmitting station include ensuring that:

1. The transmitter final amplifier is properly tuned
2. The final amplifier is not overdriven
3. A good impedance match exists between the final amplifier and the antenna system
4. All connections in the antenna system are electrically sound

Additional measures for suppressing harmonic radiation include:

5. Installing a low-pass filter in the antenna feed line
6. Using a transmatch (antenna tuner) in the antenna feed line
7. Using band-stop filters (traps) in the antenna feed line to suppress energy at specific harmonic frequencies

Spurious Emissions

Many types of electronic devices emit RF energy at frequencies that can interfere with radio and television equipment. A classic example is the color-burst signal emitted by television receivers at approximately 3.59 MHz. This frequency corresponds to a wavelength of 80 m and falls within one of the most popular amateur radio bands.

Personal computers produce emissions at RF that can cause interference to radio receiving equipment. Amateur radio operators are familiar with this phenomenon because it is common for a "ham shack" to contain a computer alongside a radio receiver. Cathode-ray-tube (CRT) monitors produce RF at many frequencies as a result of the electron-beam scanning apparatus. The central processing unit (CPU) also produces RF. The microprocessor clock frequency, or one of its harmonics, might fall within a radio band where reception is contemplated. Various integrated-circuit devices within the CPU can generate RF at many frequencies simultaneously.

Radio transmitters, when poorly designed and/or improperly operated, can produce powerful signals at frequencies other than the desired transmission frequency. Such signals can propagate for thousands of miles and cause widespread interference to other services. The most common cause of such spurious emissions is oscillation in the final amplifier circuit. Such oscillations are called *parasitics.* They can, in some cases, be eliminated by placing small inductors in series with the collector, drain, or plate leads of the final amplifier transistor, field-effect transistor (FET), or vacuum tube. However, in stubborn cases, the only effective cure is to reengineer and rebuild the amplifier.

Spurious Response

A radio receiver can sometimes pick up signals from frequencies other than the one to which it is tuned. There are several ways in which this can happen.

One common *spurious response* scenario takes place when a strong signal appears at the *image frequency* of a superheterodyne receiver. Suppose a single-conversion receiver has an IF of 9 MHz. Further suppose that it is set to receive a signal at 25 MHz. The local oscillator within the receiver is therefore tuned to 16 MHz; this will mix with the incoming 25-MHz signal to produce an output at 9 MHz ($25-16=9$), as shown in Fig. 15-8A. Now suppose that a strong signal appears at a frequency of 7 MHz. This will also mix with the 16-MHz local oscillator signal to produce an output at 9 MHz ($16-7=9$), as shown in Fig. 15-8B. There are various ways to deal with situations of this kind. The use of one or more high-selectivity, tuned RF amplifiers ahead of the mixer will enhance the desired signal and suppress image signals. In especially severe cases, for example, when a nearby broadcast station happens to be transmitting at a receiver image frequency, the use of a *band-stop filter* (also called a *trap*) in the antenna system can help. Another alternative is to use a *double-conversion receiver.*

Another type of receiver spurious response occurs with *intermodulation.* Nonlinear electronic components inside a receiver can cause unwanted mixing among external signals. Poor electrical connections

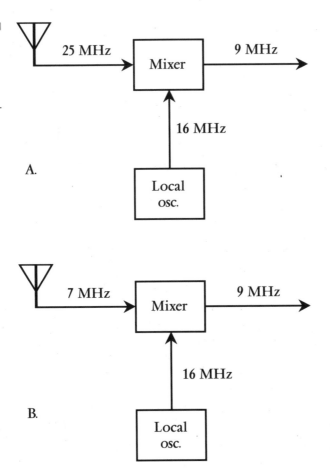

external to the receiver can also cause this problem. In the downtown areas of some large cities, there are so many radio transmitters operating simultaneously that intermodulation is observed in all but the most sophisticated radio receivers. In severe cases this can completely disable a receiver. You might have heard this phenomenon while riding in a taxicab in downtown New York, Philadelphia, or Chicago. The same problem is often encountered by radio amateurs using VHF mobile transceivers.

Chapter 7 contains in-depth information about receiving systems.

Band Congestion

The radio spectrum is a limited resource. Unlike optical fiber and cable systems, which can be manufactured and assembled to obtain ever-increasing amounts of bandwidth as the public need increases, there are only so many megahertz of EM spectrum space in existence that can be effectively used in a given region at a given time. If you have operated an amateur radio station on the 14-MHz band during a weekend, or listened to shortwave radio between 7 and 8 MHz on any evening, you are aware of the sorts of interference that can result when too many stations are crowded into too little spectrum space.

Directional antennas can reduce band-congestion interference to some extent. Even if two stations are on exactly the same frequency, their signals are not likely to arrive from the same direction. At frequencies of about 10 MHz or higher, Yagi and quad antennas are practical and can provide several decibels of unwanted-signal attenuation when they are pointed away from an interfering signal. At VHF and UHF, more attenuation is possible because antennas can be made more directional. At microwave frequencies, dish and helical antennas can be used, and these provide even more directionality.

When there are many stations on a frequency band, the receiver's IF passband should be set to the minimum that will allow comfortable reception. For CW radiotelegraphy this can be as narrow as approximately 100 Hz; for SSB it can be as narrow as about 2 kHz; for AM voice it can be as narrow as about 4 kHz (plus or minus 2 kHz from the carrier frequency); for narrowband FM it can be as narrow as about 6 kHz (plus or minus 3 kHz from the unmodulated-carrier frequency). If an interfering signal happens to be a steady carrier wave or Morse radiotelegraphy signal, an IF or audio *notch filter* can be used to provide upward of 40 dB attenuation at its frequency, while leaving all other frequencies within the passband unaffected. Notch filters in the IF chain work better than audio notch filters in most cases.

Common-Mode Hum

In a direct-conversion radio receiver, the *beat-frequency oscillator* (BFO) can be modulated by AC from the utility mains. The usual cause of this problem, known as *common-mode hum*, is one or more ground loops in the station arrangement. Common-mode hum tends to be an increasing annoyance as the operating frequency increases. It is more likely to take place with an end-fed wire that comes right into the radio room, as opposed to a center-fed antenna located well away from the station.

Common-mode hum can be reduced or avoided by making sure there are no ground loops at the station, and by using an antenna with a properly balanced feed line, locating the radiating part of the antenna at least one-fourth wavelength from the radio equipment. It might also be necessary to install RF chokes in each lead from the power supply to the radio.

Peculiar Interference

In the 1970s, a previously nonexistent signal appeared on the shortwave radio bands. Because of its constant pulse rate and the way the pulses sound, it became known as the *woodpecker.* At times, the woodpecker signal is strong and causes interference to radio communications over a wide band of wavelengths.

The woodpecker is believed to have originated in Russia during the Cold War, and was probably used as an over-the-horizon radar system for the early detection of incoming ballistic missiles. Some engineers have theorized that the woodpecker might be used to modify the ionosphere for the purpose of gaining control over EM propagation conditions at shortwave frequencies. The signal consists of regular pulses at a rate of approximately 10 Hz. The bandwidth is usually several hundred kilohertz. The transmitter power is on the order of megawatts. The center frequency changes, apparently following the *maximum usable frequency* to take advantage of optimum long-distance ionospheric propagation.

A well-designed noise blanker or noise limiter can provide some reduction in woodpecker interference. Switching off the receiver's *automatic gain control,* if that is possible, and using an audio clipper can also help to alleviate the interference. However, because the woodpecker pulses are of longer duration than those of impulse noise, they cause more severe interference, and this interference is more difficult to deal with using noise blankers and limiters. The woodpecker is a phenomenon that might be termed *peculiar interference,* having some of the characteristics of noise and some of the characteristics of an interfering signal.

One method that has been suggested to deal with peculiar interference (such as the woodpecker) is to sample the interfering signal at a certain stage in the IF chain, have a microcomputer isolate and analyze its waveform, replicate the interference waveform, and then introduce the replica at a later point in the receiver's IF chain so the replica exactly cancels out the actual interfering signal. Failing such a sophisticated method, directional antennas can be employed, and if possible, the desired communication or reception can be carried out on a frequency considerably different from the band in which the interference is taking place.

Jamming

All the forms of noise and interference discussed so far in this chapter have been assumed to occur by chance or circumstance. Solar flares, thunderstorms, faulty electrical appliances, and band overcrowding take place not as a result of some sinister design by human beings, but simply because this is an imperfect world. However, interference is sometimes caused by malicious activity. This is known as *jamming*.

In wartime, jamming has traditionally been used as a means of disrupting enemy communications, radar, radiolocation, and radionavigation. If the equipment operates on a specific, known frequency, then a powerful signal can be transmitted on that frequency, modulated in such a way that the interference completely fills the passband of the signal. For example, to wipe out an SSB communications link with a channel 3 kHz wide, an SSB transmitter is modulated with a complex waveform having a maximum frequency of 3 kHz. The suppressed carrier of the interfering transmitter is set to the same frequency as the suppressed carrier of the signal transmitter. To obliterate a CW radiotelegraphy signal, random characters are sent at *zero beat* with (on exactly the same frequency as) the CW signal. The jamming of radar usually requires the use of another radar system; this is a sophisticated art in which false targets (so-called *bogeys*) can be made to appear, and/or real targets can be made to change their apparent location or completely disappear. Wartime jamming can be considered either destructive or constructive, of course, depending on point of view.

Malicious interference occasionally takes place on amateur radio bands and citizens band (CB), although it is probably less common than most radio hams or CB'ers believe. Some people, apparently having amateur radio or CB licenses, get a thrill out of ruining other people's enjoyment of the hobby. The code-practice and bulletin transmissions of W1AW, the amateur radio station operated by the American Radio Relay League (ARRL) in Newington, Connecticut, have occasionally been targeted. A CW signal can be easily wiped out. There might be 500 people listening to an ARRL code-practice session or bulletin at any given time on, say, the 40-m amateur band. A single person might ruin reception for 100 of these people. The alleged "reason" for the interference might be that the person disagrees with ARRL policies or dislikes the idea that W1AW begins transmission on a certain frequency as a matter of course, without yielding to anyone else who is using that frequency. These jammers perhaps forget that without the ARRL, amateur radio would not exist. But such people's mental problems did not evolve according to reason, so it is unrealistic to suppose that trying to reason with such people can change their behavior.

Jamming in peacetime is, of course, not confined to amateur radio or CB. Commercial radio services experience it; sometimes it even occurs on radio and television broadcast bands.

Jamming by civilians is illegal, and the Federal Communications Commission (FCC) has prescribed penalties. But it is difficult to catch jammers. Much can be said in favor of finding and arresting people who cause malicious interference, prosecuting them, and fining them and/or throwing them in jail. But while it is possible to locate sources of malicious interference, it requires a considerable expenditure of time and money. Proving guilt in a court of law is a further expense. Perhaps there is a market for companies who will locate jammers and provide evidence of their guilt that can pass the test of "beyond a reasonable doubt." If proven guilty, a jammer could be required to pay all the costs of the prosecution, in addition to a fine, which would help cover the cost of developing more advanced anti-jamming technologies.

No matter how sophisticated the legal efforts against jamming might become, the problem will never disappear altogether. There are ways to reduce or eliminate interference caused by jamming. These measures can be applied on a case-by-case basis. They include:

1. Changing the communications frequency

2. Using spread-spectrum modes if allowed on the band in question

3. Switching to alternative communications media, for example, the Internet instead of CB radio

4. Using directional antenna systems (a jammer's signal will often arrive from a different azimuth bearing than the signal from the desired station)

Non-RF Considerations

Noise and interference are generally thought of as being associated with the radio spectrum. But these problems also limit the performance of optical, IR, and acoustic wireless systems.

Optical and IR Systems

At first thought, it might seem that any optical or IR system would work best in the dark. This is not necessarily true. Bright daylight can bias a photocell and reduce its sensitivity, but in a well-designed system, this need not be a problem. By narrowing the receiver's field of view sufficiently in a line-of-sight optical or IR communications system, most of the background illumination can be cut out, even in bright sunshine. The exception, of course, is when the desired signal happens to be coming from the same direction as rays from the sun.

A more serious type of optical or IR noise and interference is produced by modulated sources of light or IR. Lightning is one example. Perhaps you have used a modulated-light receiver, similar to the one described in Chap. 8, and heard the noise resulting from lightning flashes. This noise sounds remarkably similar to the sferics heard in a radio receiver. If you are trying to use an *optical cloudbounce* communications system such as the one described in Chap. 8, a thunderstorm can cause as much interference as if you were trying to communicate via radio on an HF amateur band. Another source of modulated visible or IR energy comes from bulbs of all kinds when they are supplied with 60-Hz utility current. The interference from these bulbs sounds like a strong 120-Hz hum (there are two pulses for every 60-Hz AC cycle). Fluorescent bulbs also produce considerable noise at harmonics of 120 Hz. Bulbs produce some output in the IR range as well

Figure 15-9
A telescope can be used to cut down on optical and IR noise and interference.

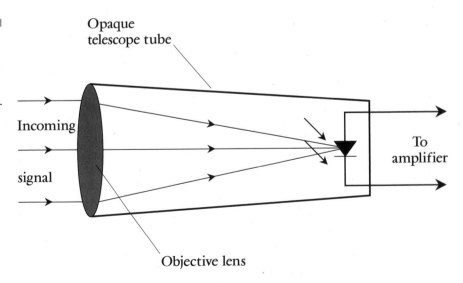

Opaque
telescope tube

Incoming

signal

Objective lens

To
amplifier

as in the visible range; incandescent lamps are the worst offenders in this respect.

Noise and interference caused by lightning or by modulated sources other than the desired signal can be reduced by the same technique as that used to minimize daylight. By cutting down the field of view of the receiver, extraneous modulated energy is reduced while the strength of the desired signal is not affected. This can only be done up to a certain point if optical cloudbounce or *optical atmospheric scatter* modes are used, because the signals in these modes arrive from a diffuse spot or region and not from a point source. With *optical line-of-sight* or *IR line-of-sight* systems, and especially if lasers are used, the receiver's aperture can be narrowed by using an astronomical telescope with the photocell placed at its focus (Fig. 15-9). The focal point will depend somewhat on the wavelength. In the drawing, the telescope tube is shortened for clarity. The field of view can be further narrowed by using an eyepiece having a very short focal length, thus maximizing the magnification. The photocell should be placed outside the eyepiece, where you would normally position your eye. The telescope can be aimed using a finder scope, which is standard equipment with most astronomical telescopes.

Another scheme that can sometimes reduce "bulb hum" is illustrated in Fig. 15-10. This circuit operates like the VLF noise-canceling circuit described earlier in this chapter. The "hum sensor" is situated so it picks

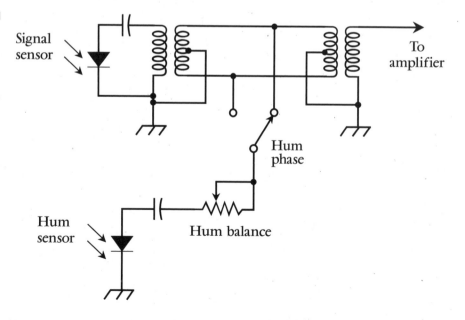

Figure 15-10
A hum-cancellation scheme for modulated-light communications.

up little or none of the desired signal, but plenty of "bulb hum." The switch and the potentiometer are adjusted by trial and error until a null or minimum occurs in the hum. This circuit will not always work, because utility AC generally consists of waves in three phases, with each wave separated by 120 degrees from the other two. Because of this, there will often be components of "bulb hum" that cannot be nulled out by a phase shift of 0 or 180 degrees. Unless all the hum is in the same phase, therefore, the scheme of Fig. 15-10 will be marginally effective at best.

Acoustic Systems

In acoustic wireless systems and devices, noise and interference occur because of physical vibration in the molecules of gases such as air, liquids such as water, and solids such as concrete or steel.

In an acoustic communications system, such as the one described in Chap. 7, noise within the human hearing range has little effect, unless the noise has harmonic content. Some noise results from air movement, and the movements of objects (such as tree leaves) caused by wind. Insects and various other small animals sometimes emit sonic and ultrasonic waves. This type of interference can often be more interesting than the intended signal, especially if the acoustic communications system was built for experimental purposes. You might find yourself connecting the output of the receiver to an oscilloscope rather than a headset, and holding the transducer near crickets, grasshoppers, ants, fruitflies, and lizards. Even more interesting results can be obtained underwater. The sounds emitted by porpoises have been likened to seagulls, squeaking doors, and other disturbances, but some scientists believe the creatures are communicating in a language as sophisticated as—and perhaps more sophisticated than— human speech.

In a computer *speech-recognition* system, acoustic interference can present a major problem. If you are in a quiet room and there are no sudden unexpected sounds, a modern speech-recognition program can produce almost perfect "copy." In fact, some systems get better and better the more a particular person uses them, because the software learns the dialect and accent of the user. Unfortunately, many offices do not present a good environment for the use of a speech-recognition system. Slamming doors, raised voices from an adjacent desk or cubicle, ringing telephones, and even shuffling papers can be interpreted by the system as sounds intended for it to translate. The results are annoying and can sometimes be amusing. In severe cases, extraneous acoustic noise can render a speech-recognition system unusable.

In ultrasonic intrusion-detection or motion-detection systems, acoustic noise rarely causes a problem in itself, although if the noise is loud enough to physically vibrate objects in the vicinity, a burglar alarm can be falsely triggered. These systems typically work by detecting a change in the phase of an acoustic wave having a very short wavelength. Ultrasonic systems can be fooled if an acoustic wave of the same or nearly the same frequency is emitted by some other device.

Sonar, which is used for underwater depth-finding and/or the location of objects, can sometimes be fooled by unexpected obstructions such as fish, sand, or mud in the water, or even an abrupt change in temperature or salinity. False "echoes" can also be produced by deliberately transmitting signals at the same wavelength and pulse frequency as the sonar. This is a form of *acoustic jamming* and can be used by submarines in warfare. It is the sonar analog of radar jamming.

16

Personal
and Hobby
Communications

Wireless technology makes it possible for anyone to be in constant touch with friends, business associates, news media, and the Internet. The term *personal communications* encompasses all these modes, including cellular telecommunications, pagers, beepers, shortwave listening, citizens band (CB) radio, and amateur radio.

Cellular Telecommunications

A *cellular telecommunications* system is a network of specialized *repeaters* that allow mobile/portable radio transceivers to be used as telephone sets. The transceivers can be used alone or in conjunction with portable computers.

Cellular transceivers resemble handheld cordless telephone units, except the range is much greater. In an ideal cellular telecommunications network, all the transceivers are always within range of at least one repeater. The range of coverage for any repeater, also called a *base station*, is known as a *cell*.

If a telephone transceiver is in a fixed location, such as a restaurant or residence, then communication generally takes place via a single base station. If the transceiver is in a moving vehicle such as a car or boat, it goes from cell to cell in the network. This is shown in Fig. 16-1A as seen from some point high above the system. The heavy curve is the path of the vehicle. The base stations (dots) "hand off" access to the mobile transceiver. The cells overlap somewhat, but the system is such that the strongest or clearest repeater is always chosen automatically.

The system is set up in such a way that no two adjacent base stations are on the same frequency. This prevents the base stations from interfering with each other, and keeps conversations from getting crossed. All the base stations are connected to the regional telephone system by wires, microwave links, or fiberoptic cables.

To gain access to a cellular telecommunications network, you must purchase or rent a transceiver, and pay a fee. The fees vary, depending on the location and the amount of time per month you use the service. When using such a system, it is important to keep in mind that the conversations are not necessarily private. It is easier for unauthorized people to eavesdrop on electromagnetic (EM) communications than to intercept wire or cable communications.

A computer can be hooked up to the telephone lines for use with online networks. For many people, getting on the Internet (including the World Wide Web) and also using *electronic mail* (e-mail) are the main

Figure 16-1
Basic cellular
telecommunications
scheme (A), and the
connection of a
laptop computer to
a cellular telephone
transceiver (B).

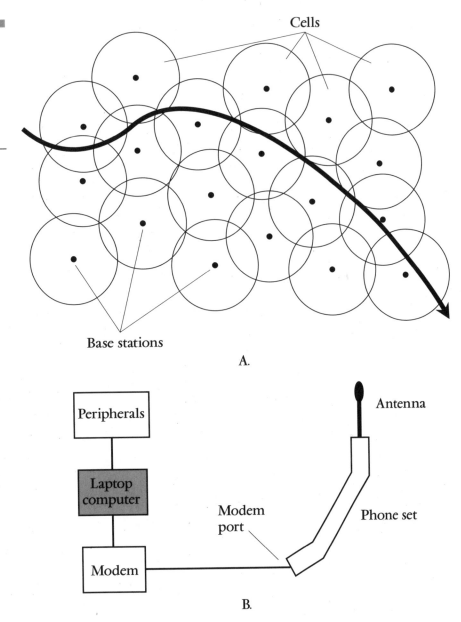

A.

B.

motivations for buying a computer. You can get online from your car, boat, or portable telephone, as well as from your house, using cellular telecommunications.

You can connect a laptop (notebook) computer to a cellular phone set with a portable modem. This will let you get online from anywhere within range of a cellular base station. A block diagram of this scheme is shown in Fig. 16-1B. Modems are built into some mobile and portable telephone units. Virtually all laptop computers have provision for easy installation of a modem. The modem comes in the form of a so-called *PC card* that slides into a slot on the side of the computer. Some laptop computers have modems built in.

Many commercial aircraft have telephones at each row of seats, complete with jacks into which you can plug a modem. If you plan to get online from an aircraft, it is important that you use the phones provided by the airline, and not your own cellular telephone units, because the latter can cause interference to flight communications and navigation equipment. You should also be sure that the airline allows you to use portable computers while the aircraft is in flight.

Cell phones have proved themselves useful in emergency situations. An increasing number of cellular systems employ location-finding equipment, so when a call is placed to the emergency 911 number, the location of the cell phone can be determined to within a few hundred feet.

Suppose you are driving in a heavy snowstorm. It is dusk, and darkness comes quickly. Visibility drops to near zero. Snow blows across the road; the wind speed increases and the temperature drops. Finally you wind up in the ditch, having lost sight of the road. It's 7:00 P.M. on a Tuesday. You know you're a couple of hours north of Minneapolis, but you didn't see any mile marker signs along the highway. The temperature is near zero Fahrenheit and the snow is beginning to drift. You have one-fourth of a tank of gas. There are no other vehicles within sight. You have not seen another vehicle for several minutes. This is a life-threatening situation. But if you have a cell phone, you can dial 911 and request help. If you are in a region where location-finding equipment has been deployed, rescue personnel will immediately know where you are. Your additional description of the situation (the highway number, town you are near, type of car you are driving, etc.) can be of help, but rescue personnel would probably be able to find you even if you could do nothing more than say, "Help! I'm stranded!"

Cell phones can also be of use in various other situations that might not be emergencies but can nevertheless be inconvenient without communications. One example is having your car break down. Even if you're in a city, a cell phone can be used to call a tow truck. This saves you having to walk in an unfamiliar neighborhood to a convenience store or telephone booth.

The main limitation of any wireless telecommunications, including cellular, is a relative lack of privacy. Aspects of this are discussed in the next section.

The Wireless Local Loop

A system similar to cellular telecommunications makes use of wireless base stations in certain areas, with a specified range of coverage for each base. This is called *wireless local loop* (WLL). Instead of running copper wires to each and every subscriber in the network, wires run only as far as the base stations. Individual telephone sets and data terminals are linked into the system by radio transceivers. An illustration of this scheme is shown in Fig. 16-2. Heavy lines represent wire connections; thin lines represent wireless links. The ellipses and circles represent the individual subscribers in the system.

There are some differences between WLL and cellular systems. First, the WLL network need not provide continuous coverage over vast areas, as must a cellular system. A given base station need only provide coverage to subscribers within the range specified by the local exchange. Second, while a cellular system must be capable of handling massive amounts of communications traffic during certain times of the day or week (for example, weekdays from 8:00 to 10:00 A.M. and 4:00 to 6:00 P.M.), a WLL base station can be designed to serve a certain maximum number of subscribers without fear of overload in case of some unusual occurrence such as a traffic accident on the freeway during rush hour. Third, while the majority of users in a cellular network are mobile, most (or all) of the subscribers in a WLL system are at fixed locations. The antenna for each subscriber can be set up in an optimal position (found by trial and error) and then left in that position, so there will be little variation in signal quality. A directional antenna, such as a Yagi or dish, can be used at fringe locations where antenna gain can provide an advantage.

The individual subscriber to a WLL can use a telephone set similar to an ordinary home phone. Several telephone sets might be connected together by a system of wires confined to the house or business. The radio transceiver can be placed on an upper floor or in the attic, and the antenna can be located on the roof. Some or all of the telephone sets might be replaced by personal computers or data terminals. In the immediate vicinity of the base station, handheld telephone transceivers, similar to cell phones, might be used. The range of the base station

Figure 16-2
Wireless local loop
telephone systems.
Heavy lines are wire
links; thin lines are
wireless links. Circles
are telephone sets;
ellipses are data
terminals.

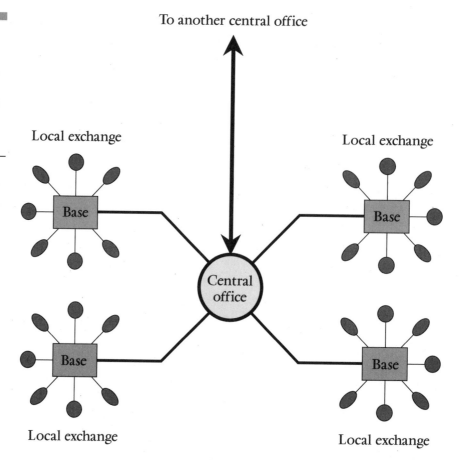

To another central office

Local exchange

Local exchange

Base

Base

Central office

Base

Base

Local exchange

Local exchange

would be much more limited for such sets, because the tiny antennas and low transmitter power reduce their effectiveness compared with full-sized, fixed sets.

One of the biggest challenges to WLL systems involves security. A radio link is much easier to "tap" than a copper wire. *Encryption* of signals can be used, but this is costly and there has been controversy over the legality of *strong encryption* (which is needed to fully ensure privacy). A typical WLL system is no more secure than a cellular system. Perhaps the best policy is not to say anything, or send or receive any data, that could cause harm to anyone should the information be intercepted and the information be misused. You would certainly not want to order merchandise using a credit card, for example. This sort of vigilance is easier to talk about than to maintain, however. One slip of the tongue or one accidentally downloaded sensitive piece of e-mail could cause trouble. As technology advances and governments become more receptive to the demands of businesses and individual citizens for communications security, this potential risk might diminish, but it will never go away entirely.

Another security issue involves the potential for unauthorized use of the system by nonpaying people. While it is possible to register each subscriber and restrict access to a given base station to the set of paid and registered subscribers, it is possible for an intelligent hacker to "clone" a registration and gain entry to the system at the expense of a paid subscriber. Technological advances should gradually reduce this threat, but in the meantime, plenty of lawyers are ready and willing to make money from this type of security breach.

Another possible limitation of WLL systems is the fact that radio links are more susceptible to noise and interference than wire links. *Electromagnetic interference* (EMI) from appliances such as television (TV) sets and personal computers might occasionally cause problems. Amateur or CB radio equipment might interfere with system performance under some circumstances. Heavy local thunderstorms will produce some noise in the form of sferics. Lightning can also be a direct physical hazard to people and equipment when outdoor antennas are used.

Wireless Pagers

Pagers, also called *beepers,* are as common as cellular telephones. Millions of pagers are in use in the United States alone. They are extensively used for business and personal applications.

Simple System

The simplest type of paging system employs a small, battery-powered radio receiver that picks up signals in the very-high-frequency (VHF) or ultra-high-frequency (UHF) portion of the radio spectrum. Transmitters are located in various places throughout a city, county, or telephone area code district. Ideally, there are enough transmitters, with well-placed antennas and sufficient output power, to ensure coverage over the entire region without dead zones. Several frequency bands are assigned for use by pagers; they differ from one country to another. The receiver picks up a simple encoded signal that causes the unit to display a series of numerals and/or letters of the alphabet.

To send a signal to a pager, you dial a certain telephone number. Any telephone set can be used for this; it can be hard-wired or cellular. After a "ready" signal, usually a series of beeps, you can enter any combination of characters up to a certain maximum. Usually, this is your local phone number. The pager receiver then emits a loud series of high-pitched tones, and a liquid-crystal display (LCD) shows the same string of numbers that you dialed into your telephone. The party being paged then knows which number to call in response to the signal.

Some people have various numeral and/or letter combinations that they have agreed upon in advance to convey specific messages. For example, "252-1700" might mean "Meet me at Joe's at 5:00 P.M." Or, "777-0830" might mean "I'll be in the office at 8:30 A.M." This eliminates the need for the paged party to call back merely to receive a simple, one-sentence message.

Advanced Features

Obviously, codes such as the fictitious ones denoted above have limitations. For one thing, they're inconvenient. They're easy to forget. Perhaps their sole advantage is that they can offer some measure of privacy. Organized criminal groups, such as drug dealers, took advantage of this almost from the very day the pager was invented. For the ordinary citizen and business person, however, more advanced pager features can make a big difference in ease of use.

A pager equipped with *voice mail* allows the sender to leave a brief spoken message following the beep. Then, rather than punching in the digits "252-1700," the sender might say "Meet me at Joe's at 5:00 P.M." Any number of other short messages can be enunciated, obviously a much more powerful system than the "secret code" scheme. A small speaker in the pager receiver allows the listener to hear the voice. One problem with this scheme is that the voice often comes out with less than optimal fidelity, because the speaker is physically tiny. Also, the power required to generate the audio consumes current and shortens the life of the battery in the pager receiver.

Pagers and cellular telephones naturally go together. Suppose you're driving on the freeway and your pager beeps. A voice mail message says, "This is Fred. Call me as soon as possible." You probably won't want to pull off the freeway at the next exit, hunt for a pay phone at a convenience store, and then call Fred while the cashier looks suspiciously at you. You'd rather call him right from your car. This is the height of wireless convenience. However, you might still encounter problems if the driving is difficult, because operating a cellular telephone increases the risk of getting into a traffic accident.

Diminishing Returns

In recent years, younger and younger people have been using pagers and cellular telephone systems, partly because the hardware has become more affordable. These devices also serve as a status symbol. They can make a young person think that he or she looks important. But often, the unnecessary use of wireless devices is merely a curiosity, especially to the older generation who, despite the lack of such advanced systems in their time, managed to survive and prosper.

Have you ever been at the beach, or riding along a bike trail, and seen someone talking on a cellular telephone? Didn't it seem a little bit silly? Have you ever been in a movie theater or at a funeral and heard someone's cellular telephone ring, or heard someone's pager go off? Didn't this strike you as rude? There are times and places where wireless paging and telephone devices serve useful purposes, and once in awhile such places include beaches, bike trails, movie theaters, and churches. But pagers, in particular, are carried around more often than necessary.

Wireless E-mail

Electronic mail is a convenient alternative to conventional paging. A pager equipped to receive e-mail has a larger LCD than a conventional pager receiver. The unit resembles a handheld computer or small calculator. When the unit emits its characteristic sequence of beeps, the user can pull the unit out of a pocket or unclip it from the belt, and look at the screen to scroll through and read messages. Messages can then be stored for later retrieval or transfer to a laptop or desktop computer. An advantage of e-mail over voice communications is that it provides written data that can be referenced at a later time. Another asset is the fact that the LCD, and storage of the messages in memory, requires almost no battery power.

Transceivers

An increasing number of portable e-mail pagers can send messages as well as receive them. This is done via a system similar to a cellular telephone network. The pager contains a small radio transmitter with an attachable antenna. Coverage is reasonably good in areas well served but nonexistent in other regions. "Dead spots" often exist with such systems. For example, you might be able to access and send e-mail from one side of a motel room but not from the other side. Wireless e-mail transmitters are generally forbidden for use on aircraft in flight, as are cellular telephones, because the signals can interfere with aircraft navigation and communications equipment.

Some traveling businesspeople find wireless e-mail to be a great boon. This is especially true in cities that have well-planned wireless communications networks. When visiting a client, the salesperson can simply pull the e-mail transceiver from a coat pocket or briefcase and be connected within seconds to the main office. No longer is it necessary to carry a laptop computer, a modem, and all the cables and cords necessary to connect it to a power source and a hard-wired telephone line. New wireless networks are constantly being developed with traveling salespeople specifically in mind.

A potential inconvenience (as well as a convenience) inherent in e-mail is the fact that the user must look at, and read, the data on the screen. This takes some time, because the message consists of more than a simple, instantly recognizable sequence of characters. The composition of e-mail requires the use of a keyboard. Wireless e-mail transceivers are known for having keyboards too small to use in the same way a computer keyboard is operated. More often than not, you'll end up poking at the buttons with the point of a pen or pencil.

Caution!

Reading and composing e-mail is dangerous while driving a motor vehicle, operating some kinds of machinery, or even walking down the street. People are aware of the risk in trying to use wireless e-mail transceivers while doing other things, but when a message arrives, it is human nature to read it and perhaps respond, no matter where the user is or what he or she is doing. Such impatience can lead to injury or death. If you must read or compose wireless e-mail messages in a car or truck, pull off the road or exit the freeway, find a safe and legal place to stop, and do your work while the vehicle is not in motion.

Wireless Fax

Fax is the common expression for *facsimile,* a method of sending and receiving nonmoving images over telephone lines or radio. Fax was originally developed in the early 1900s, years before the first experimental TV signals were transmitted. Today, fax machines are almost as commonplace as telephone sets. This communications mode is used extensively by businesses, government agencies, and individuals.

How It Works

For the past few decades, sending and receiving of fax messages has been getting progressively cheaper, while mailing letters has been getting more expensive. For overseas mail, or when you need to get a document someplace right away, fax is now less expensive than any mail carrier. Fax is more reliable than the mail. It is basically impossible for a fax to get lost. You always know that a fax has reached its intended recipient, because you can have them acknowledge receipt by sending you a return fax. If you receive no return fax, you can send your fax again. Fax is also faster than regular mail. It gets to its destination with the speed of a phone call. In today's fast-paced society, that can be critical. The saving of two days here and two days there, multiplied many times over, translates into a big profit increase. In business, time is money.

A fax machine has two parts: an *image transmitter* and an *image receiver.* Figure 16-3 shows the transmission of a wireless fax signal from one place, called the *source,* to someplace else, called the *destination.* This type of arrangement has been in use by radio amateurs since World War II. In Fig. 16-3 only the sending part of the fax machine is shown at the source, and only the receiving part is shown at the destination. A complete fax installation generally has both a transmitter and a receiver, allowing for two-way exchange of fax messages.

To send a fax, you place a page of printed material in an *optical scanner.* This device converts the image into a series of binary digital pulses, that is, high and low signals (also called 1 and 0). Any image can be

Figure 16-3
Transmission of a
wireless fax signal.

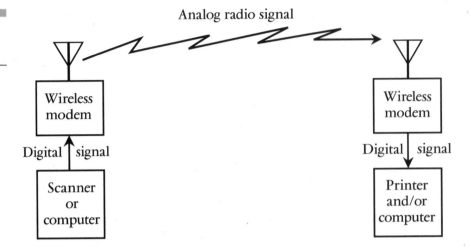

converted in this way, although generally the output will be in shades of gray. (Color fax is available, but it costs more than standard fax.) The output of the scanner is sent to a *modem* that converts the binary digital pulses into a signal suitable for transmission over a telephone line or, as shown in the drawing, via a wireless link. In some cases the link is partly via wireless and partly via the conventional telephone circuits. An example is the use of a fax machine with a cellular telephone unit.

At the destination, the analog signals from the telephone line or radio are converted back into digital pulses like those produced by the optical scanner at the source. These pulses are then routed either to a printing device or to a computer or terminal. A standard fax has a printing device similar to a photocopying machine. The *image resolution* (level of detail) of high-quality fax is good enough to allow reproduction of gray-scale photographs. Fax has been used by the news networks for many years for the transmission of such photographs, which are known as *wirephotos.*

A computer can function as a paperless fax machine. You can receive faxes and display them on the screen. This is especially popular among radio amateurs, who receive the fax images from weather satellites, Space Shuttles, and other radio "hams." A personal computer can convert its data into fax format and send this over the phone lines or radio. To do this, you need a *fax/modem.* Anything on your screen consists of digital data, which can be transmitted via fax/modem. New computers have fax/modems built in. External fax/modems are available for use with older computers.

You cannot simply take a piece of paper and send its images over computer fax. You need a separate, "outboard" optical scanner to do that. (Conventional fax machines have the scanner built in.) If you intend to send hard copy via fax, a standard fax machine is the cheapest way to go. If you plan to use a computer scanner for other applications, such as digitizing photographs or printed documents, it can be a good investment.

Wireless Fax/Modems

A *wireless fax/modem* is an expansion board or card that can be installed in a personal computer. Using a wireless fax/modem, it is possible to send anything on a computer display via wireless link to some other computer or fax machine. With low-end units, color will generally not be reproduced; the image will be transmitted in gray scale whether it consists of pure text, pure graphics, or a combination of text and graphics. It is impossible to fax hard copies (this page, for example) using a wireless fax/modem. This is the main limitation of these devices. To send faxes of hard copy, you must add an optical scanner to your computer system.

You can receive a fax and store it using a personal computer. The data can be stored on a hard disk or on a diskette. This is a big asset of a fax board or fax/modem. It is convenient to keep images and text on disk, because this practice minimizes the amount of paperwork you have to file or leave lying around. But fax images consume considerable disk storage space. One full-screen bitmap occupies several hundred kilobytes. Text faxes, however, can be converted into ASCII format using *optical character recognition* software, greatly reducing the storage space required.

If you want to make a hard copy of a fax that you have received using a fax board or fax/modem, you can usually print the data using a conventional computer printer. The quality of the image will vary depending on the kind of printer you use. Laser printers produce the best fax hard copies. Inkjet printers are quite good also. The newest high-resolution laser printers can produce hard copies as good as, and sometimes better than, those from a dedicated fax machine.

Amateur-Radio Fax

It is possible to send fax images by means of narrowband amateur-radio transmissions, in a manner very much like the way it is done over telephone channels. Amateur fax transmissions have image resolution comparable to that of any commercial transmission. The only problems occur because the image can be degraded because of fading, noise, or interference from other stations. These problems are most likely on the high-frequency (HF) amateur bands, such as 3.5, 7, or 14 megahertz (MHz). They are less severe or frequent at VHF and UHF.

An amateur-radio fax transmission requires several minutes to send. The longer the time for sending the image, the better the resolution, in general. The image is sent in lines, similar to the way in which a television raster is scanned. In amateur work, 120 or 240 lines per minute (two or four lines per second) are scanned.

Amplitude modulation (AM) is generally used, and it can be either of two general types. In *positive modulation,* the instantaneous signal level is directly proportional to the brightness at any given point on the image. In *negative modulation,* the instantaneous signal level is inversely proportional to the brightness at any given point on the image.

Radio amateurs can use surplus commercial fax apparatus to build their fax stations. Often the equipment must be modified. This is a challenge to the hobbyist who likes to tinker with hardware. Most commercial units are designed to work with AM, but they can be converted to work with *frequency modulation* (FM). Some radio amateurs like to build all of their fax equipment from scratch.

In recent years, several manufacturers have come out with interface units that allow personal computers to be used to receive, and also to send, amateur wireless fax transmissions. The images can be displayed on the monitor screen with excellent resolution. It is even possible to use these interfaces for color fax. There are numerous functions that enhance the versatility of the system, such as a zoom feature for viewing a portion of the image close-up.

One especially popular use of amateur-radio fax is in the receipt and retransmission of weather-satellite data. The satellites send pictures to the surface via high-resolution fax. Radio "hams" then send the images to each other. Two-way amateur-radio contacts are sometimes carried out on fax; this is somewhat like *slow-scan television* (SSTV), except the pictures take longer to send, and they show more detail.

Hardware Miniaturization

Most personal communications devices are meant to be portable, and this means they must be small and lightweight. Few people would like the idea of a cellular telephone set or pager as large and heavy as a TV set, even a portable one. Perhaps you remember the brief heyday of the so-called *luggable computer*, which was about the size and mass of a brief-case full of books. It was difficult to carry around and practically impossible to use on an aircraft, train, or bus. Some people used lug-gables simply because some form of portable computer was essential to their work. When *laptop computers*, also called *notebook computers*, became widely available, the luggable lost its market.

Hardware miniaturization involves making the size and weight of portable communications devices as small as possible, to a point. But a device must have at least a certain size and mass to present an acceptable user interface. Some cellular telephones can be folded up so they are hardly any larger than a credit card. But when they are in use, they are about the size of an unfolded wallet. The earpiece and mouthpiece must be separated by a physical distance not less than the distance between the user's ear and mouth; in an adult this is 5 to 7 inches. The antenna must be at least a certain minimum length if it is to provide reception and transmission of signals over the necessary distance. The keypad must be large enough so the keys can be actuated individually, without difficulty, by adult fingertips.

In portable electronic equipment of all kinds, the component that has received the most attention in recent years, in regard to miniaturization, is the *battery*. An entire computer can be fabricated onto a single inte-grated circuit (IC), and displays and keypads have reached their smallest practical size. New battery technologies are constantly being tested. The batteries in most portable wireless devices are rechargeable. This saves money in the long term, and it is also more convenient than the use of batteries that must be replaced every few hours. Lithium-ion batteries offer good stored-energy-to-volume ratio, and are popular in portable wireless devices. Nickel-metal-hydride batteries are also employed. In some devices, solar batteries can supplement the power pack, and can even be used in place of electrochemical batteries in direct sunlight. The main problem with solar batteries is the fact that they are physically large. Also, if they are to generate their maximum rated power output, they must be positioned so sunlight will shine directly on them.

Physical size is not the only consideration in hardware miniaturiza-tion. Mass and density are important as well. In general, devices become less heavy as they get smaller, but not always in the same proportion. Assuming the density (mass per unit volume) remains constant, cutting each dimension (height, width, depth) of a device in half will reduce the volume, and therefore the weight, to one-eighth of its former value. But density can vary. Generally it tends to increase somewhat as a device is made smaller; this is mainly because engineers make every effort to utilize every cubic millimeter of available volume within the device, so air takes up as little space as possible. Some devices have been built that have density so low that they will float on water. This low density must be achieved by sacrificing some volume miniaturization, but in some instances the tradeoff can be worth it in terms of sales appeal. Consider, for example, the appeal to boating enthusiasts of a cellular telephone receiver or pager that will float if it is accidentally dropped. The device would of course have to be water-resistant, as well has having a specific gravity less than 1.

Despite problems with superminiaturization, there will be demand for some wireless devices that can be worn on the wrist, kept in a pocket, inserted in the ear like a hearing aid, or surgically implanted in the body. Consider a hypothetical *wrist communicator*, a combination pager and cell phone that can be worn on the wrist. A usable alphanumeric computer keyboard would be impossible to fit onto such a device, and a telephone keypad would be very difficult to manipulate. The user would have to use the point of a sharp pencil or pen to actuate the keys. The display would be limited to a few letters, numbers, and punctuation marks. Perhaps *speech recognition* and *speech synthesis* could provide a user-friendly interface for a superminiature device. But there are some kinds of data that are best sent and received in written form. E-mail, for example, would be nearly impossible to send and receive using a screen and keypad less than 1 inch square.

One of the less popular (at least among their wearers) forms of superminiature wireless personal communications device is used by law enforcement. The device is a simple transmitter or transponder similar to those attached to wildlife to monitor their whereabouts. If a person is confined to house arrest, for example, except during business hours, compliance can be checked from a central station. A possible violation causes the system to notify the police, who can come to the user's house and check to be sure he or she is there. No user interface is necessary for this device, so it can be extremely small.

Shortwave Listening

The HF radio band, ranging in frequency from 3 to 30 MHz [wavelengths from 100 down to 10 meters (m)], is sometimes called the *short-*

wave radio band. The EM waves are actually longer than those employed in most radio communications, but they are short compared to the wavelengths that were commonly used when the term was coined.

In the early twentieth century, practically all communication and broadcasting was done at frequencies below 1.5 MHz (wavelengths of more than 200 m). It was thought that the higher frequencies were useless and that energy at such frequencies could not propagate for more than a few miles. The vast, supposedly worthless region of the radio spectrum consisting of wavelengths "200 meters and down" (frequencies of 1.5 MHz and above) was given to amateur radio operators. In fact, amateurs were restricted to these frequencies. But the amateurs discovered their frequencies were capable of supporting long-distance communications. So-called shortwave transmissions often produced superior results compared with transmissions at longer wavelengths, allowing reliable contacts spanning thousands of miles using transmitters with only a few watts of radio-frequency (RF) output power. Soon, commercial entities and governments took keen interest in the short waves, and amateurs lost most of the HF band.

The high frequencies are still used today, mainly for international broadcasting and, ironically, amateur radio operation. To a large extent, especially in the developed countries, the great economic powers have moved their operations to wavelengths shorter still, where spectrum space is more abundant and communication is not affected by sunspots and the earth's ionosphere.

An HF radio communications receiver, especially one that offers continuous coverage of the HF spectrum, is sometimes called a *shortwave receiver.* Most general-coverage shortwave receivers function at all frequencies from 1.5 through 30 MHz. Some also work in the standard AM broadcast band at 535 kilohertz (kHz) to 1.605 MHz. A few shortwave receivers can function below 535 kHz in the so-called *longwave radio band.*

Anyone can build or obtain a shortwave or general-coverage receiver, install a modest outdoor wire antenna, and listen to signals from all around the world. This hobby is called *shortwave listening* (SWLing). Millions of people in the world enjoy it. In the United States, the proliferation of computers and online communications has, to some extent, overshadowed SWLing, and many young people grow up today ignorant of a realm of broadcasting and communications that still dominates in much of the world.

There are various commercially manufactured shortwave receivers on the market today, some for quite low prices. A simple end-fed wire receiving antenna costs next to nothing. Most electronics stores carry one or more models of shortwave receiver, along with antenna equipment, for a complete installation. One problem with such receivers is that they sometimes lack the mode flexibility, selectivity, and sensitivity necessary to engage in serious SWLing. If you are interested in this hobby and want to obtain high-end equipment, shop around in the consumer electronics and amateur radio magazines. Most electronics stores and book stores carry periodicals and books for the beginner as well as the experienced SWLer. A library can also be a good source of information, especially if you are interested in older shortwave receivers, some of which can still be found at amateur-radio conventions and flea markets.

A shortwave listener in the United States need not obtain a license to receive signals. In general, a license is required if shortwave transmission is contemplated. Shortwave listeners often get interested enough in communications to obtain amateur radio licenses, so they can engage in two-way communications with other radio amateurs throughout the world.

Citizens Radio Service

The *Citizens Radio Service,* also known as *citizens band,* is a radio communications/control service. The most familiar form, called *Class D Citizens Band,* operates on 40 discrete channels near 27 MHz (11 m) in the HF radio spectrum. Other classes include *Class A Citizens Band,* used by the General Mobile Radio Service, and *Class C Citizens Band,* used for radio remote-control purposes.

History

The frequencies near 27 MHz were originally selected for CB allocation by the Federal Communications Commission (FCC) because, at the time CB was first conceived, it was too expensive to mass-produce two-way equipment for use at higher frequencies. Normally, the maximum communications range at 11 m is 10 to 20 miles. However, at times of peak sunspot activity, ionospheric *F-layer propagation* (sometimes called *skip*) makes worldwide communication possible on this band.

The original 11-m band had 23 channels, and AM was used. The maximum power limit was 4 watts (W). Unlicensed operation was allowed if the power was 100 milliwatts (0.1 W) or less. Because of the explosion in

CB use during the 1970s, the number of channels was increased to 40, and *single-sideband* (SSB) was introduced to conserve spectrum space and improve communications reliability. The peak-envelope power (PEP) limit was set at 12 W for SSB.

The Citizens Radio Service has been periodically plagued by so-called *CB bootleggers*, who use illegal amplifiers and/or operate outside the limits of the 11-m band. This activity reaches its peak during times of maximum sunspot activity, when long-distance propagation becomes commonplace at 27 MHz. Some CB operators, legitimate or otherwise, obtain amateur radio licenses to explore the wider horizons offered by that service.

Uses

In emergencies, CB radio can save lives. Motorists, hikers, boaters, and small-aircraft pilots use these radios to call for help when stranded. Small CB radio sets, with reduced-size magnetic-mount antennas, are available for motorists, intended expressly for use in emergencies. Full-sized quarter-wave mobile antennas provide superior range and are the optimum choice for people who want CB for mobile emergency preparedness.

Some CB operators enjoy the recreational aspect of radio communication. This is especially true for children, who use low-power "walkie-talkies" to communicate over distances of up to a mile or two. Some people use CB radio to communicate between one fixed (base) station and one or more mobile stations, as a convenience.

During the 1970s, CB radio became immensely popular among truckers. The long, lonely hours of highway driving are more endurable when two-way radios allow communication with people in various parts of the country. The "CB boom" took place largely as a result of CB activity among truckers. They developed a slang, in addition to the *10 code* that was already in use among CB-ers, that became the trademark of a popular subculture.

REACT and GMRS

The acronym REACT stands for *Radio Emergency Associated Communications Teams,* a worldwide organization of radio communications operators. REACT operators commonly provide assistance to authorities in areas devastated by earthquakes, hurricanes, floods, and other disasters. Messages are relayed into and out of affected regions. REACT operators provide assistance to motorists and truckers on highways and freeways, calling for help to aid stranded people, and advising of traffic conditions and problems. REACT operators work with amateur radio operators to provide communications in emergencies. On the Class D band, the emergency channel at 27.065 MHz (channel 9) is monitored by REACT operators.

The *General Mobile Radio Service* (GMRS) operates at frequencies between 460 and 470 MHz using Class A Citizens Band. In Class A, transmitters are allowed to operate with up to 50 W output. There are 15 channels; FM and some digital communications modes are used. Because the Class A band lies in the UHF part of the radio spectrum, ionospheric propagation does not occur. The maximum communications range is 20 to 40 miles. Some tropospheric propagation occasionally takes place, resulting in communications over distances of hundreds of miles, but most communication beyond 40 miles is accomplished with the help of *repeaters.* Numerous GMRS repeaters are operated by REACT at 462.675 MHz. Class A operators must be licensed by the FCC.

Equipment

Class D citizens band radio equipment is not expensive. The hardware you'll need depends on whether you plan to install a fixed station or a mobile station.

A 40-channel, 12-W SSB transceiver, designed for desktop use, is the basic radio for fixed-station CB operation. These radios are entirely self-contained and are available in various designs to suit different tastes. The antenna system normally uses a coaxial-cable feed line and a quarter-wave or 5/8-wave ground-plane radiator cut for 27 MHz. Vertical polarization is standard because this facilitates communication with mobile stations. The antenna should never be installed where it might fall or blow onto utility power lines or where such lines might fall or blow onto the antenna. The antenna should be disconnected and grounded when it is not in use. The radio should never be used during thunder-showers. If you have any concerns about the safety of installing or using a CB antenna or equipment, consult a professional engineer.

Mobile transceivers typically run 12 W SSB and operate from the vehicle battery. Connection should be made directly to the battery. Small, inductively loaded, magnetic-mount whip antennas can be used, but the most effective antennas are a quarter wavelength high (about 9 feet). If you have concerns about installing a CB radio in your car or truck, consult a professional. Some automotive sound system dealers will install CB radios for a nominal fee. As with base stations, mobile CB stations should never be operated during thunderstorms. Caution should be exercised to be sure that the antenna does not come into contact with low-hanging power lines, tree limbs, or other obstructions. Ignition noise, alternator noise, and other forms of interference are common with mobile two-way radio installations. For information on how to deal with these problems, consult an electronics engineer or automotive electronics dealer. Some amateur radio publications, notably *The ARRL Handbook for Radio Amateurs* (published annually by the American Radio Relay League, Inc., Newington, Connecticut), contain information on mobile radio installation and operation.

Amateur Radio

In most countries of the world, people can obtain government-issued licenses to send and receive messages via radio for nonprofessional purposes. In America, this hobby is called *amateur radio* or *ham radio*. There are hundreds of thousands of radio "hams" in the United States.

Who Uses It?

Anyone can use ham radio. Radio hams communicate by talking, sending Morse code, or typing on computer terminals. This last method, typing the text on a computer, is similar to using the Internet. In fact, hams can set up their own radio networks, and some amateur radio clubs have limited wireless Internet hookups.

Some radio hams chat about anything they can think of (except business matters, which are illegal to discuss via ham radio). Others like to practice their emergency-communications skills, so they can be of public service during crises such as hurricanes, earthquakes, or floods. Still others like to go out into the wilderness and talk to people thousands of miles away while sitting out under the stars. Hams talk on the radio from cars, boats, and even bicycles.

Amateur radio has special appeal to electronics and communications experimenters. Many new discoveries in wireless communications have been made by radio hams who were tinkering with hardware, building prototype circuits even when the conventional wisdom seemed to say they could never work, and not necessarily thinking about making a financial profit. Some of these ideas eventually found their way into general public use. While some radio amateurs have made money from their inventions, patenting some devices and/or selling their ideas to industry, most consider their experimentation a hobby. The experimenter often wants only to see if something will work, and money is not an issue. This creates a "think tank" atmosphere that is a proven breeding ground for technological breakthroughs.

Licensing

You need to get a license to transmit on the amateur frequencies. The transmission of radio signals without a license can result in fines and/or imprisonment. There are several levels, or classes, of ham radio licenses available in the United States, all of which are issued by the FCC.

Amateur radio is an electronics-oriented hobby, as is personal computing. Radio hams are more likely to own computers than are non-hams. Computer users are more likely to be interested in amateur radio than people who avoid computers. If you use a computer very much, and especially if you're interested in hardware, you should not have trouble obtaining an amateur-radio license and getting on the air.

Fixed-Station Equipment

The components of a basic ham radio station are shown in Fig. 16-4. The personal computer (PC) can be used to network via packet radio with other hams who own computers. The station can be equipped for online telephone (landline) services. The PC can control the antennas for the station and can keep a log of all stations that have been contacted. Most modern transceivers can be operated by computer, either locally or by remote control over the radio or landline.

Safety   The most important consideration is operator safety. There should be minimal risk of electric shock from high-voltage wires. Station wiring must always conform to good electrical standards. If in doubt, a professional electrician should be consulted. Any ham radio station should be able to pass a fire inspection. Antenna feed lines should

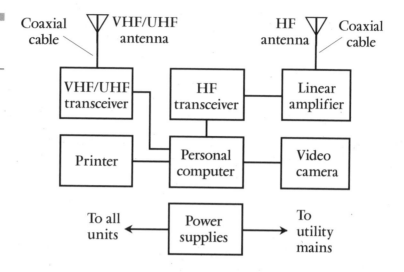

Figure 16-4
A simple amateur radio station.

be disconnected from the radio equipment, and connected to a ground rod, ideally outside of the house or apartment building, whenever the station is not in use. This minimizes the danger of equipment damage or personal injury from lightning. Lightning arresters should not be relied on to give effective protection.

Grounding   For good station performance, you should always try to get the best possible ground system. This is important not only for safety reasons but to minimize EMI. In recent years, the proliferation of personal computers and other sophisticated, solid-state devices and accessories has made it much more likely that different pieces of equipment in an amateur radio station will interfere with one another. All equipment should be grounded to a common bus that is connected to an earth ground. The best *electrical ground* is a long rod driven into highly conductive soil. This might not, however, be a good *RF ground*. A ground plane, or a system of radials, might be needed in addition to the electrical ground. Coaxial cable, or "coax," is best for equipment interconnection. The cables should have good *shield continuity*. A high-quality transmission-line type of coax, such as RG-58/U, is sufficient for most interconnection purposes. When multiple conductors are used, the manufacturer usually supplies the appropriate type of cable.

Heart of the Station   The operating table or desk should be sturdy. You don't want it to collapse under the weight of the station equipment. The combined mass of the radios and accessories can easily exceed 100 pounds, and in large stations, might be 200 pounds or more. The desk should be at least 28 inches deep, and preferably 30 to 36 inches. This allows plenty of room for your arms and for papers such as logs and data sheets. Some operators place the equipment on a shelf elevated 6 to 10 inches above the main desk surface. A good illumination scheme should be employed. Some hams like indirect lighting; others prefer direct, close-in illumination. The main electronic component of a ham station is, of course, the "rig" or radio. A station might have several rigs. These are usually transceivers, although separate transmitters and receivers are sometimes used. This unit, or these units, should be placed at the most accessible position(s) in the station. Each piece of equipment should have enough space around it to allow for proper ventilation.

Linear Amplifiers   If you have one or more linear amplifiers, it/they should be placed where the controls can be reached easily. Many linear amplifiers have "outboard" power supplies. These can be placed out of sight and out of the way of the operating position, such as on the floor under the operating table. Linear amplifiers have large current requirements. The electric circuit must be adequate to provide this current. Most bedroom-type circuits supply 15 amperes (A) at 117 volts (V). This is not sufficient to power a full-legal-limit amplifier. Ideally, 234-V circuits, such as those intended for laundry machines or kitchen ranges, should be used. It might be necessary to have a special circuit installed by a professional electrician to provide adequate power for your linear amplifier. High power increases the chances for EMI, both to station accessories and to home entertainment equipment such as hi-fi sound systems, TV receivers, videocassette recorders, and telephones.

Antennas   A station can only work as well as its antenna system will allow. There are many different types of antennas, both commercially manufactured and "homebrew." Basic RF antennas are discussed in Chap. 4.

Mobile Equipment

Mobile amateur-radio equipment is operated in a moving vehicle such as a car, truck, train, boat, or airplane. Mobile equipment includes all attendant devices, including antennas, microphones, computers, and power supplies.

The design and operation of mobile electronic equipment is much simpler today than it was a few decades ago. The main reason for the improvement is the development of solid-state technology. Most mobile power supplies provide low-voltage direct-current (DC) power; 13.8 V is by far the most common. A tube-type circuit requires a *power inverter* for operation from low-voltage DC; the inverter changes the DC to 117 V alternating current (AC). Solid-state equipment can usually be operated directly from a mobile DC supply.

Mobile equipment is generally more compact than fixed-station apparatus. In addition, mobile equipment must be designed to withstand larger changes in temperature and humidity, as well as severe mechanical vibration.

Most two-way mobile communications installations consist of a vertically polarized omnidirectional antenna and transceiver. The transceiver is a transmitter-receiver combination housed in a single cabinet. This saves significant space and cost. Most mobile operation is done at VHF and UHF, where the antennas are manageable in size. But some mobile operation is carried out on the HF amateur bands, also known as the "low bands." Inductively loaded vertical antennas are used in virtually all low-band mobile ham installations.

Portable Equipment

Portable amateur-radio equipment is any apparatus that is designed for operation in remote locations. Portable equipment is usually battery-operated and can be set up and dismantled in a minimum time. Some portable equipment can be operated while being carried; an example is the *handy-talkie* (HT) or *walkie-talkie*.

With the improvement in miniaturization and semiconductor technology, portable equipment can have complex and sophisticated features. A few decades ago, the mention of a microcomputer-controlled HT would have elicited ridicule. Today, such devices are commonplace.

Portable equipment must be compact and light in weight. For this reason, such equipment always has modest power requirements, because a high-power battery is bulky and heavy. In addition to being small and light, portable equipment must be designed to withstand physical abuse such as vibration, temperature and humidity extremes, and prolonged use without complicated servicing.

Portable equipment is sometimes used in mobile applications. For example, a handheld unit can be connected to a mobile antenna and used in a car or truck.

Traffic Handling

An important justification for the existence of amateur radio is the public-service communications that hams provide. As people gain experience in the hobby, they tend to focus most of their attention on one or two specialties within the hobby. Messages sent via ham radio are called *traffic. Traffic handling* consists of originating, passing, and delivering messages. Some hams do almost all of their operating in this vein of public service.

Nets   Most traffic handling is carried on in an organized manner by groups of stations that meet regularly on specific frequencies. These groups are known as *traffic nets.* Most of the messages on these nets are of minor importance. The activity is done largely to sharpen individual operating skills for use in emergencies. Some traffic nets operate using continuous-wave (CW) radiotelegraphy. Others use voice communications, primarily SSB on the bands at 1.8 through 29.7 MHz, and FM on the bands at 50 MHz and above. Much traffic-net operation is found on the 80-m (3.5-MHz) CW band, the 75-m (3.8-MHz) SSB band, and the 2-m (144-MHz) FM band. In recent years, an increasing amount of traffic handling has been done in digital modes at all frequencies. *Packet radio,* discussed in Chap. 9, lends itself perfectly to high-volume, high-speed, accurate, and efficient traffic handling.

Radiogram Format   A radio message or *radiogram* is sent according to a specific format. The first part of the radiogram is the *preamble.* It contains a message number, a *precedence* (see below), handling instructions (if any apply), the call sign of the originating station, the *check* (number of words or groups in the message), and the date and time (optional). The second part of a radiogram is the *address* of the person to whom the message is to be delivered. The third part of the message is the *body* or *text.* The fourth and final part of a radiogram is the *signature* (the name and/or call sign of the originating person).

Precedence of Message   Precedence is the degree of importance attached to a message. There are three levels of precedence. A *routine* *message,* abbreviated R on CW, is by far the most common. It is usually a simple greeting, such as "Happy new year." It has the lowest priority. A *priority message,* abbreviated P on CW, is a message that must be delivered within a certain time frame and/or that has significance over and above routine. Traffic in an emergency situation, but not itself of great urgency, falls into this category. An example is "We all survived the earthquake." Priority messages supersede routine messages. An *emergency message,* always spelled out fully on CW, is any message that is a matter of life or death to a person or persons. An example is "Send more medication immediately." Emergency messages always supersede priority or routine messages.

Contesting

One of the most popular ham radio activities is *contesting.* There are many different contests during the calendar year. Most are on weekends. They take place on a variety of different bands and in any or all of the various modes. In most contests, the objective is to make as many contacts as possible during the time period. Contests can be recognized by the presence of many stations on a band, carrying on short, rapid-fire, repetitious contacts.

Some radio hams participate in several contests each year, and some never participate in any. Contests serve the purpose of keeping operating skill sharp, and this can be invaluable in times of emergency, when large volumes of traffic must be handled accurately and efficiently. Contests are also great fun for those with a fondness for challenge and for those who like to place themselves and their stations in competition with others. Most contests have certificates for the winners in various geographical areas.

The following paragraphs are short descriptions of a few contests in which hams can operate. This is by no means a complete list. Amateur radio magazines carry announcements, schedules, and rules for contests well in advance of the actual event dates.

Field Day   On the last full weekend every June, ham operators "go portable" and set up stations using emergency power. Home stations can take part in this contest, but the objective is to simulate a nationwide emergency in which commercial power would not be generally available. Contacts are simple, including only the number of transmitters, the type of power arrangement, and the ARRL section. Other activities on

Field Day include testing insect repellants, learning about *Murphy's laws* (expressions of the limitless and inconvenient ways in which hardware can malfunction or fail), and eating hamburgers, pizza, and barbecued chicken in the middle of the night.

Sweepstakes Every autumn, a major contest takes place, known as the *November Sweepstakes*. One weekend is for phone (voice), and the other is for CW (radiotelegraphy). Most of the activity takes place on the HF bands. This is a major event for many hams. The contact exchanges are comprehensive, and the competition is keen, requiring skill and endurance. It is mainly a U.S. contest.

DX Contests Many hams love the challenge of DX, or long-distance contacts. There are several DX contests during the course of the year. Such contests present an excellent opportunity for radio hams to add new and/or rare countries to their list of places they have "worked." In a DX contest, a rare station will command a *pileup* in which many stations call every time he or she listens. High power and large antennas are an advantage to U.S. hams in these situations. But operating skill can often allow a modest station to break through pileups. There is a special thrill in doing this; it is akin to making a hole-in-one at golf.

160-m Contest Usually held on the first full weekend each December, the ARRL 160-m contest is a special challenge because of the large antennas needed for good signals on this band. The band is open for long-distance communications only during the hours of twilight and darkness. Another 160-m contest, the *CQ Worldwide*, takes place in January. This is part of a much larger contest sponsored by *CQ* magazine.

QSO Parties In ham radio jargon, "QSO" stands for "radio contact." The various ARRL sections (usually states, and sometimes parts of states) sponsor local, scaled-down contests. These attract fewer hams than the large contests but can still be great fun. QSO parties are announced in the ham magazines.

DXing

One of the most interesting, challenging, and exciting kinds of ham radio operation involves making contacts with stations in foreign lands. Ham radio signals know no geographic bounds, and ham radio friendships recognize no political lines. A contact with a station in another country is called *DX*. The pursuit of contacts in many foreign countries is *DXing*. A ham who likes to work DX, or primarily DX, is a *DXer*.

Station Design It is not necessary to have a huge antenna or high power in order to work DX, even rare countries. A simple wire antenna, strung between two trees and fed with coaxial cable or open-wire line, can produce plenty of DX contacts even with a low-power transmitter. This is particularly true on the 10-, 12-, and 15-m amateur bands. The single most important piece of hardware for working DX is a good receiver. It should be sensitive, stable, and selective, and should have good dynamic range. Most commercially manufactured amateur receivers and transceivers are satisfactory for working DX. In older receivers, a *preamplifier* at the antenna input can sometimes make a big difference in sensitivity, especially on the bands above 14 MHz. The second most important thing for a DX station is a good antenna. A *quad antenna* or *Yagi antenna* is preferred for 10, 12, 15, 17, and 20 m, and also for 30 and 40 m if space and money are available. Such antennas should be at a height of at least one-half wavelength, and preferably a full wavelength or more, above the ground and surrounding obstructions. For the lower bands (40, 80, and 160 m), a *vertical antenna* will perform well if it is provided with a good RF ground. A *dipole antenna* is satisfactory if it is placed at a height of at least one-fourth wavelength, but preferably one-half wavelength or more. The *inverted-V antenna* is noted for its good DX performance on 80 and 160 m if the center is at least one-fourth wavelength above ground. The third, and optional, ingredient for working DX is a *linear amplifier* that can deliver the maximum legal power, or close to it.

Operating DX Not mentioned as station equipment, but as important as the hardware, is a good, patient, courteous operator. DX stations are often easy to find on the ham bands, not because their signals sound unique, but because of the presence of pileups, or groups of stations calling. You might hear one of these and wonder how you can ever hope to be heard by the DX station, when you are just one of dozens or even hundreds of stations calling. It is easy to imagine that the combined power of all these stations will drown you out. It's not hard to envision gigantic yagis and quads, and big linears run by hardened DXers, making it impossible for you to get through. This writer has faced all of these challenges using a wire antenna thrown over a sapling and a mere 85 W output, and snared rare DX on the very first call. One method of being heard is to call the DX station on a frequency no one else appears to be using. This is easier on CW than on SSB. But sometimes, especially in a heavy pileup or when the DX operator is experienced, the DX station will give specific instructions concerning the calling frequency. You should always follow these instructions, making sure, of course, that your license class allows you to use that frequency. Another method of being heard is to try to stand out because of your operating courtesy. Keep calls short. Never call while the DX station is in QSO with someone else.

QSLs Some DXers like to collect *QSL (confirmation) cards* from their DX contacts. This is necessary in order to demonstrate that you have had contacts with the stations; you will need them for the DX Century Club, for example. The easiest way to send, receive, and keep track of QSL cards is via a *QSL bureau.* Some hams don't care about QSLing, but you should send one if you receive one from a DX station and the operator specifically requests it by checking the box "PSE QSL."

For More Information There are books available that discuss amateur radio DXing. Any of these make good reading. They are advertised in all the ham-radio-oriented magazines. You can also learn about operating DX from a local ham who is into this aspect of amateur radio. Most DXers like to show off their stations and their accomplishments.

Awards

In order to make ham radio operating fun and challenging, various organizations give out award certificates to hams. Most of the awards are administered by the ARRL. Smaller groups of hams, and other organizations, give awards also. Here are some of the more popular awards.

DX Century Club For contacting 100 or more foreign countries, and for collecting the QSL cards from those contacts, the ARRL will give you a certificate of membership in the DX Century Club (DXCC). Some hams have contacted more than 300 countries. The call signs of these hams are regularly listed in *QST,* the official magazine of the ARRL. Competition is keen among top DXers.

Code Proficiency The ARRL regularly holds on-the-air tests, transmitted by station W1AW, for Morse-code copying ability. To obtain a certificate for a certain speed, you must get 1 minute of solid (perfect) copy. Once you get a certificate at a certain speed, such as 15 words per minute (wpm), you can obtain endorsements later for higher speeds in 5-wpm increments. The times and frequencies for these tests, sometimes called "qualifying runs," are published in *QST.*

WAS and WAC   The ARRL will give you a certificate for submitting QSL cards for contacts with each of the 50 states in the United States. This is called the Worked All States (WAS) award. A certificate is also available for having contacted stations in each of the continents: North America, South America, Europe, Asia, Africa, and Australia. QSL cards are required for verification. This is called the Worked All Continents (WAC) award.

A-1 Operators' Club   If you show good operating skill, efficiency, and courtesy while in QSO with two or more members of the A-1 Operators' Club, you will be invited to join the club.

Other Awards   If you carry on a contact for more than 30 minutes with a ham who is a member of the *Rag Chewers' Club* (RCC), you might be invited to join the club. This indicates you can have a long conversation, as well as a short QSO, on the radio. If you have been a ham for 20 years or more, and can prove it to ARRL (such as with an old license or Callbook listing), you can become a member of the *Old Timers' Club* (OTC). For carrying on a CW conversation at a speed of 80 wpm or more, copying in your head but understanding what is said, a group of hams will welcome you into the *Five Star Operators' Club.* Numerous other awards exist, and they all carry attractive (and sometimes humorous) certificates. You'll learn about these as you gain experience, and make friends, as a radio ham.

Handi-Ham System

At the Courage Center in Golden Valley, Minnesota, there is a program to help handicapped people learn about ham radio and become hams. Once these people get licensed, the Handi-Ham System provides technical assistance, station equipment, physical assistance, and upgrade tutoring. The *Handi-Ham System* is a worldwide organization. It is represented in all 50 states in the United States.

Ham radio is a link to the world for anyone. But for disabled people, ham radio has an added, special meaning. With amateur radio, one can talk to the whole world without leaving the comfort and convenience of home. Some of the best ham operators in the world are handicapped.

The Handi-Ham idea was conceived by the late Ned Carman of the Rochester (Minnesota) Amateur Radio Club, and several other Minnesota hams during the late 1960s and early 1970s. Ned, not himself handicapped, was a fine operator and a public servant. His preferred activities were traffic handling and helping new hams. The Handi-Ham club station bears Ned's call sign.

Information about the Handi-Ham System can be obtained from the Courage Handi-Ham System, 3916 Golden Valley Road, Golden Valley, MN 55422. It has a Web site at:

http://www.mtn.org/handiham/

QRP

The abbreviation *QRP* is one of the ham radio *Q signals*, used mainly in radiotelegraphy. It means "Reduce your power." The abbreviation has, in recent years, come to mean "low power" and/or "low-power operation." This is a relative thing, of course; to some hams, 100 W output is QRP, while to others, only levels less than a few milliwatts are true QRP. The most accepted figure is 5 W output, or less, on the bands below 30 MHz.

Operating QRP is a challenge that appeals to many hams. To the QRP enthusiast, there's nothing quite like working some rare DX station using less than 5 W. This is especially true if the contact is made when numerous other, high-powered stations are also calling.

One of the biggest advantages of QRP is that it allows the use of battery power. This makes it possible to go practically anywhere and carry a complete ham radio station in a small suitcase. A QRP station can also be powered by solar panels or other alternative energy sources. Another advantage of QRP is that it reduces the tendency for problems to occur with EMI.

The easiest bands on which to make DX contacts using QRP are 6, 10, 12, and 15 m. This is because of the propagation characteristics on these bands. This writer has had the experience of working a New Zealand station on 28 MHz CW, using 3 W output to a three-element Yagi antenna on the roof of a three-story apartment house. The signal report was good; I was told my signals were perfectly readable and moderately strong. With this same station, contacts in Europe and South America were made in large numbers during peak times of day.

QRP can yield exciting and fascinating contacts on any band. Some hams have worked thousands of miles using less than 5 W on 160 m.

QRQ

The signal QRQ means "Increase your CW sending speed." It also refers to high-speed CW in general. Keyboard keyers allow an operator to send CW at a speed limited only by typing ability. Character buffers (small memory chips) eliminate any need to worry about spacing. A good typist can send CW at 40 wpm or more with minimal effort.

Machines can "copy" CW and display the text on a screen. This, too, can be carried out at speeds considerably greater than most people can read in their heads or take down on paper. However, for high-speed printed communications, *radioteletype*, using Baudot or ASCII codes and *frequency-shift keying*, are preferred because these methods reduce the number of teleprinter errors.

There are numerous hams who enjoy communicating via QRQ Morse code, sending with keyboards but reading the code in their heads. Some of these operators can send and read at speeds of 60 wpm or more. The best can send and read up to about 100 wpm. You can find these hams in the lower parts of the General/Advanced Class CW band segments. They send CQ at 40 to 45 wpm, increasing speed once a contact gets under way.

Rag Chewing

Many hams like to carry on long conversations. There are almost as many subjects to talk about in an amateur contact as there are in any ordinary conversation. Sometimes, a contact can go on for hours. This pastime is called *rag chewing*. When three or more hams get involved in a rag chew, each one makes a contribution in a rotating sequence. This is done often in informal nets. It is called a *round table*. It is similar to a chat room on the Internet.

Rag chewing is an integral, and important, part of ham radio. Lifetime friends can be made that way. It can be fun. Confirmed rag chewers like to take some time getting to know the person on the other radio, rather than being concerned only with equipment, propagation conditions, and contact volume and efficiency. There are frequencies on which rag chewing is generally accepted, and those on which it is not. In some cases, rag chewing is frowned upon; for example, it should not be done on a repeater. Common rules of courtesy and etiquette apply to rag chewing, as to all ham radio communications. If you might get in the way of someone else's need for a frequency, your rag chew should be carried out on an unoccupied or low-priority frequency. Rag chewing with rare DX will earn you many enemies in a very short time (if the DX station will even engage you). This is because whenever a rare DX station appears on the air, dozens or even hundreds of stations want to get a contact to confirm a new country worked.

There are some modes on which rag chewing is difficult or impossible for physical reasons. These include *meteor scatter*, low-orbit *satellite communications* (when there is only one satellite), and *moonbounce*.

For Further Information

A good way to learn about ham radio is to contact the headquarters of the American Radio Relay League, 225 Main Street, Newington, CT 06111. Its Web site is at:

http://www.arrl.org

The ARRL publishes books on all subjects in ham radio, as well as training materials and the monthly magazine *QST* (which means "Calling all radio amateurs"). The people at ARRL headquarters can tell you the location of the nearest amateur radio club, where you can meet local hams and find out if this hobby is right for you.

Lightning

Lightning is a special hazard to radio amateurs, CB operators, and short-wave listeners because of the outdoor antennas commonly used in these personal radio systems. Lightning kills more people every year in the United States than hurricanes or tornadoes. The main danger to people is from electrocution; in the case of property, damage can result from fires, induced currents, and explosions. This section suggests precautions to protect radio users and equipment from the dangers of lightning. But this section cannot cover every scenario. Sometimes people are injured, or electronic equipment is damaged, despite precautions. The best one can do is statistically minimize the danger.

The Nature of Lightning

There is a constant potential difference between the earth's surface and the ionosphere. The lower atmosphere acts as a dielectric, so the earth and ionosphere form a giant capacitor. The charge in this atmospheric capacitor is stupendous, and the dielectric frequently breaks down with discharges of lightning. The discharges are almost always concentrated in areas of precipitation, particularly in heavy rainstorms.

A lightning surge is called a *stroke*. It lasts for only a small fraction of a second, but in this time it can start fires, cause explosions, destroy electrical and electronic equipment, and electrocute people and animals. There are four types of lightning that are common in and near thunderstorms. These are:

A. *Intracloud lightning* (within a single cloud)

B. *Cloud-to-ground lightning*

C. *Intercloud lightning* (between different clouds)

D. *Ground-to-cloud lightning*

These four types of stroke are illustrated in Fig. 16-5, and are labeled A through D, respectively. Types B and D present the greatest danger to electronics hobbyists and wireless equipment, although types A and C can sometimes cause an *electromagnetic pulse* (EMP) sufficient to damage sensitive apparatus.

A stroke begins when the charge between a cloud and the ground, or between a cloud and some other part of the cloud, gets so large that electrons begin to advance through the air from the negative charge pole. A small current, called a *stepped leader,* finds the path of least resistance through the atmosphere via "trial and error." This can take from a few milliseconds to several tenths of a second. Once the stepped leader has established the circuit by ionizing a channel in the air, one or more *return strokes* immediately follow, attended by current that can peak at more than 100,000 A. This current is responsible for most of the destructive effects of lightning.

If you happen to be operating a CB radio, ham radio, or shortwave station and lightning strikes nearby, you might be seriously injured or even killed. And of course, a direct hit is much worse than the induced charge from a nearby stroke. Fortunately, direct hits don't occur very often, but (as the adage warns) it only takes one. The way to avoid or minimize the personal danger to yourself is to stay away from all electrical and electronic equipment, especially antenna systems, whenever

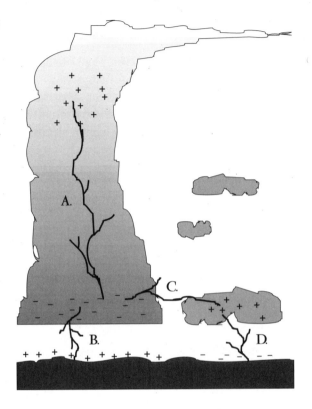

Figure 16-5
Four types of light-
ning stroke:
intracloud (A), cloud-
to-ground (B), inter-
cloud (C), and
ground-to-cloud (D).

there is a thundershower in the area, or whenever there is any risk of lightning.

Lightning can cause hardware damage from the EMP, which induces large currents in antennas, antenna feed lines, and utility power lines even if there is no direct hit. Electrostatic-sensitive components are most easily destroyed. These include metal-oxide-semiconductor (MOS) transistors and integrated circuits. The front-end circuitry of a receiver is extremely vulnerable. Semiconductor components in general, especially those near the antenna terminals or in the power supply, are also susceptible to damage from the EMP.

Protecting Yourself

Lightning can take place at any time, but it is most common in or near areas of precipitation. If thunder can be heard, lightning is occurring. Do not assume that the noise is coming from some other source. Heavy precipitation at or near a place (including snow and sleet), dark clouds, a sand storm, a dust storm, or an erupting or smoldering volcano can be a source of lightning. The only necessary ingredient is a large potential difference between two regions of the atmosphere that are close enough together, or between some parcel of air and the ground.

The following precautions are recommended if lightning is taking place near you.

1. Stay indoors, or inside a metal enclosure such as a car, bus, or train. Stay away from windows.

2. If it is not possible to get indoors, find a low-lying spot on the ground, such as a ditch or ravine, and squat down with feet close together until the threat has passed.

3. Avoid lone trees or other isolated, tall objects such as utility poles or flagpoles.

4. Avoid electric appliances or electronic equipment that makes use of the utility power lines or that has an outdoor antenna.

5. Stay out of the shower or bathtub.

6. Avoid swimming pools, either indoors or outdoors.

7. Do not use the telephone.

Equipment Protection

The antenna is the most dangerous part of any radio station when lightning is taking place nearby. A direct hit is not necessary for destructive and dangerous currents to be induced in the antenna element(s) and feed line. The power lines can also get high-voltage surges that can damage equipment plugged into wall outlets. Precautions to minimize the risk are as follows.

1. Never operate, or experiment with, a radio station when lightning is occurring anywhere near your location.

2. When the station is not in use, disconnect all antennas and ground all feed-line conductors to a good DC electrical ground other than the utility power-line ground. Preferably the lines should be left entirely outside the building and connected to an earth ground at least several feet from the building.

3. When the station is not in use, unplug all equipment from the wall outlets.

4. When the station is not in use, disconnect and ground all rotator cables and other wiring that leads outdoors.

5. *Lightning arresters* provide some protection from charge buildup, but they cannot offer complete safety and should not be relied upon for routine protection.

6. *Lightning rods* reduce the chance of a direct hit but should not be used as an excuse to neglect the other precautions.

7. Power-line *transient suppressors* (also called "surge protectors") reduce computer "glitches" and can sometimes protect sensitive components in a power supply, but these should not be used as an excuse to neglect the other precautions.

8. The antenna mast or tower should be connected to an excellent DC earth ground using heavy-gauge wire or braid. Several parallel lengths of American Wire Gauge (AWG) No. 8 aluminum ground wire, run in a straight line from the mast or tower to ground, form an adequate conductor. This conductor might be called upon to carry a momentary current of thousands of amperes. The conductor must be able to survive in case of another strike.

9. Other secondary protection devices are available and can be found in electronics-related and radio-related magazines. None of these should be relied upon for complete protection; the primary methods 1 through 4 above are mandatory to minimize the risk to equipment.

For More Information

Refer to the Lightning Protection Code, published by the National Fire Protection Association, Batterymarch Park, Quincy, MA 02269. Hobby electronics magazines often carry articles on the topic of lightning protection. Various books also have information, but it is important to be sure it is recent, because new, improved protection methods are occasionally found.

The Role of the Machine

Do you think electronic gadgets are incompatible with your brain? Do you envision such devices as alien or dangerous? Many people suffer from *technophobia*, an unreasonable fear of machines, especially electronic systems.

Necessity

Maybe you don't need a cell phone or pager. Some people get along fine without TV sets; some even make do without cars. There is no law that says you must buy something simply because it exists. If you have no need for a cell phone and do not want one, then why buy one?

If you have been reading this book, you probably believe that wireless technology can enhance the quality of your life. Maybe you've bought a cell phone and pager, but don't use them much. Perhaps you have not yet chosen from the vast selection of wireless devices available today. Take plenty of time selecting hardware. Once you've bought equipment, don't try to rush the learning curve. In the beginning, treat the new machine as a toy. Have fun with it, and you'll forget your fears.

Building Confidence

Some instruction manuals are written at a level too advanced for the neophyte. Manufacturers and distributors have begun to recognize this problem and are writing documentation in a breezy style. Some manufacturers provide two or more instruction manuals, with varying degrees of technical complexity, filling the needs of users at all levels of experience. With computers, detailed instructions are provided in "help" files that can be read from the display. Most manufacturers have technical-support phone numbers that you can call as much as you need. An increasing number of vendors provide Web sites for the same purpose.

Some electronics "experts" seem to enjoy overwhelming novices with jargon. If you run into such a person when looking for advice, get help from somebody else. Plenty of dedicated experts are willing and able to help the newcomer. Their enthusiasm shows. Some will tell you that they were once technophobic.

Servant or Master?

Another manifestation of technophobia is a fear that machines have too much power in society and that they are "taking over." Science-fiction writers love this theme, and exploit it endlessly. They point out that a blind trust and belief in machines is no better than blind fear, and might in fact be worse.

People are starting to question the desirability of automation, especially in the workplace, where technological advancement can mean the loss of jobs. Technology unquestionably has a dark side. A healthy skepticism can ensure that it enriches, rather than impoverishes, our lives. Your cell phone, pager, wireless e-mail system, or other equipment is meant to serve you, not rule you. When irresponsibly used, either individually or collectively, electronic devices and systems can create or exacerbate many kinds of problems. When responsibly used, they can enhance your life by giving you more free time, increasing your earning power, and widening your career opportunities. This is especially true for small-business owners and freelancers.

SUGGESTED ADDITIONAL READING

American Radio Relay League, Inc., *The ARRL Handbook for Radio Amateurs* (American Radio Relay League, Inc., published annually).

Dorf, Richard, *The Electrical Engineering Handbook* (CRC Press, 1993).

Garg, Vijay, and Wilkes, Joseph, *Wireless and Personal Communications Systems* (Prentice-Hall, 1996).

Gibilisco, Stan, *The Illustrated Dictionary of Electronics,* 7th ed. (McGraw-Hill, 1997).

Gibilisco, Stan, *TAB Encyclopedia of Electronics for Technicians and Hobbyists* (McGraw-Hill, 1997).

Schneiderman, Ron, *Future Talk: The Changing Wireless Game* (IEEE, 1997).

Van Valkenburg, Mac, *Reference Data for Engineers* (Sams Publishing, 1993).

INDEX

B

About the Author

Stan Gibilisco is a professional technical writer who specializes in books on electronics and science topics. He is the author of *The Encyclopedia of Electronics, The McGraw-Hill Encyclopedia of Personal Computing,* and *The Illustrated Dictionary of Electronics,* as well as over 20 other technical books. His published works have won numerous awards. *The Encyclopedia of Electronics* was chosen a "Best Reference Book of the 1980s" by the American Library Association, which also named his *McGraw-Hill Encyclopedia of Personal Computing* a "Best Reference of 1996." Stan Gibilisco's Web site is

http://members.aol.com/stangib